STUART FISHER

Canals of Britain

A COMPREHENSIVE GUIDE

Canals of Britain

Contents

Published by Adlard Coles Nautical
an imprint of A&C Black Publishers Ltd
36 Soho Square, London W1D 3QY
www.adlardcoles.com

Copyright © Stuart Fisher 2009

First edition 2009
Reprinted 2010

ISBN 978-1-4081-0517-7

This book is produced using paper that is made from wood grown in managed, sustainable forests. It is natural, renewable and recyclable. The logging and manufacturing processes conform to the environmental regulations of the country of origin.

Project editor: Julie Bailey

Typesetting, layout and maps by the author

Typeset in 9pt Bembo

Printed and bound in China

Note: reference to waters do not necessarily imply that access and passage on those waters or their banks is legally permitted or that they are safe in all conditions. The author and publishers cannot be held responsible for any omissions of references to hazards on these waters. Circumstances can change without warning. The user should assess the situation using all information available at the time and act appropriately.

Foreword

It is surprising that, other than Hugh McKnight's *The Shell Book of Inland Waterways*, there have been very few complete works describing the canals of Britain. Adding to this surprise is that Hugh first published his book in 1975 and it has not been updated since the 1980s.

Considering so many narrowboaters have navigated the complete canal system, it is nothing short of amazing that this new, highly informative work is the result of 'paddling' the waterways by kayak. As one of those who has, by narrowboat, navigated just about all the inland waterways and made some 2,000 miles of coastal connections from and to rivers and canals, I have only the utmost admiration for Stuart Fisher who, before writing the *Canals of Britain*, authored another equally informative production after paddling the entire coastline of Britain! THAT IS an experienced canoeist!

Stuart's expertise as a 'paddler' is matched by his writing talent. If you are looking for an informative reference to British waterways here it is. It even includes landlocked waterways that we narrowboaters cannot get to or, in many cases, could not navigate on even if we could, except perhaps with the ship's dinghy!

Chris Coburn MBE

Introduction

We are fortunate to have, in this country, a canal network like no other in the world. It was the first commercial canal system, leading the way for the Industrial Revolution, but has remained largely as built. The canals are mostly small and intimate. Restoration in recent years, supported by lottery and other funding, has outstripped the pace of construction even during the Canal Mania years.

Overseas, where canals have been enlarged to take modern commercial craft, you can look at the distant bank and wonder whether you will be run down by something the size of an office block or a multi-storey carpark, complete with cars. In this country the biggest risk is running aground and you could usually wade through the mud to either bank. It is a safe environment with limited scope for getting into serious trouble. For the walker and cyclist, canals provide routes which are mostly flat. As far as possible, descriptions are given downhill and with the flow for those who have a choice of direction.

We have canals with scenery which changes frequently, open countryside, wildlife, heritage industrial buildings, canalside public houses, modern city centres, wild moorland and coastal harbours, all mixed up. Anywhere on the system fantastic engineering structures can be found.

Sixty canals are described, all but the shortest ones. I have excluded the river navigations, by which many people mean joined up waters deep enough to take at least a narrowboat. However, for those with suitable boats there are 65,000km of navigable river in England and Wales alone, scope for a whole library of books. Not all the canals described in this book are linked to the rest of the system and not all are physically passable for many boats or have towpaths you can or may use. You may need a spirit of adventure, like the earlier recreational boaters.

I have been fortunate enough to have travelled all the canals in this book with a kayak. That is not to say that I was always in it. A 7km portage through Failsworth and Miles Platting on the Rochdale Canal, before the concrete cap was removed during restoration, was memorable for the wrong reasons. Some of the canals I have also travelled by narrowboat, by bike or on foot, including a happier run down the Rochdale after restoration.

Who uses the canals? If you look at canal magazines you will see smiling couples or families busy in the summer sunshine. In practice, you may find the picture rather different. Often the canals are deserted, except for wildlife that finds the canals an ideal environment, without needing all humans and boats to be banned. My unscientific survey suggests that cyclists might be the commonest humans, followed by dog walkers. Joggers, local walkers, ramblers and boaters are less frequent users, even though it is boaters who are asked to foot the licence bill.

The sun isn't always shining, either. Rain, graffiti and barbed wire are a few of the less attractive elements that appear in the photos, hopefully presenting an honest balance of what will be met around the system.

The intention is that the book should be engaging to all who travel the canals. I do not usually give navigation instructions, depths and headrooms, portage routes or what to do when the towpath runs out. If the present state of a canal is such that it is limited to one kind of user, usually someone able to undertake portages, I may refer to that category of user, otherwise I talk more generally. I draw attention to features near the canal, especially in heritage cities such as Bath and Chester, because most canal travellers will not want to pass through without stopping.

I raise a couple of concerns for purists. There may be some who take issue with the use of metric units.

Metrication was supposed to have been completed in 1965. It is the system which was taught in school to most people still below retirement age. Canals are not just for an ageing population. We need to attract younger users. Ah, you protest, the canals were built to imperial units. If this was a book about the pyramids, would you expect me to give all the measurements in cubits and spans?

I make one concession, small value coinage. This has no more sense of its contemporary value in post decimal figures than if left in shillings and pence.

The other confession is that I have often referred to bridges by the numbers of the roads they are carrying. Canal company bridge numbers (or even just names) are less accessible to those without detailed canal maps.

If something is shown on a map it is within 4km of the featured canal. On the canal or towpath you are going to need the listed OS 1:50,000 maps or something comparable to support your spirit of adventure.

This book has been developed from a series of guides first published in *Canoeist*. Revisiting canals, it is surprising how much additional greenery there is, often meaning that old pictures could not be retaken from previous positions. At the same time, much canalside grass is now kept much tidier by British Waterways than in the past. The canals are changing constantly. You may find things which have changed since the book went to print or, heaven forbid, actual errors. If so, feel free to protest and slap my wrists at canalsofbritain@canoeist.co.uk or via Adlard Coles Nautical.

Stuart Fisher – November 2008

Acknowledgements

I am extremely grateful to Julie Arnold, Malcolm Bower, Graeme Bridge, Peter Brown, Tony Collins, Bob Gough, Di Harris, Ian Hunter, Peter Keen, Tommy Lawton, John Lower, Chris Morgan, Trevor Morgan, John Roddis, Peter Stone and Andy Talbot for checking content. They sometimes undertook masterly studies in considerable detail. On exceptional occasions I have not taken all the offered advice. The blame stays with me for any errors.

Generally, I have failed to give adequate credit for the restoration work of the Waterway Recovery Group and the many canal societies. I list the relevant websites at the end of each chapter. Do visit these to get a fuller picture of what the societies have been and are doing to restore the canals to full use.

Within Adlard Coles Nautical I am particularly grateful to my commissioning editor, Janet Murphy, and project editor, Julie Bailey.

Last but not least, this book would not have happened without the support of my wife, Becky, and sons, Brendan and Ross, who have also toured the canal network widely.

Photographs

Chris Coburn p4
Becky Fisher p150 bottom, p162 centre left, p168 bottom left
Ross Fisher p140 bottom right
Chris Jones p34 right two, p35 bottom two, p36.
All other photographs by the author.

By same author
Inshore Britain

Legend for maps

— Featured canal
— Other canal or river
— Motorway
— Other road
— Railway

Open water or sea

Inter-tidal zone

Built-up area

Woodland

Scale 1:200,000.
North is always at the top.

1 Birmingham Canal Navigations, Old Main Loop Line

Birmingham's oldest canal

Of Britain's operational canals, only the Fossdyke and the Bridgewater predate the Birmingham. The Birmingham Canal Company were authorised to build their line in 1768 and it was opened as far as Wednesbury as a contour canal by Brindley the following year, picking up many industrial premises. The whole 36km was completed in 1772. Because of the minerals and industry it was immediately highly successful. In 1790 the Smethwick summit was lowered from the 150m level to the 144m Wolverhampton level with the removal of three locks on each side, saving time and reducing water supply problems.

Many of the loops of the old line have been cut off or filled in but the section between **Tipton** and Smethwick remains, a contour canal with more features of interest than the more efficient New Main Line.

The old line divides from the new in front of the Old Bush at Tipton Factory Junction, just a stone's throw from the top lock on the New Main Line. With a westerly wind there is the acrid smell from nearby metalworking industries. At this end the water is not too bad but the quality deteriorates towards Birmingham.

The Malthouse Stables have been restored as a community recreation centre.

Tipton Green has been pleasantly landscaped to make the most of the canal, with assorted new housing standing around the Fountain Inn. This was formerly the home of 19th-century canalman and prize fighter William Perry, the Tipton Slasher.

The Tipton Green & Toll End Communication used to lead from Tipton Green Junction. From Tipton Junction there is a spur, the Dudley Canal No. 1 Line, which leads to the Black Country Museum and the **Dudley** Tunnel.

Housing, with gardens behind greater reedmace, gives way to old brick factories. The view opens out at Burnt Tree with new housing around a marina. The Silurian limestone ridge behind is topped by a couple of masts on Darby's Hill. The canal is wider here as it approaches Dudley Port. A container yard on the left is an update of the area's former purpose.

The canal crosses the Netherton Tunnel Branch Canal on an aqueduct and it is worth looking over the parapet at the imposing cutting leading to the mouth of this 3km long tunnel.

At Brades Hall Junction there are two arches on the left, the first abandoned but the second leading past an ivy-covered wall to the first of the Brades Locks on the

Gower Branch Canal. This drops down as a midway connection to the New Main Line. Once again it is an area of industry. Factories vent acrid gases over the canal as oil begins to appear on the surface but this doesn't last for long.

The character of the canal changes sharply at **Oldbury** as the M5 is in close proximity to the canal for the next 2km, mostly overhead. As with the River Tame in Birmingham, an elevated route over a waterway has proved to be the most acceptable line for a motorway to be squeezed through the city.

This does not indicate any lessening in canal complexity, however. North Junction and South Junction of the closed Oldbury Loop Line are still visible and the Houghton Branch Canal leads away under the motorway. Once under what can be a useful canopy for a rainy day, Oldbury Locks Junction accepts the Titford Canal which descends through six locks from the Crow, a feeder from Rotton Park Reservoir.

The concrete jungle intensifies as the A457 passes between canal and motorway and the canal comes out into the open air for a breather. At one point there is a traditional brick-arched bridge over the canal, noticeably out of place among all the vast concrete columns and walls. After passing Cadbury's chocolate factory on the right, the complexity reaches its zenith. The M5 passes over the main railway line, which passes over the Old Main Loop Line, which passes over the New Main Line on the Steward Aqueduct. The aqueduct would probably have been more of an honour to Birmingham Canal Navigations committee member Stewart if the name had

The Gower Branch Canal leads away from Brades Hall Junction through the Brades Locks.

The Malthouse Stables centre at Tipton Factory Junction. Only the tall part is original.

Contrasting scales of bridge; a footbridge and the M5 viaduct.

been spelled correctly. Telford's 2.1m iron-trough aqueduct of 1826–8 is now a listed building.

Spon Lane Wharf and Junction are now beneath the motorway. Spon Lane Locks Branch provides a connection down to the New Main Line through Spon Lane Locks, the remaining bottom three from the six that descended from the original summit. These three locks are probably the oldest working locks in Britain. Top Lock has a split cantilevered bridge to pass ropes through without unhitching towing horses.

The cutting between Sandwell and Smethwick has been designated the Galton Valley Canal Heritage Area. It has one of Britain's greatest concentrations of canal architecture.

Chance's former glassworks on the right were founded in 1824 and include a number of listed buildings. Chance pioneered sheet glass, produced optical glass for lighthouses after 1838 and made the glass for the Crystal Palace in 1851, going on to manufacture microscope lenses, rangefinders, telescopes and searchlights. Pilkingtons subsequently took over the firm.

Looking back on the left side the prominent feature is the listed small timber belltower of 1847 on the seven-storey offices of Archibald Kenrick & Sons, who were making ironmongery from 1791. The foundry beside Top Lock is that of George Salter, manufacturers of such things as spring balances, weighing equipment and steam locomotive safety valves.

The motorway was built in 1969 but suffered badly from chloride attack by road de-icing salts to the extent that late in 1991 a start was made on replacing complete crosshead beams. A 33m x 1.7m x 1.1m beam was the first to be removed and replaced as the entire six lanes of motorway plus hard shoulders were jacked up, the first time this operation had been undertaken in Britain. Further crossheads followed although there was considerable delay when deck movements were found to be four times greater than anticipated.

Eventually the M5 turns away as the canal enters a deep cutting below the original summit, lined with bulrushes. Notable on the left bank were colliery loading chutes, built about 1930. These were fed by a narrow tramway from the Sandwell Park and Jubilee collieries, later replaced by conveyors. While boats were being loaded, boatmen and horses were able to shelter in a brick building, the ruins of which still stand on the opposite bank. The chutes were dismantled in 2006.

A scheduled ancient monument, Summit Bridge of 1791 is a great brick arch with an unusual sloping parapet wall. Concentric arches reduce towards Galton Tunnel, which has a towpath and passes under the A4168 Telford Way. Samuel Galton was a Birmingham Canal Committee member, self-educated in the sciences and the owner of a gun foundry. Galton House was built on the right before the tunnel, its gardens continuing beyond.

On the left side is a marshy bank with a flight of steps leading up to the Night Inn. Tiers on the south side beyond the New Main Line support the railway line, the A457 with some very gaudily painted buildings and then a slim church spire.

Smethwick New Pumping Station of 1892 had two steam engines to pump water from the New Main Line to the Old Main Loop Line to replace water lost by boats locking down at Spon Lane and Smethwick. Built to supersede the Smethwick Engine, it ceased operation in the 1920s although a diesel engine was installed for fire fighting during the Second World War. It is a restored listed building.

Smethwick Brasshouse of 1790, later the District Iron & Steelworks, had its own canal wharf at Brasshouse Lane. The top three locks and the existing summit were discarded by Smeaton in 1790 and the present level used.

Smethwick New Pumping Station. The New Main Line is in the deeper cut which lies to the right.

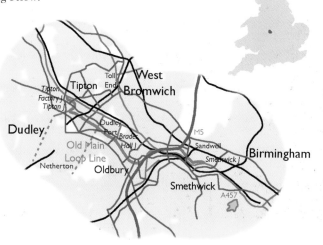

Smethwick Lock, formerly duplicated. The chamber on the left side has been filled in.

The Engine Arm was a feeder to the summit level from Rotton Park Reservoir via the Boulton & Watt engine which operated for 120 years. It leads across Telford's magnificent Engine Arm Aqueduct of 1825, a scheduled ancient monument, past the Galton Valley Centre. The Old Main Loop Line towpath is taken over the Engine Arm Canal on a brick footbridge with indented honeycomb stone quoins.

The three locks down to the lower level are the bottom ones of the original six, all listed buildings. In 1789 Smeaton duplicated these three locks. Brindley's originals were filled in during the 1960s. There was a Toll House between the upper pair. The layout of Pope's Bridge, carrying Bridge Street, shows the alignment of the two lock flights. The Old Navigation to the north of Pope's Bridge is part of the Navigation Hotel, dating from at least 1832.

The two main lines meet at Smethwick Junction, Soho, notable features being two cast-iron footbridges installed in 1828 after being prefabricated at the Horseley Ironworks in Tipton. Once again these are listed buildings. Their semi elliptical shape gives an advantage over segmental curves by allowing greater headroom for horses passing below.

Distance
10km from Tipton Factory Junction to Smethwick Junction

Navigation Authority
British Waterways

Canal Society
Birmingham Canal Navigations Society
www.bcn-society.co.uk

OS 1:50,000 Sheet
139 Birmingham & Wolverhampton

2 Birmingham Canal Navigations, New Main Line

Telford gets straight to business

The major improvement to the Old Main Line came between 1825 and 1838 when Telford engineered the New Main Line between Deepfields and Birmingham. He introduced bold cuttings and embankments, producing extra water space to ease traffic congestion and shortening the route by 11km at the lower 138m Birmingham level.

Possibly because Birmingham is the only major city not located on a large river, it has had to rely on its manmade waterways, having more canals than Venice. The whole canal network spreads out from Birmingham and it is to the Birmingham Canal Navigations that all the loose ends connect. It is, therefore, intensely complex, completely built-up and industrial. Its past and present commercial influence reduces its potential as a cruising waterway, resulting in lighter traffic.

The canal starts from Gas Street Basin at the end of the Worcester & Birmingham Canal in the centre of **Birmingham**. The Tap & Spile and a canal exhibition precede Broad Street Bridge, a vast tunnel with at least two increases in cross section along its short length and buildings on top. Much of the New Main Line has twin footpaths, partly because of the complexity of the system and partly to reduce congestion.

The Pitcher & Piano and the Handmade Burger Company are among the amenities at the much restored Brindleyplace opposite the International Convention Centre and Symphony Hall, a former working environment where business suits are more at home these days.

Much of the blue brickwork on towpath bridges has been renovated to convert the towpaths to attractive canalside walks, now fully surfaced with brickwork.

Farmer's Bridge Junction is at the start of the Birmingham & Fazeley Canal, opposite Sherborne Wharf on the Oozells Street Loop. Three of the four corners of the junction are occupied by the Malt House hostelry, the National Sea Life Centre and the National Indoor Arena.

Buildings by the canal further west were derelict, their windows broken and brick walls breached and broken down. These are being replaced by the amenity of modern residential buildings, positive action that is to be welcomed.

Three loops show where the contour canal used to run before being straightened, Oozells Street Loop, Icknield Port Loop via Rotton Park and Soho Loop which winds its way past Winson Green prison. The Soho Loop rejoins at the first of several toll islands on the canal. By this stage it has already been joined by the West Coast Main Line, remaining in close proximity for most of the distance. Another arrival which is a feature of this part of the canal system is the presence of purple lupins growing wild on the embankments.

Bridges over side arms have low lattice parapets, manufactured in the local Horseley works. Metal products factories are a constant feature of the canal with their attendant odours.

At **Smethwick** the Old Main Line diverges to the right, rising through three locks from the Birmingham Level to the Wolverhampton Level. A feeder, Engine Branch, crosses the New Main Line on Telford Aqueduct. It is a magnificent cast-iron structure, highly decorated, its dark brown paintwork highlighted with red and white detailing. The feeder name came from the Boulton & Watt steam pumping engine which fed the Birmingham Canal Navigations summit level for 120 years.

The New Main Line turns into its boldest cutting at Sandwell. On the north bank is Galton Valley Canal Park. There was formerly a view straight down the 21m deep cutting to Telford's elegant iron Galton Bridge of 1829. A fitting end to the straight cut, it has a 46m span, the world's longest canal span when built, and is 23m high. In 1974 this changed with the construction of a 112m long tunnel next to it with an embankment over the top to carry the A4168. Galton Tunnel has a towpath.

The Broad Street Bridge in a brick panorama.

The International Convention Centre at Brindleyplace.

Farmer's Bridge Junction with the Birmingham & Fazeley Canal leaving to the right under the iron bridge between the Malt House and the National Indoor Arena.

With a railway bridge just beyond, Galton Bridge can no longer be seen at its best from either direction.

Again, a long straight leads under another interesting group of bridges. A brick road bridge has been widened with a concrete arch which provides a striking facade. The Old Main Line crosses to the higher ground on the south on the Steward Aqueduct. The M5 viaduct is supported on uncompromising nodes in the centre of the canal.

A slip road drops through the three Spon Lane Locks to join the New Main Line at Bromford Junction.

Pudding Green Junction leads off northwards to the Walsall Canal in an area of small but bustling works. By now the New Main Line is on a dead straight 4km run through to Tipton. At Albion Junction the Gower Branch, with its deep locks, connects with the Old Main Line.

The New Main Line now runs onto an embankment, and the only extensive views of the route are to be seen in the form of the hills around **Dudley**.

From Dudley Port Junction the Netherton Tunnel Branch runs parallel to the Gower Branch and, in the distance, it can be seen passing under the Old Main Line and up to the mouth of Netherton Tunnel, the largest cross section tunnel in the country.

Residential properties close in on the south side of the canal, and a canal cottage sits on top of the embankment near a couple of aqueducts. The Ryland Aqueduct of 1968 clears the A461 in a single 24m concrete span.

Two branches which have been lost from the north side here were Dixon's Branch and the Toll End Communication, which linked with the Walsall Canal.

Beyond the Noah's Ark the three Factory Locks bring the New Main Line up to rejoin the Old Main Line at Factory Junction. Noteworthy are a split

The New Main Line crosses the picture while the Oozells Steet Loop is ahead with Sherborne Wharf next to a National Sea Life Centre.

bridge over the bottom lock, a boatman's chapel now converted into a factory, a large warehouse and a Birmingham Canal Navigations cast-iron boundary post. The Factory Bridge of 1825 has now been removed to the Black Country Museum in Dudley.

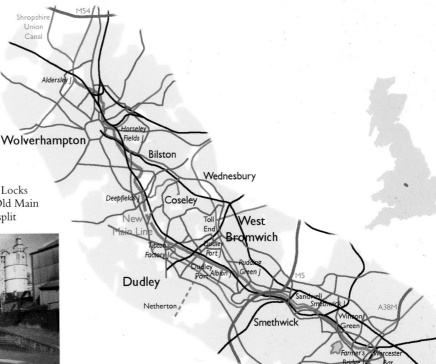

Smethwick Junction with the Old Main Loop Line leaving to the right.

Telford Aqueduct carries the Engine Arm over.

Country. These days the area seems to be used for storing plant. It is not all construction plant, though. Water plants are joined by increasing amounts of algae. Yellow iris is also present along the banks.

The canal approaches a gas holder in a large loop before returning to a straighter line.

As the canal edges into **Wolverhampton**, capital of the Black Country, the atmosphere deteriorates. Factory owners have given up the battle to keep up with vandals and hardly a pane of glass is intact. One works has rags and plastic bags pushed through the holes to lessen the

A striking widened road bridge, the Steward Aqueduct, carrying the Old Main Line and the M5 bridge.

Bloomfield Junction formerly took the Ocker Hill Branch away to connect with the head of the Tame Valley Canal.

The 329m **Coseley** Tunnel has twin towpaths with visible white handrails.

The Boat is just before Deepfields Junction and the Wednesbury Oak Loop, a remainder of the contour canal. A nearby tip has been landscaped, but factories lie in various stages of dereliction. **Bilston** Steelworks was, until 1981, the last surviving blast furnace in the Black

airflow. In quick succession come factories of corrugated-iron as often as of brick. Outlet pipes discharge foul looking liquids. Horseley Fields Junction is the end of the Wyrley & Essington Canal.

A large tunnel with twin towpaths supports a multi-storey carpark on top. Beyond it is a British Waterways depot with barge entry doors in the side, once again in Broad Street. The basin beyond it is the jewel of the canal, an attractively laid out garden area with benches and narrowboats moored at the top of the 21 lock Wolverhampton Flight. This is the proposed site of the major Wolverhampton Interchange Project with office and residential buildings. The locks carry on for over 2km right down to Aldersley Junction, with never more than 300m between locks.

The flight begins beside a traditional lock keeper's cottage. M&B's Springfield brewery, a municipal refuse disposal plant, a landscaped former railway line and a network of existing railways are other features. The West Coast Main Line leaves on a viaduct which crosses over a railway bridge as it, in turn, is crossing the canal. The flight drops under another viaduct, this time taking the Shrewsbury to Wolverhampton line. Beyond the viaduct the transformation is dramatic. The city is left behind. Wolverhampton's horse race course is on the left and beyond it are only trees and a quiet towpath down to Aldersley Junction, for the pleasant appearance of which Friends of the Earth claimed the credit. Although relatively rural in appearance, the Staffordshire & Worcestershire Canal has been extremely busy here in the past, being only 900m from Autherley Junction at the head of the Shropshire Union Canal.

Distance
24km from Gas Street Basin to Aldersley Junction

Navigation Authority
British Waterways

Canal Society
Birmingham Canal Navigations Society
www.bcn-society.co.uk

OS 1:50,000 Sheet
139 Birmingham & Wolverhampton

British Waterways boat in Coseley Tunnel.

3 Wyrley & Essington Canal

The main line with the Wyrley & Essington leaving left at Horseley Fields Junction.

The Wyrley & Essington Canal connected the Birmingham Canal Navigations' Main Line at **Wolverhampton** with the Coventry Canal at Huddlesford, forming the most northerly loop of the Birmingham Canal Navigations and today displaying its most attractive scenery. Almost all of the remaining part is on one level. The canal follows a tortuous line around the contours with all the locks positioned on the numerous branches.

The line was constructed to Wyrley in 1795 and extended to Brownhills.

It leaves the Birmingham Canal Navigations Main Line at Horseley Fields Junction. Light traffic on it means that it is remarkably clear except for the water-weeds and a significant amount of flotsam.

The West Coast Main Line crosses immediately and then the canal goes back a couple of centuries as it passes a constriction caused by what was once a toll island for coal barges in the centre of the canal. Tunnels under the towpath carry former spurs to serve adjacent factories with a distinctly derelict air. The contrasts come spasmodically, a few cottages backing onto the canal, tower blocks, a plant yard and the blackened spire at Heath Town.

Near the Jolly Collier, a half-dismantled railway bridge has an outer beam truncated at mid span over the canal.

After the Bull's Head, wasteland opens up on the right but the left bank is in good order with playing fields and grassed areas, a modern

church with striking green roofs and a hospital. The old brick bridge leading up to the hospital has been repaired very obviously and carries a nameplate, as do the other bridges on this canal.

Beyond it, a substantial lattice bridge, of a pattern to be seen again later, carries the towpath at Wednesfield Junction across the Bentley Canal. This was abandoned as a through route to Bentley in 1961 with the buildings of **Willenhall** having spread over it in recent years. **Wednesfield** Junction also had an island with one of the octagonal tollbooths.

A red sandstone towered church lends dignity to an area where the Boat, Royal Tiger and Spread Eagle public houses flank the canal, together with a school. This seems to be a popular exercise area for dogs.

Houses have their gardens backing onto the left bank, often managing to make special features out of their canalside locations. A modern one has shutters, bull's eye windows and great attention to detail, while another has a large aviary with masses of ironwork making it unapproachable and invisible from the road. Attempts to improve the environment at New Invention include a picnic table on the bank opposite the Boat Tavern. At Short Heath, the only wooded area on the whole canal is enfolded in a large sweep as the route turns northwards.

Remains of the toll island.

Bridges carry the towpath over factory feeders near Horseley Fields Junction.

The canal sweeps round near Birchalls Junction.

Gradually, the roar of traffic becomes audible as the M6, here one of the busiest pieces of road in the country, pulls alongside and crosses over.

The wasteland-scape throws up car scrapyards, tethered ponies and compounds of geese, all looking slightly forlorn. In this no man's land the Sneyd maintenance depot forms an oasis of interest, the well cared for traditional brick buildings being a delight to discover. Next to them, the Sneyd & Wyrley Bank Branch climbs up from Sneyd Junction towards Great Wyrley.

The main line doubles back on itself to cut down the other side of the valley, passing a range of industrial buildings which stand back from the canal so that they are imposing in their entirety rather than dominating by proximity.

Birchills Junction takes the **Walsall** Branch Canal away as the main line turns north, once again. Ill feeling and rivalry prevented the connection from being made until 1840 when the Wyrley & Essington Canal and Birmingham Canal Navigations were amalgamated and came under railway control. Unusually, the railway developed traffic on this canal, building interchange basins. Consequently, the Wyrley & Essington saw some of the last commercial traffic on the Birmingham Canal Navigations. The end came suddenly in 1966, when coal transport ceased.

Moving away past playing fields and a church with a strange little square green spire, the canal again returns to residential area. It passes clumps of watercress which indicate that the water is fairly clean. The concrete footbridge has been hit, nearly shearing it in two and requiring it to be heavily braced.

Harden is the only place where the canal comes up noticeably above the surrounding land, here a large grassed field. Houses on the left include a couple with large complexes of homing pigeon lofts, that peculiarly English hobby.

Cat-proof pigeon lofts at Harden.

Rural conditions at Pelsall.

Cast-iron bridge over the canal at Ogley Junction with Brownhills behind.

The higher level section slips past tree-lined hospital grounds, moves round a fortified works and scrapyard and then clears the housing of Little Bloxwich between the Barley Mow and the Bridgewater. It then breaks out into open country, the most attractive stretch of the whole canal, between Little Bloxwich and Pelsall. Having come down from Wolverhampton, it is immensely satisfying to pause by the farm at Fishley, the late afternoon sun shining over one shoulder, and listen to just the distant rumble of traffic and the closer cawing of rooks, the only sounds to break the silence.

The Lord Hay's Branch to Newtown has now been lost but over 2km of the Cannock Extension can be seen running away from Pelsall Junction in a dead straight line. The importance of the canal hinged on the branches serving the Cannock and **Brownhills** coalfields when the Black Country pits declined. This branch, which was vital to the exploitation of the Cannock field, was completed in 1858. The northern end was abandoned in 1963.

After the Royal Oak and an estate of recent houses alongside the canal, the tortuous nature of the canal is emphasised, the bends now being replaced with corners, tight for narrowboats. Various spurs remain as pointers to former industrial activity but now there is little to show, just a car scrapyard, a pylon bearing floodlights to illuminate a concrete batching plant and a storage area for portable cabins.

The Swan stands away behind a house with a prominent barrel-shaped dovecote alongside the canal.

Brownhills is dominated by several brown tower blocks. Next to the canal are several supermarkets, together with an outdoor market. A hip-roofed church is just one more style to contrast with the many others along this canal.

Catshill Junction brings in the Hay Head or Daw End Branch, the most important of all as it linked up with the Rushall Canal to provide access to the industry of Birmingham.

As the Wyrley & Essington loops round Brownhills, it passes the Anchor and then reaches open country on the right, with extensive views in the direction of Shenstone. On the left, however, the houses continue and the local youth population is evident, partaking in activities including rafting on piles of foam sheets.

Beyond the houses, the square brick chimney of a casting works at Ogley Junction is a prominent landmark. Here, the Anglesey Branch joins, bringing water supplies down from Chasewater. A lattice bridge takes the towpath across and the Wyrley & Essington turns right into a basin which is now the effective terminus of the canal. The canal was extended to Huddlesford in 1797 via the 30 locks of the Ogley flight but this was abandoned in 1954 and now only a garden centre stands on what was once a significant engineering feature.

The canal has lost 10km from here. The route formerly ran to the north of the prominent church spire at Wall and past the more famous spires of **Lichfield**. It is hoped to restore this as the Lichfield Canal.

Now just 500m remains at the far end, the south-west end of which is a marina with entry barred although a public bridleway crosses the stump near its southwest end. The Wyrley & Essington Canal joins the Coventry Canal by the Plough at Huddlesford, a small community cut in two by the railway.

Foot of the Anglesey Branch.

Distance
34km from Horseley Fields Junction to Huddlesford Junction

Navigation Authority
British Waterways

Canal Societies
Birmingham Canal Navigations Society www.bcn-society.co.uk, Lichfield & Hatherton Canals Restoration Trust www.lhcrt.org.uk

OS 1:50,000 Sheets
128 Derby & Burton upon Trent
139 Birmingham & Wolverhampton

4 Walsall Canal

Spare a copper

The Walsall Canal links the Wednesbury Old Canal at **West Bromwich** with Walsall in the West Midlands. Construction began in 1786 although the centre of Walsall was not reached until 1799, after this relatively short and straightforward canal had curved its way round Wednesbury.

The canal has eight locks, all of which come immediately as the Ryders Green Flight. At the southern end of the canal it leaves the **Wednesbury** Old Canal at Ryders Green Junction and passes the Eight Locks public house by the top lock.

This canal is more free of weeds than are some others. The marshy area to the right by the bottom lock is predominantly greater reedmace but toadflax gets a grip, with its pale yellow flowers, where dry ground is available. Opposite is the point where the Haines Branch entered.

A railway bridge crosses at Toll End, carrying the Midland Metro light rail rapid transit system. The canal is a route of modern and former transport links. The junction with the former Danks Branch can be seen. Soon after, the Lower Ocker Hill Branch, or the remaining 300m of it, leaves on the left next to tennis courts and other sports facilities.

The Tame Valley Canal leaves on the right between a couple of the lattice cast-iron bridges which are among the attractive features of the Black Country canals. After it is the Ocker Hill power station with two prominent concrete chimneys.

A canal cross exists at the point where the former Gospel Oak Branch joins. Presumably this must have given some logistical problems with the movement of horses and towlines at busy times.

Another railway line crosses and this was proposed as a line of the Midland Metro. The Monway Branch has disappeared under the earth but the banks of the Walsall Canal itself remain untouched and lupins and orchids may be found. The water is ochre-coloured but remains relatively clear because it is only lightly used and, like sections with other less than inviting colours of water, minnows can be seen swimming about in significant numbers.

Powerlines now follow the canal all the way to Darlaston Green. From Bilston to Darlaston Green the Black Country Route flanks the canal.

The Bradley Branch Canal, now heavily overgrown with reeds, leaves to the left at Moorcroft Junction. The reeds provide a haven for moorhens, here and elsewhere along the canal.

The Moxley Stop was near the red sandstone spire of the church in Moxley. Today, a children's playground is more prominent.

As the canal moves on past the lines of the former **Bilston** and Willenhall Branches, the countryside is one

Unusual cemetery entrance off the towpath.

Sikh temple overlooking the canal at Pleck.

The Walsall Branch Canal leaves via Walsall Locks beyond Walsall Junction.

of derelict fields, occasionally occupied by horses. A notable feature of the Black Country canals is that the people always seem friendly whereas the usual experience elsewhere is that country people are more sociable. Black Country friendliness is experienced on this canal and must be counted as one of its assets.

After a school and then Bug Hole Wharf in the vicinity of an electricity substation and more schools, the canal runs through a section with factories but also with newer canalside houses which, from their leaded windows, appear to be aimed at the higher end of the market.

The powerlines leave at **Darlaston** Green Wharf and the canal curves round to cross an aqueduct over the railway, immediately before the weed-choked remains of the Anson Branch. Over the next reach the canal is on low embankment with a cemetery to the south around a church with a small spire. A little further on are two very large gas holders next to the M6 but it is the busy motorway itself which is now prominent. Aqueducts take the canal over a minor road and the fledgling River Tame, protected by a stop gate, before the canal passes under the M6 immediately south of Junction 10.

In this vicinity, the whole area is riddled with old coal mines. Heavy metal pollution from former copper refining on the derelict site to the left of the canal has been a major problem. Drainage from polluted ground went into nearby headwaters of the River Tame. The former factory here was one of two which were said to account for 18 per cent of the copper and 17 per cent of the nickel in the water at the Tame/Trent confluence.

The canal passed moves on into Pleck where a Sikh temple stands next to the canal.

After a public house by a bridge the canal passes a former canal wharf building with the hoist points and doors at various levels, still obvious above the water. High fences and industrial premises front another cemetery, this one with an interesting circular entrance gateway.

After some recent housing by the canal, the **Walsall** Branch Canal leaves up Walsall Locks from Walsall Junction. This important link with the Wyrley & Essington Canal was not made until 1841 because of company rivalry but now cuts out a significant detour via Wolverhampton for boats going north.

Walsall Public Wharf is now a grassed area where a bar has waterside seating facing a canal terminus building which has been restored. High blocks and glass walls rise behind. Walsall has changed much since it took its name from the Old English for Walh's valley, 'Walh' meaning a Welshman.

Distance
11km from Wednesbury Old Canal to Walsall

Navigation Authority
British Waterways

Canal Society
Birmingham Canal Navigations Society www.bcn-society.co.uk

OS 1:50,000 Sheet
139 Birmingham & Wolverhampton

The restored canal terminus building at Walsall.

5 Tame Valley Canal

Bypassing Birmingham's congestion

The Tame Valley Canal was one of the last canals to be built, running eastwards across the West Midlands from Tame Valley Junction, on the Walsall Canal, to Salford Junction, on the Birmingham & Fazeley Canal, which is at the head of the Grand Union Canal. Although planned earlier, it was not built until 1844, by which time the Farmer's Bridge to Salford Junction section of the Birmingham & Fazeley Canal was hopelessly congested with 24 hour a day, 7 day a week working. The Tame Valley Canal was constructed to bypass this section and was very sophisticated with long, bold, straight sections that made use of high embankments and deep, wide cuttings, twin towpaths and brick-lined banks.

Other than tennis courts and other sports facilities beyond the Lower Ocker Hill Branch, the Tame Valley Junction area is inauspicious. Lattice-arched footbridges are an attractive feature of the Black Country canals.

The first reach of the canal is crossed by the Black Country Spine Road.

Just before Golds Hill Wharf the canal is crossed by a railway bridge with a pier on a central island in the canal. This route was earmarked for a possible line of the Midland Metro light rail transit system. The following bridge carries the track of the first line.

This reach has factories which smell of cellulose paint and sawdust. Fans vent across the towpath. Long plastic drainpipes project through walls and attempt to shoot their contents beyond the towpath. There is the usual graffiti but there are also plenty of flowers to brighten up things. Birdsfoot trefoil accompanies blackberries and there are no less than three colours of clover – the usual white, the less common pink and a much less frequently seen deep magenta variety – adding a splash of colour in summer. Algae and other weed in the canal shows it is not used too often. Some patches grow right across with lengths of wood and other floating debris among them.

A bridge arch carries the towpath over the entrance to Holloway Bank Wharf after a slight double bend. This comes before a 3km straight which is closely followed by powerlines. Houses on the outskirts of West Bromwich are close on the right beyond long grass. More wild vegetation on the left stands at the top of a high embankment with a stream, pond, college and school at the foot. The land gradually rises again to Church Hill at **Wednesbury**. A short aqueduct carries the canal over a minor road before the canal runs into a deep cutting. This ends just before the large A4031 bridge. The canal emerges to cross an aqueduct over a minor road, with a public house below on the left, and then crosses another aqueduct over the railway.

An island divides the canal as it prepares to turn onto its most dramatic section, the kilometre long embankment which runs high across the Tame Valley. It is now accompanied by the Midland Links motorway, the triangle where the end of the M5 meets the M6, a unique canal environment. The canal first crosses the western arm of the M5 on a concrete aqueduct. The height of the embankment can be judged by looking at the elevated section of the M6. This runs beside the canal but can be seen moving away to the north-west on spindly columns. A stop gate provides protection as the canal approaches the aqueduct that is to carry it across the River Tame. It then passes an electricity substation.

At the far side of the valley the motorway climbs away up to **Great Barr**. The canal divides round another island as it approaches Rushall Junction, where it was joined by the Rushall Canal to win the coal trade from

The canal is carried over the western spur of the M5 motorway on a modern aqueduct.

the Cannock mines. This is now a popular cruising link. The water quality in the Tame Valley Canal improves markedly from here with the only visible wild plants being small clumps of watercress along the edges.

The canal passes under the eastern spur of the M5 and then leaves the noise and traffic behind. It dives into a deep wooded cutting from where there is little evidence that the canal is passing between the housing estates of Grove Vale, the atmosphere being so pleasantly rural. Two bridges pass high across the cutting. The first carries the dual carriageway A4041 with a street light balanced precariously above the centre of the canal. The other is a slim footbridge carried on the piers of what looks like a former railway viaduct.

The kilometre long cutting emerges onto a longer embankment, with extensive views over the Tame valley and back towards **West Bromwich**. After another island, one side blocked off by Hamstead Wharf, there are two aqueducts. Housing estates lie below the canal on both sides although there is a deceptive amount of greenery about.

Passing TS Leopard with its anti-aircraft gun, the canal plunges into another deep cutting. There are frequent outcrops of red sandstone. These are sufficiently soft for them to have needed to be reinforced with sections of red brickwork in many places. Meanwhile, the canal is edged with black bricks laid with black mortar which provides a very smart finish to the canal.

Roses between thorns: the lock cottage at Witton surrounded by industrial premises and debris.

The first few of the many bridges that make up the Gravelly Hill Interchange complex through which the canal passes.

This is said to be one of the best angling waters in the Midlands.

The smartness is also present in the Perry Barr First Flight of locks, which were refurbished with attractive brickwork. Barr Top Lock Wharf lies beyond the A34 bridge. The first seven locks come as a 700m flight, on the way passing the **Birmingham** Alexander Stadium in Perry Park. It has an artificial football pitch, running track and full athletics facilities in a large, modern site.

The M6 crosses on the Thornbridge Viaduct, a viaduct which was extensively repaired using a very powerful water lance to cut out deteriorated concrete, a German tool being used in Britain for the first time. Four more locks follow, forming Perry Barr Second Flight. Lock 11 is set between a container terminal and a sports ground lined with poplar trees, which shed their downy white seed onto the canal profusely in the autumn. The area around Perry Barr Wharf has several industrial areas. Among these are two significantly older churches at Upper Witton, the first near the canal, with a rather short spire, and the other set amidst a cemetery, although there is little peace here these days as the M6 crosses back over.

The canal has moved into a heavily industrial area for the final two locks, Perry Barr Third Flight. Beside the second of these is an old lock cottage covered with roses,

defiantly holding on in its surroundings of chimneys, water towers, barbed wire, corrugated iron and half-dismantled vehicles. All is not derelict, however. As well as the lock cottage, there are various other places where the towpath can offer a colourful assortment of vetch, convolvulus and purple lupins, the latter seeming to thrive in the vicinity of Salford Junction.

The junction lies at the far end of Junction 6 of the M6 – Gravelly Hill Interchange, better known as Spaghetti Junction – the most complicated motorway interchange in Britain, offering freeflow routes between the M6, A38M and A5127 and also linking various local roads. Most of this happens in three dimensions and is best seen from this canal as roads snake through the air in all directions. It begins gently at first, a pipe bridge, a lattice canal footbridge and an arched cable bridge. Then it develops in earnest as slip roads flail through the air, the M6 crosses, the Birmingham to Lichfield railway passes through and ever more slip roads arc over, until the canal passes into a large box tunnel with two tow-paths below the A38M Aston Expressway. It emerges into the relative simplicity of Salford Junction below the M6. The interchange deteriorated badly after its completion and repair work involved jacking up the deck of sections of slip road still in use to extract the decaying crosshead beams for replacement.

Distance
14km from Tame Valley Junction to Salford Junction

Navigation Authority
British Waterways

Canal Society
Birmingham Canal Navigations Society
www.bcn-society.co.uk

OS 1:50,000 Sheet
139 Birmingham & Wolverhampton

The Tame Valley Canal finishes at Salford Junction, meeting the Birmingham & Fazeley Canal which crosses the picture beneath the M6 motorway.

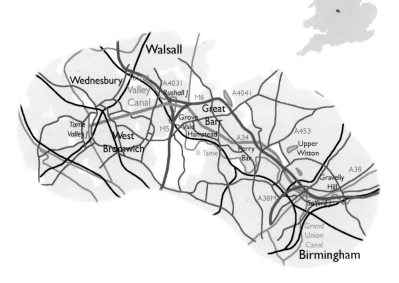

6 Birmingham & Fazeley Canal

Birmingham's 'Bottom Road'

Decorative ironwork beneath Snow Hill station.

The Birmingham & Fazeley Canal was designed as a connector linking a number of other canals despite opposition from the rival Birmingham Canal Company, which later merged its interests.

The Birmingham & Fazeley leaves the New Main Line of the **Birmingham** Canal Navigations at Farmer's Bridge Junction. This was one of the most overcrowded points on the industrial canal network and, for this reason, lights were installed on the Farmer's Bridge lock flight which went over to 24-hour working, rare in Britain.

The canal immediately passes Cambrian Wharf, a set of 18th-century cottages and a toll office in front of the DVLA building. The Firkin Brewery replaces former wharves and has one of the two wharf cranes incorporated into its outside seating area.

Bridges in Birmingham generally have red doors in parapets to allow firemen to drop hoses through to the water source. This idea has been mirrored on a couple of recent bridges with railings, the railings having opening panels also painted red.

The Farmer's Bridge Flight, the Birmingham 13, is filled with interest for anyone with a liking for industrial and canal architecture. The top lock is one of a couple with tow rope guide pins. The brick towpath is ribbed to allow horses to grip. Side ponds serve wharves, some of which clank, hiss, whirr or echo to the sound of radios as businesses are carried on inside the adjacent factories. Overflow weirs are of the bellmouth type, most discharging into side ponds below. Strenuous efforts have been made to produce an amenity out of the canal corridor. Walkways make a feature of the canal and one promenade has been built upon a series of new brick arches. These extend an existing network of arches in two planes over a side pond. They are further complemented by a series of arches within arches, of different heights and radii, carrying a road over a lock. Owing to the lack of space, dividing barriers have been erected between the canal and the towpath below some of the locks. At lock 13, on the edge of the Jewellery Quarter, is a five-storey mural on a Mowlem office wall. It is best seen from the offside of the lock or less well from the Livery Street bridge, buildings obscuring other views to it.

All around are tower blocks. The dominant feature, however, is Birmingham's BT Tower, which stands

The Birmingham & Fazeley Canal leaves Farmer's Bridge Junction.

Modern design on a new bridge over the canal.

Farmer's Bridge Flight falls away from Cambrian Wharf.

A subterranean lock – not for the claustrophobic.

The Farmer's Bridge flight canyon. A dividing rail gives safety.

beside the canal. The concrete legs of the accompanying tower block stand in a lockside pond and all around the lock. It is a curious juxtaposition – the intricate detail of the locks contrasts with the grey slabs of concrete, the former well worn and mellow, the latter becoming chipped by passing boats.

In contrast, the huge brick cavern under Snow Hill station stands stark and empty but for some decorative ironwork. At one time the canal was heavily lined with dark brick walls, hemming it in and giving it the local name of the Bottom Road. These walls are now going, opening the towpath up as a pleasant place to walk. The local people are noticeably friendly.

Just beyond the Aston Expressway, Aston Junction allows the Digbeth Branch to peel off to the right to meet up with the Grand Union Canal. Aston Junction has been renovated in the best canal tradition. Dereliction is then greater and there are numerous bricked-up side arms that once served thriving industry. HP Sauce moved to Belgium in 2007 after over a century of production. Most factories remain anonymous. The flight is not without its interesting features, too, such as a recessed alcove in the brickwork of one bridge, the radius just large enough to take a lock balance beam and the height just enough for a man to stand.

Approaching Gravelly Hill Interchange of the M6, plant life begins to increase. Strangely, the most common plant encountered is purple lupin which adds a welcome splash of colour. Blackberry and elder bushes and reeds also help to bring the canal to life.

One end of the Mowlem mural beside lock 13.

Lupins have a foothold on the aqueduct over the River Tame, too. The canal now closely follows the river for most of its length.

Salford Junction, in the middle of Gravelly Hill Interchange, connects the Birmingham & Fazeley with the Tame Valley Canal and the northern end of Britain's longest canal, the Grand Union. As well as the usual scrapyards, timberyard and old industrial premises, there are some newer factory units.

Although the locks are narrow with single top and bottom gates, the canal itself is usually quite wide, no more so than where one factory completely

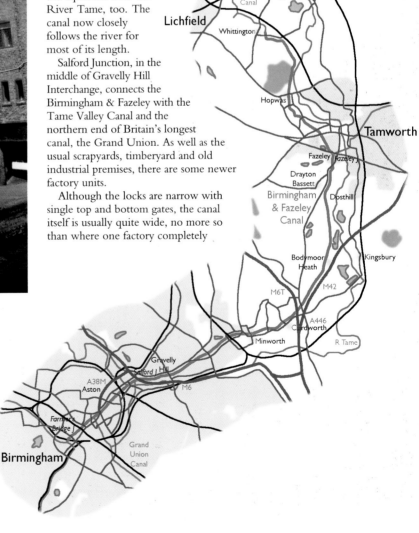

The great cavern under Snow Hill station.

Salford Junction, now below Gravelly Hill interchange of the M6. The Grand Union Canal begins under a traditional bridge.

spans the canal and towpath for 200m. Light enters from the side so it does not seem claustrophobic.

Before Fort Dunlop, one pylon of the line on the right is cylindrical, tapered and silver, more typical of Continental designs than of British. As aircraft pass overhead, landing at Birmingham Airport, the canal passes a storage yard for new cars. There is also a Boots warehouse. Tower blocks rise on the right as the canal reaches the first of the Minworth locks. The transport theme is revived here by two giant cycle wheels outside a factory building. Painted light blue, they appear more like colliery winding wheels.

Minworth top lock burns or skips away a continual stream of flotsam. From here the water is covered by chopped weed rather than timber and plastic.

A factory which showed how to make a feature of the canal was the Cincinnati works with lawns leading to the top of a bank covered with a thick growth of ivy. The site was acquired for housing in 2007. Beyond, the rusty fence style is revived.

Daylight is visible through the short Curdworth tunnel.

Minworth, too, has an industrial park, one shed even being constructed out of concrete pipes. The cooling towers of the Hams Hall power station can be seen in the distance. Nearer, the giant Minworth sewage works are not seen but a southerly wind makes their presence known.

Unusually, the bridges on the Birmingham & Fazeley are named rather than numbered. By now they are becoming the more traditional rounded arch type. One between the Boat and a steak bar displays a peculiarity which is to be seen regularly now, a small arched hole leading back into one side of the bridge. This was for stop-logs, to be dropped into the canal in the case of a breach, although there are no logs in place now.

New housing estates front the canal before it passes a scrapyard with a specialism in military vehicles.

Now the canal breaks into open country for the rest of its route, with views over farmland to the left. The square tower of the squat St Nicholas' church at Curdworth, with its finely carved font, is seen over the hedgerow. Yellow flags become a regular feature, complemented later by the pink flowers of arrowhead.

Curdworth tunnel is only 52m long but the trees that overhang the cutting approaching it add to its gloom. The towpath goes through; it carried notices asking boats to keep right, a fanciful idea as only the smallest boats would have much choice in the matter.

The A446 bridge just after the first lock of the 11 lock Curdworth flight shows three different sizes of arch. The bridge has been progressively widened to take the present dual carriageway. As Bodymoor Heath is approached, the M42 moves alongside the canal before turning away through Kingsbury Water Park, a set of former gravel pits.

The tank farm at Kingsbury's oil storage depot is seen with higher land rising behind for the first time.

The unique bridge crossing the canal at Drayton Bassett.

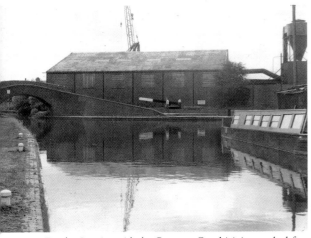

Fazeley Junction with the Coventry Canal joining on the left and the lock gate seat.

Two locks beyond the Dog & Doublet, a swing bridge precedes the 38th and final lock on the canal. The following swing bridge at Drayton Bassett is accompanied by a fixed footbridge quite unique to the British canal system. At each end is a Gothic arch doorway in a castellated brick tower with spiral staircase. Dosthill House is seen away to the right but it is probably the proximity of Drayton Manor, unseen on the left, which

Peaceful willow-fringed meadows in the Tame valley near Tamworth.

has more to do with the design. Formerly the home of Sir Robert Peel – who invented the policeman – the manor now has a zoo, golf course, cafeteria, ballroom and other features designed to pack in the masses.

Passing hosiery mills, the canal arrives at Fazeley Junction. This should have been the end of the canal. John Smeaton completed it in 1789 but it had not been started until the owners of the Oxford, Coventry and Trent & Mersey Canals had all agreed to complete their own projects. Unfortunately, the Coventry Canal ran out of funds, even though it was to become one of the most profitable canals. The Trent & Mersey company built down as far as Whittington Brook, this section being purchased by the Coventry Canal, but the intervening section was completed by the Birmingham & Fazeley, this section following the contours for economy, rather than being straight to give speed.

Fazeley Junction has the top of a lock gate sited as an unusual seat, backed by a large crane with a mountain of tree trunks in a timberyard and by a double-fronted brick canal cottage. Beyond the Three Tuns and a canal shop, a large and venerable tree was rather larger until part of it dropped across the canal in a 1984 storm.

The line of the canal onwards from Fazeley Junction was the setting for the ghost story *Three Miles Up*.

The canal winds away along the left bank of the Tame valley, past willow-fringed meadows of grazing cows, heading towards the mast above Hopwas. In this village the canal is flanked by the Red Lion and the Chequers on opposite banks.

Hopwas Hays Wood carries notices that it is near firing ranges and the public are warned to keep out while red flags or lights are displayed. At this point the canal comes right alongside the river for a while, although it is not seen at the foot of the hillside. In turn, the river is replaced by the railway as a neighbour.

An IWA plaque indicates the change of ownership at Whittington Brook.

Distance
33km from Farmer's Bridge Junction to Whittington Brook

Navigation Authority
British Waterways

Canal Society
Birmingham Canal Navigations Society
www.bcn-society.co.uk

OS 1:50,000 Sheets
128 Derby & Burton upon Trent
139 Birmingham & Wolverhampton

7 Coventry Canal

The canal that wasn't built fast enough

The Coventry Canal is an early contour canal that begins in the West Midlands and heads north-west across Warwickshire to the east of a ridge of higher land, into Staffordshire. It was promoted by Bedworth mine owners who planned to take Bedworth coal to the Trent & Mersey and Oxford Canals. It was started in 1768 and boats brought out coal from 1769. It took a further 22 years to build the rest, during which time engineer James Brindley was sacked for spending too much of his time on other canal projects although his statue stands in the basin at Coventry. Once built, it was one of the most prosperous canals, originally carrying heavy coal traffic from the Warwickshire coalfield but also making much profit from selling water. It was still making a profit when nationalised in 1947, although narrowboat traffic ceased after the Second World War.

As far as Lichfield, it is followed by the railway, initially the Coventry to Nuneaton link and then the West Coast Main Line.

Coventry takes its name from Cofa's tree, Cofa being a tribal leader of about 600–650. Coventry's most famous resident was Lady Godiva, who rode naked through the city as part of a challenge by her husband, Loefric, Earl of Mercia, to lower taxes. This resulted in the repeal of the Heregeld tax in 1051. While there is a statue of her in the Broadgate, there are no fewer than three statues of the original Peeping Tom, the only person to look, an interesting reflection on priorities.

The 14th-century St John's church was used as a prison when Cromwell defeated the Scots, hence the expression about being sent to Coventry. The church of St Michael was formerly a priory, the bishop's seat from 1095–1129, a parish church from 1373 to about 1450 and a cathedral from 1918. Although the 90m steeple still stands, the rest was reduced to ruins in the 1940 Blitz which destroyed so much of Coventry. The city centre inside the ring road has been largely rebuilt since the war. The new cathedral was designed by Sir Basil Spence, completed in 1962, and has a nave with Graham Sutherland's 23m tapestry, *Christ in Glory*, John Piper's stained glass, Epstein's bronzes and a font from a boulder found near Bethlehem.

The Museum of British Road Transport houses 150 cars and commercial vehicles, 75 motorcycles and 200 bicycles, spanning over 180 years of history, including the first safety cycle, which was built in Coventry in 1885.

The Bishop Street Basin of 1769 has two parallel arms, the western one being obstructed by a low swing bridge. The whole complex has been completely restored and is a well signposted asset to the city. At the entrance, opposite the Admiral Codrington, is the original triangular weighbridge office, possibly from 1787. It is now a canal society information point. The slate-roofed warehouses on the north-west side, which contained grain, sugar and cement between 1787 and 1914, were restored in 1986 and converted to workshops and art studios. On the east side are wooden awnings perforated with dovecote entrances in what were coal wharves. These were built below water level to allow coal to be unloaded easily into carts. On the north side were 18m long arched coal vaults. The Canal House was re-sited in 1852 and housed successive canal managers.

Draper's Field Bridge has a small arch and no towpath. This unusual design was to allow the basin to be kept free of unwanted intruders at night.

The canal bends round to the left to face the Daimler

The old weighbridge office at Coventry.

The former coal wharves complete with dovecote holes in Bishop Street Basin in Coventry.

The northern end of Bishop Street Basin with the anti-intruder Draper's Field Bridge.

Power House. Originally it was a pair of three-storey Coventry Cotton Factory mills until a fire in 1891. Henry Lawson purchased them for the Daimler Motor Company in 1896 and they became the Coventry Motor Mills, producing the first British car, the 2 cylinder Daimler, which had tube ignition, solid tyres and tiller steering. Second World War bombing rather reduced the size of the operation.

A striking footbridge suspended from a diagonal arch marks Electric Wharf.

After Cash's Lane Bridge are a bank of impressive houses. These are Cash's Hundred Houses, although only 48 were actually built, of which 37 remain. They were owned by Joseph Cash of woven nametapes fame, clearly a forward thinking man. Built in 1857, workers' families lived on the first two floors and work was undertaken on the top floor which features large windows. A lineshaft ran the length of the building and provided power for 70 years to assist individual workers to be competitive with factory looms.

The elegance declines with Courtald's Pridmore Road artificial fibre factory, set up before the First World War. Here reedbeds develop in the canal. After Prince William Henry's Bridge there is a timber yard and then a scrapyard.

Hawkesbury Junction with the Coventry Canal on the left and the Oxford Canal on the right. The building with the high chimney was a pumping station to feed the canal.

wharf, with canal cottages in Hollybush Lane. From here the two canals ran side-by-side for nearly 2km to Hawkesbury until the dispute was finally settled in 1836 and a more sensible attitude was applied.

The canal passes under powerlines and the M6, from the West Midlands to Warwickshire and a mineworking area.

Hawkesbury Junction was built in 1802. It is also known as the Sutton Stop after the first lock keeper, whose family operated the lock for over half the 19th Century. There is a 150mm rise stop lock where the Oxford Canal enters. A southward pointing spur shows the original proximity of the Oxford to the Coventry down to Longford. Across the junction is a fine 1827 cast-iron roving bridge with X-lattice sides, its

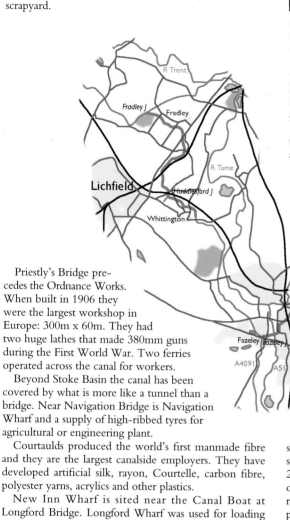

Priestly's Bridge precedes the Ordnance Works. When built in 1906 they were the largest workshop in Europe: 300m x 60m. They had two huge lathes that made 380mm guns during the First World War. Two ferries operated across the canal for workers.

Beyond Stoke Basin the canal has been covered by what is more like a tunnel than a bridge. Near Navigation Bridge is Navigation Wharf and a supply of high-ribbed tyres for agricultural or engineering plant.

Courtaulds produced the world's first manmade fibre and they are the largest canalside employers. They have developed artificial silk, rayon, Courtelle, carbon fibre, polyester yarns, acrylics and other plastics.

New Inn Wharf is sited near the Canal Boat at Longford Bridge. Longford Wharf was used for loading coal. Between the two are some landmark gas holders.

Signs along the towpath from the Coventry basin indicate that it forms the Oxford Canal Walk. Longford Junction is where the Oxford Canal originally joined the Coventry Canal, the woodyard being the Oxford Canal

semi-elliptical arch spanning 18m. Its 2.3m width is more of a problem for narrowboats than for horses, providing a block on sight for them, most of which have to make a tight 180° turn. Between the Second World War and 1970 this was the site of a Salvation Army mission for boat people at one of the busiest points on the canal network. It was also the site of Sephton's boatyard. The Greyhound is still popular. Now this is a landscaped conservation area and includes the Sowe Valley Footpath and Centenary Way.

The route becomes much busier as it forms part of a route from the Oxford Canal to the north-west which avoids Birmingham's built-up areas, part of the Warwickshire Ring. The water is more churned up but there is less floating debris.

The tall chimney just beyond the junction is on an 1837 steam pumphouse that raised

Cash's Hundred Houses – a forward-thinking industrial and residential development.

Young cattle enjoy life at Hartshill.

water 35m from a well with one engine and from a stream underneath into the canal with another. The *Earl of Mercia* Newcomen atmospheric engine, which had worked in Griff Colliery for nearly a century, was transferred here to replace the original *Lady Godiva* engine.

A series of lakes precede a spur on the left before **Bedworth** which is the Newdigate Colliery arm, what remains of 8km of private canals built from 1764 through Sir Roger Newdigate's Arbury Hall estate. It included 13 locks, one of which was a unique three-way lock. Navigation had mostly ceased by 1819.

Bedworth was a mining town which also established hat and ribbon making industries. It has an 1890 church, by Bodley and Garner, and fine almshouses. The built-up area is passed largely unseen in a deep cutting overgrown with hawthorn bushes, finishing by the Navigation at Coalpit Field. Beyond this the canal wall is draped with periwinkle. A drydock undertakes narrowboat building and repair in the former Charity coalmine basin.

The Ashby-de-la-Zouch Canal joins at Marston Junction, where there was a stop lock.

Griff has a quarry where unique volcanic rock has been graded chemically into layers of different colours. The colliery arm was in use until 1961. The hollows are said to be the original of the Red Deeps in George Eliot's *Mill on the Floss*, while the 18th-century Gothic mansion of Arbury Hall became Cheveral Manor. A canal bridge to the north of the quarry has settled as a result of the mining.

The Boot Inn stands at Boot Wharf. Nearby places of worship include a mosque and a 1946 church built by German prisoners of war at Chilvers Coton.

Nuneaton had a priory and the 'nun ea tun' was the nun's river estate. With its redbrick houses it has been a mining town for 500 years, has been a textile producer since the 1800s and was the birthplace of George Eliot in 1819. The library has a collection of George Eliot photographs and memorabilia and the museum also features her, together with local history, archaeology, geology and mining.

A canalside racing pigeon loft seems to fit the environment.

The Birmingham to Nuneaton railway crosses and follows the canal and a battery of telephone lines runs beside the towpath.

Tuttle Hill roadstone quarries, Midland Quarry, Judkin Quarry with Mount Judd and Boon's Quarry follow. In 1971 *Civilia: the End of Sub Urban Man* proposed that the red wasteheaps could be landscaped into an urban site. A brick kiln hides in the undergrowth by the canal beyond the B4114 bridge.

The canal now follows the River Anker to Tamworth, while a windmill and an aerial stand at the top edge of its valley. The ridge, occupied by Hartshill Hayes Country Park, 55ha of wooded hillside, the remains of a castle and red heaps from the Hartshill quartzite quarries, become progressively more attractive. There are extensive views across the river valley.

The Anchor Inn has not so much a children's playhouse as an estate in which to occupy their time.

Oldbury Camp Iron Age hill fort is well preserved with 2.8ha enclosed in a 1.8m bank. It is the reported site of the defeat of Boudicca by the Romans in 60 AD. The Romans were heavily outnumbered but killed 80,000 Britons. Boudicca poisoned herself to avoid capture.

One of the gems of the canal system is Hartshill Wharf with a covered drydock, restored blacksmith's shop in a waterways maintenance yard, refined clocktower with weathervane and a derrick crane. There is an air of Victorian Gothic with rounded windows, arches

The covered drydock at Hartshill Wharf.

The derrick crane and the clocktower with its weathervane are just two of the features that give Hartshill maintenance yard its character.

and curved walls to allow horses to pass with loads of wood for narrowboats. The wharf is an oasis in the countryside.

Oak trees and gorse cloak the hillside while there is a mix of shelducks, magpies, black-headed gulls and terns. The cuckoo can be heard in the spring. Across the valley is the site of the Manduessedum Roman settlement. Rather nearer to Mancetter Wharf are a manor house, a 13th-century church with 18th-century slate tombstones and Victorian Gothic almshouses.

Atherstone, with its Georgian buildings along the main street, claims the distinction of being 100 miles from London, Liverpool and Lincoln. On Shrove Tuesday, hundreds of people play medieval football in the streets. There are no rules and the shopkeepers take the precaution of boarding their windows. Rather gentler sport is provided by a golf course on the other side of the canal, near the canal shop. The town has been a millinery centre for 200 years, using wool felt. Three firms were located by the canal.

Beyond the Barge & Bridge is a coalyard with an old lorry and coal-carrying narrowboat, nicely turned out, still trading and an example of the way things used to be.

This brings the Coventry Canal to the end of its long top pound. There is a new housing estate on the east side of the canal.

Atherstone was the end of the Coventry Canal for a decade when the promoters ran out of money. In 1778 the Oxford Canal was opened from Banbury, exerting pressure for the Coventry Canal to be completed through to the Trent & Mersey Canal. The expensive but necessary flight of 11 locks lowered the canal 25m. The locks are close together at first but more widely spaced lower down. They have side ponds, emptying quickly and filling slowly, including what is probably the only remaining example of a type of side pond that was formerly used frequently. It is now closed to use.

On the offside of the top of the lock flight is a tyre dump. The towpath side lists local church services for the benefit of canal users. Back from the canal is Merevale Hall, a 19th-century Tudor mansion, and the remains of the 12th-century Merevale Abbey, where a church has been based on an Argentinian friary.

After lock 5 there is a bridge on the line of the Roman Watling Street, followed immediately by the A5 dual carriageway at the end of its diversion past Atherstone town centre. After two more locks the Trent Valley railway crosses by Baddesley Colliery Basin site.

Clear of the lock flight, the canal passes through yellow sandstone cuttings at Grendon.

The railway crosses back. Next to it is an obelisk to a chapel destroyed in 1538 in Henry VIII's Dissolution of the Monasteries. Rather nearer are the remains of a swing bridge, set in an area of gorse and bluebells.

Polesworth has a Norman church with a 15th-century abbey gateway, the Tudor-style Nethercote School and a Victorian vicarage, incorporating part of an Elizabethan school where Drayton and possibly Shakespeare were taught. On the far side of the village, Pooley was a boatbuilding centre until the 1950s especially known for

The rural canal reaches away from the bottom lock of the flight at Grendon.

its traditional decoration. Circular pipe firing kilns were also built here. Pooley Hall, among the trees, is a Tudor brick mansion of 1509, possibly the oldest occupied building in Warwickshire. Partly fortified, it has its own chapel inside. The gatehouse and clerestory are from a 10th-century abbey. Until his death it was owned by Edwin Starr.

A nearby memorial is to Pooley Hall Colliery men killed in the First World War. Pooley Hall, Alvecote and Amington pits combined to form the North Warwick Colliery, closing in 1965. At this time coal was still being taken away by narrowboat.

Beyond the M42 are slag heaps with motorcycle trails up and down them. This area formerly had wharves where coal was transferred to the railway.

Alvecote Priory picnic area is in a ruined dovecote and Benedictine priory of 1159, by the Alvecote Colliery site.

Alongside a golf course the canal passes from Warwickshire into Staffordshire. On the other side of the canal are Alvecote Pools, flooded flashes resulting from mining

A microcosm of the past embodied in a coalyard at Atherstone.

subsidence. A nature trail has 100 species of bird and 250 flower species.

The village church has a weatherboarded bellcote and Norman work inside. Amington's more recent church of 1864, by Street, has a Burne-Jones window.

Housing follows the south side of the canal, together with the Gate Inn, until the west side also becomes built-up at Bolehall. The Anchor, by a canal spur, precedes the final two locks at Glascote, the upper with posts heavily worn by ropes. They are accompanied by a modern steelworks and the Reliant car factory. The A5 has been diverted through here. The Burton upon Trent to Birmingham railway passes over and the A51 also crosses, with the Park Inn slotted between them.

The first of two aqueducts, protected by a pill box, takes the canal over the River Tame, giving views down to Tamworth.

Tamworth, formerly Tomtun, was originally Saxon, mostly grey stone. It is built at the confluence of the River Anker with the River Tame. It was razed to the ground by the Danes in 874 and again in 953. The castle has a rare Norman shell keep and tower on a motte, a herringbone curtain wall, a medieval gatehouse, a Tudor chapel and great banqueting hall, Jacobean state apartments, Victorian suite and a bedroom haunted by the Black Lady, St Editha, the 9th-century founder of Polesworth Abbey. The castle has been occupied for nearly 800 years. Offa, king of Mercia, had his capital and palace here in 757 and it was probably the stronghold of Ethelfleda, daughter of Alfred the Great. The Norman and Tamworth Story museum has coins from the Tamworth mint of Saxon and Norman times. The 1701 red brick town hall, with Jacobean windows, was built by Thomas Guy, although it is less well known than his hospital in London. St Editha's church has a unique double spiral staircase, in the corner of the tower, and monuments and stained glass by William

Morris. One of the town's sons was Sir Robert Peel, twice Prime Minister and founder of the Police Force.

The canal runs straight for 800m, to arrive at Fazeley Junction where it meets the Birmingham & Fazeley Canal, the Bottom Road to Birmingham.

Fazeley Junction has seen restoration work. A roving bridge and a mill building are to found near the junction. The Coventry Canal reached here in 1790, by which time the Birmingham & Fazeley Canal had already built the next section to Whittington Brook, a section that remained in their ownership. The final section had also been built but the Trent & Mersey Canal sold their section to the Coventry Canal as a disconnected piece, the boundary between the two different ownerships being marked by a stone at Whittington Brook and bridge 78, 13 bridges after bridge 77 at Fazeley Junction.

Whittington has a 19th-century church, the Swan, Whittington Lock (set off the canal in a garden) and a heron or two.

Huddlesford Junction used to bring in the Wyrley & Essington Canal, now just a spur as a result of being abandoned in 1954 but with restoration underway. The toll island is difficult for narrowboats. The junction house is Lichfield Cruising Club's headquarters.

The Plough is tucked into a corner next to a minor road at Huddlesford, a small community cut in two by the Trent Valley Railway, which passes over for the last time.

To the left lies **Lichfield**. Between 1195 and 1310 the only medieval cathedral with three spires was built, the Ladies of the Vale, sited at the shrine of St Chad and built in local red sandstone. It has a 7th-century manuscript of the gospels and a lady chapel with Flemish Herckenrode windows of 1802. It was badly damaged in the Civil War but the Victorian-restored west front has over a hundred carved figures.

The city was the birthplace of Shakespearean actor David Garrick, 18th-century lexicographer Samuel Johnson and Elias Ashmole, who left a collection of antiquities to Oxford University in the world's oldest museum, the Ashmolean. The city also has associations with Erasmus Darwin, founder of the Lunar Society.

Down by the canal, life is less erudite. Powerlines cross over and a windsock and an aerial stand on a hillside.

A railway line from Lichfield to Burton upon Trent crosses over and then follows the canal for a while. The canal turns by a boatyard to run alongside the A38, built by the Romans as Ryknild Street but now a busy dual carriageway trunk road.

For a while the canal moves away from the road, passing a tree with a nesting box the size of a coffin and opposite a Victorian pumping station, before returning to pass under the A38.

A former airfield stands beside the canal, its large hangars still dominating. Fradley developed because of the airfield.

Trees and reeds line the canal and pillboxes protect the minor road at the end of the airfield.

The Coventry Canal joins the Trent & Mersey Canal at Fradley Junction, one of the busiest points on the canal system, in front of the Swan in the middle of a lock flight. Before that, there is a swing footbridge across the Coventry Canal.

Distance
43km from Coventry to Fazeley Junction and 9km from Whittington Brook to Fradley Junction

Navigation Authority
British Waterways

Canal Society
Coventry Canal Society www. covcanalsoc.org.uk

OS 1:50,000 Sheets
*128 Derby & Burton upon Trent
139 Birmingham & Wolverhampton
140 Leicester*

The swing footbridge before the Trent & Mersey Canal at Fradley Junction.

8 Ashby-de-la-Zouch Canal

The Ashby-de-la-Zouch Canal, Ashby Canal or Moira Cut was intended to be a major link in our canal system, extending to the Trent valley to join the Commercial Canal and then serving to Liverpool or Manchester. Running north across Warwickshire and Leicestershire from the Coventry Canal, it never got any further north than Overseal and never even reached Ashby-de-la-Zouch itself. To get down to the Trent would have required heavy and expensive locking. The canal, as it remains, has no locks at all except at Moira on the disconnected section, being part of one of the longest level pounds in Britain as it follows the 91m contour. Extensions northwards were as plateways or tramroads and the canal carried heavy coal traffic although it was not very profitable. In 1804, the year it was opened, good quality coal was found at Moira and this was supplied by canal to Oxford colleges, among other customers. Medicinal springs were found at the canal head in the 19th Century, resulting in passenger boat traffic, and the canal company also tried exporting cheese. Something they have successfully exported is Measham ware, highly prized decorated brown canal earthenware, made in Church Gresley.

The canal was bought out in 1846 by the LMS railway. In 1856 there were experiments with the steam-powered barges *Pioneer* and *Volunteer*. Traffic peaked at 153,000t in 1870. In 1918 there was extensive mining subsidence in the Measham area and by the 1940s much of the canal around Donisthorpe was on embankment or low bridges. The LMS offered to give the canal to the Coventry Canal Company but they declined the offer. In 1944 the canal was shortened to Donisthorpe, abandoned north of Measham in 1957 and shortened again to its present terminus north of Snarestone in 1963. The last regular coal carriage from the North Leicestershire Coalfield took place in 1970 by the Ashby Canal Associations' trading section. The closures not only avoided the cost of repairs but also allowed further extraction of coal from beneath the line of the canal.

There are plans to reopen the canal to Measham, the Ashby Canal Association owning the canal bed as far as Gilwiskan Aqueduct. Around Moira Furnace in Donisthorpe Country Park 2km has been restored. Full restoration is intended by 2012.

Farmland is rarely more than gently undulating and views are frequently extensive, hedges often being absent beside the canal. Views are better from the water before the reedmace grows too much in the summer. Bridges across the canal are mostly arches of stone or blue engineering brick, sometimes clearly distressed by mining settlement.

These days the canal is almost entirely rural and, being a dead end, is not heavily used by canal traffic although its lack of locks and its rural feel encourage enough boats up to keep the water stirred up.

Although the Ashby-de-la-Zouch is mostly on one level, the lock-free opportunities continue from Coventry or from Atherstone.

The canal leaves the Coventry Canal at Marston Junction on the edge of the old mining town of **Bedworth**. A stop lock was built here to control traffic. The siting of the junction is the result of a late change of plan, the original intention being to join at a Griff Colliery arm to the north.

The first two bridges are Grade II listed structures of Attleborough sandstone. At Marston Jabbett the canal is also crossed by the Trent Valley Railway line. Three groups of powerlines pass over the initial reaches of the canal, diverging from Hawkesbury.

After the bridge at Whitestone, a collection of chalet buildings stand on top of the steep canal embankment. One of the owners has laid out his section of bank with a plethora of little statues from gnomes to carefully painted mermaids. As these cannot be seen from the chalets it appears they are for the pleasure of passing canal users.

Bramcote's hospital is next to the

The canal they couldn't give away

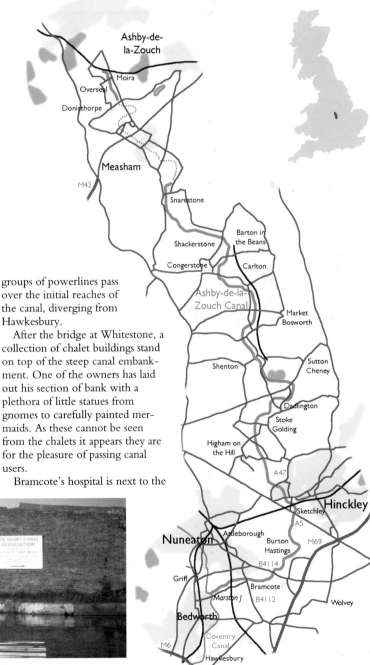

Looking along the line from the Coventry Canal at Marston Junction.

canal and to the fast B4114, where vehicles have to be encouraged to slow down to the point where they will not become airborne from the canal bridge.

On the hillside is Burton Hastings, where the font of St Botulph's church dates from 1300. Two limestone bridges across the canal are also Grade II listed structures. A distinctive powerline flyover beside the cut, with lines along and across the canal, is hard to ignore. The hillside, where Stretton Baskerville Village was lost to the field enclosures of the 16th Century, now serves to eliminate any noise from the M69, 900m away.

Next to the Lime Kilns public house, the A5, built as the Roman Watling Street, passes overhead and the canal moves from Warwickshire into Leicestershire.

A sewage works at Sketchley cleans the local effluent, as indicated by mussel shells beside the canal. The Leicester to Nuneaton railway crosses. The next bridge carries a road that runs beside a canal arm, on the other side of which is a pool which was a quarry and where there were brick and tile works.

The A47 feeds the only built-up part of the canal, **Hinckley**. The centre of the town is marked with a transmission mast which dwarfs the church spire. Hinckley has some fine timber-framed cottages and its industrial background includes having the stocking frame in 1640, long before Leicester did. It was also the home of architect Joseph Hansom, designer of the Hansom Cab, and of Triumph motorcycles. The town is left past the back of a large Tesco warehouse on the edge of an industrial estate.

1938 Triumph Speed Twin two cylinder innovation

Powerlines cross the canal near Higham Grange hospital and kennels make their presence heard. An ancient fishpond site follows a field of Shetland ponies.

The line of the branch railway from Hinckley runs beside the canal. Although the branch was laid with track and signalled, it never carried traffic and was dismantled in 1900. It is joined by the route of the railway from Nuneaton which approaches past the Motor Industry Research Association test track complex.

From the picnic area at Stoke Golding, signs point to the George & Dragon, while a striking spire draws attention to the 13th-century church of St Margaret with its decorated arcade and delicate window tracery. The village is known for the Stoke Golding Country Dance, a more complex version of the Scottish Strip the Willow.

Dadlington's church dates from the 13th Century but the Dog & Hedgehog, with its green and yellow illuminated signs, looks horribly modern. A narrowboat appears to be parked in a field but there is a slot cut at a right angle to the canal just large enough to hold it.

Many other narrowboats are moored along the canal. The canal crosses the line of the Roman road from Mancetter to Leicester before Sutton Wharf Bridge. Tea rooms and trip boats at the wharf make this a busy area.

Beyond the woods to the east of the canal Ambion Hill – the scene of one of the turning points in British history – can be seen. On 22 August 1485 Richard III, with 9,000 Platagenet forces, had control over the hill. Henry Tudor, the Earl of Richmond, faced him at the bottom with only half as many troops. Sitting on the sidelines with a ringside seat and 4,000 men was Lord Stanley. Henry went to ask him for help and Richard chose to attack in his absence. The Stanley party agreed to join Henry and Richard found himself without the advantage of height and facing an army that had suddenly doubled in size. Richard was killed and Henry was crowned on the battlefield as Henry VII, the first Tudor king, thus ending 30 years of the Wars of the Roses. It was the last time the British crown was to change hands on a battlefield. Richard's crown was found in a haw-

Woods stand between the canal and the battlefield hill.

thorn bush on Crown Hill. These days there is a visitor centre with battle re-enactments, jousting, Morris and tea dancing and falconry among the attractions at various times. There is a birdwatching hide at the edge of the wood, looking down towards the canal, which is edged with gorse bushes here.

The current terminus of the Battlefield Line, which uses ex-industrial steam locomotives, is at Shenton. The abandoned route of the railway could see future extension following the canal. The ticket office was built in Leicester but moved to Shenton in 1993 as part of a road-widening scheme.

Near the terminus is a very amateur-looking jetty but all its wooden handrail posts are topped with beautifully turned wooden knobs. A high brick aqueduct carries the canal over the road from Shenton, which has a Victorian church with a 17th-century monument to the Battle of Bosworth Field. Shenton Hall dates from 1629 and was much rebuilt in the 19th Century.

Market Bosworth has some fine thatched cottages. It was where Dr Johnson taught for a time. A timber yard is located next to a road bridge. The canal society members placed mileposts along the canal and one here shows it will be the midpoint of the canal when fully restored.

Rooks circle above their rookery in a wood in the

The aqueduct and listed bridge over the River Sence at Shackerstone station.

Snarestone tunnel is short and the bends are not obvious.

The pumping station at the present foot of the main section of the canal.

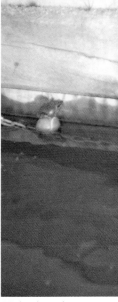

spring. A windsock on the outskirts of the town marks a landing strip. Beyond is the town with the slender spire of the partly 14th-century church.

From Carlton Bridge, where it passes over the Sence Brook, the canal dances past Congerstone with its kingfishers, wrens and water voles.

A distinctive brick tower rises above Barton in the Beans. There is a railway museum at Shackerstone. A Grade II listed viaduct with delicate cast-iron parapets carries a road over from the station. The canal follows the River Sence, with its otters, before turning sharply to cross it on a stone aqueduct.

A line of narrowboat moorings curves round a medieval pond in front of the village, which dates from Saxon times and from which Saxon jewellery has been unearthed. Prominent signing makes the Rising Sun a difficult public house to miss.

Some stands of woodland are found alongside this section of canal, oaks, hawthorns, willows and, appropriately, ashes, with coltsfoot and butterbur among them along the banks in the spring. The woods hide Gopsall Park, which formed the grounds of the former Gopsall House where Handel may have composed the *Messiah*.

A picnic table at Gopsall Wharf is one of a number that have been placed along the length of the canal.

A rabbit warren has been dug in the embankment between the canal and the line of the former railway, which used to cross to the south of Snarestone.

The canal's major structure is the 230m long tunnel at Snarestone. It is unusual for a canal to end beyond a tunnel rather than before a collapsed bore. The line of the canal snakes through the tunnel but the full bore of the tunnel is visible all the way through. As the tunnel has no towpath, the line of the former horse path leads from before the Globe Inn over the top of the low hill,

on which the 18th-century farming village of Snarestone is built.

The canal currently ends just above a Victorian Gothic pumping station, built in 1892 to supply water to Hinckley, its chimney conspicuous in the local countryside. The canal's dead end makes it susceptible to collecting floating debris and an oil film. The nearest access point is a gate, kept locked by the canal society.

The present line ends just outside the National Forest and is mostly in a Conservation Area which boasts nine species of dragonfly (including the rare red eyed damsel), the flat stalked pondweed (which is rare elsewhere) and a fair sample of herons, coots, moorhens, mallards, kingfishers and a few swans.

Beginning at Donisthorpe is a 2km section of canal which is wide, fully restored and in excellent condition, home to a Canal Society trip boat and a maintenance craft.

Initially it runs between a minor road and woodland planted on colliery spoil. After limekilns it is crossed by an arched bridge that feeds directly into Moira Furnace which is now partially restored as a museum. It was an 1805 blast furnace producing pig-iron with the compressed air blast produced by beam engine.

Beyond a swing bridge the canal passes teasels and a coach depot as it swings into the back of Moira. The tower of a fire station locates the restored Moira Lock.

Beyond the road crossing there is low ground on the right towards Sarah's Wood. The canal turns under the Marquis footbridge, named after the local colliery, supported by a pyramid of poles, to enter the terminus at Bath Yard Basin. Conkers Waterside has room for future moorings but is presently more appreciated by families using the slides and other facilities. The canal stops just short of the Leicester to Burton upon Trent railway and 8km from its intended target of the River Trent.

The bridge at the end of Bath Yard Basin.

Distance
39km from Marston Junction to Bath Yard Basin

Navigation Authority
British Waterways

Canal Society
Ashby Canal Association www. ashbycanal.org.uk

OS 1:50,000 Sheets
128 Derby & Burton upon Trent 140 Leicester

The furnace at Moira — the main reason for the canal being built to this point.

9 Oxford Canal

The country's second most popular canal also has its most winding section

The canal formed part of the original Grand Cross scheme to link the Thames, Mersey, Trent and Severn. From 1769 until the time of his death in 1772, James Brindley took the Oxford Canal from Coventry as far as Brinklow. Samuel Simcock took over, reaching Banbury in 1778, then Robert Whitworth continued to the River Thames by 1790. A contour canal, it had many meanders, especially on the northern section. It lost traffic after 1805 when the Grand Union Canal provided competition. In response, it was shortened between 1828 and 1834, cutting 22km off its length. From 1840, it lost further traffic to the railways. Even so, it remained profitable and paid a dividend right up to nationalisation in 1947. The last commercial traffic was carried in the 1950s but it is now one of the busiest canals in the country – second only to the Llangollen Canal – particularly the northern section. Thus, it is well churned up.

The stop lock and roving bridge at Hawkesbury Junction.

The Oxford Canal leaves the **Coventry** Canal at Hawkesbury Junction. It is approached under a cast-iron bridge, one of a number of these graceful structures, particularly on the northern section.

The current discussion is over whether the stop lock with its small drop should be retained or eliminated, which would reduce the depth upstream. An option may be to leave it open except when a deeper boat requires the extra draught. The left bank is where working boats waited for orders.

Powerlines diverge from an electricity substation, all that remains of Coventry Power Station, the site having been landscaped in 1979. Powerlines follow the canal to Newbold on Avon.

The canal moves from Warwickshire into the West Midlands as far as Sowe Common. At Tushes Bridge, beside the Elephant & Castle, the B4109 crosses and is in turn crossed by the M6, which moves alongside for 2km but is surprisingly well screened. On the other side, however, there are extensive views in the direction of Barnacle, the views being a feature of this contour canal. Hedges are frequently of hawthorn and the craft of hedging is shown to be in good hands with skilled hedging work in evidence.

The Wyken Old Colliery Branch is used for moorings by the Coventry Canal Society. At Sowe Common there used to be a cast-iron towpath bridge but it was removed during construction of the M6. It now crosses the River Sherbourne at Spon End in Coventry.

Playing fields at Sowe Common give way to farmland and the city is quickly left behind. Throughout the length of this canal sheep grazing is the predominant livestock activity. On the canal the inevitable mallards and Canada geese patrol the water.

The canal turns away from the M6 but is crossed by the M69. Up the hill from Ansty Pottery is an 18th-century Georgian hall, now a hotel hidden in the trees next to a small church with a delicate spire. Also in the vicinity is the Rose & Castle, a public house with a suitably canal-inspired name. There is a medieval ridge and furrow system, the first of a number that have survived along the line of this canal. Between the golf course and an aerial on the ridge above, the West Coast Main Line arrives to follow the canal as far as Hillmorton.

There are extensive views to the south-east towards Nettle Hill. Where once the canal made a loop up the Hopsford valley, it now crosses on an aqueduct. A riveted viaduct passes high over the canal and railway, its sides supported with scrolled stays. The canal heads towards the Upper Smite Village site. The M6 also passes over and heads east, opening up extensive views back across the fields to the south-west to the Coventry skyline.

Among moorings at Coombe Fields Farm is a hawthorn bush. From the top of it a heron watches still as a statue, exactly what it is. The swallows and moorhens are real enough, though. Mussel shells show the canal to be clean despite the stirred up sediment. Teasels add to the plant life.

Stretton Stop lock and toll office are approached by a plank footbridge, often swung across the water but easily swung back. The Stretton Wharf Branch goes north and is used by Rose Narrowboats, based here beside the Foss Way Roman road. At Brinklow there is a motte and bailey, built in the time of King Stephen to defend the road.

The canal was barred from entering the park at Newbold Revell by the landowner so the 12 Brinklow Arches were built over the Smite Brook. Some were used as stables, forge, hay store and dwellings although 11 were filled in when the canal was widened in 1834.

Elliptical cast-iron bridges cross the ends of the Brinklow Wharf and Brinklow Branches. On the other side, there are fine views eastwards before the canal cuts through Brinklow Hill, on top of which is a church with a sloping floor.

At Hall Oaks Corner the canal turns to pass along the edge of the more appropriately named All Oaks Wood.

Running through All Oaks Wood.

The start of the Fennis Field Lime Works Branch.

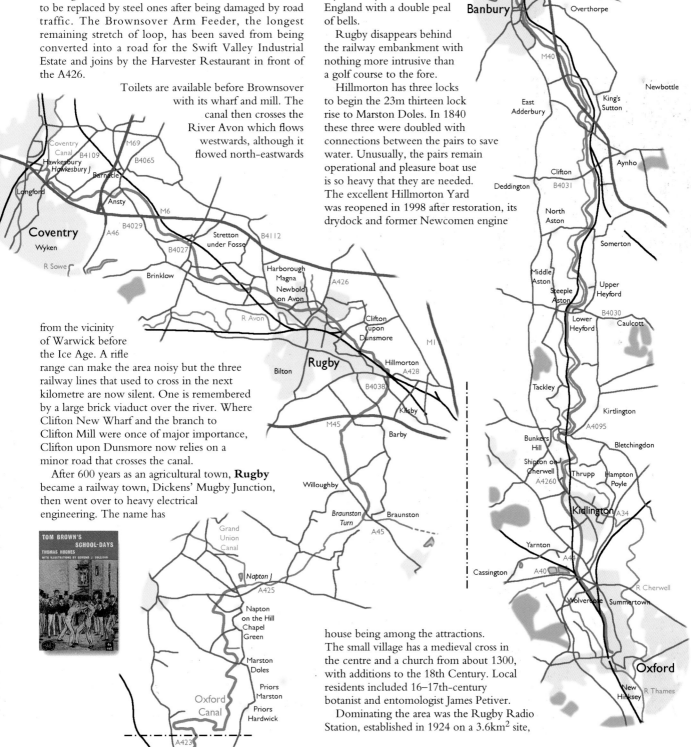

Another cast-iron bridge crosses the end of the Fennis Fields Lime Works branch.

The railway crosses at Cathiron, opening up a view to the church at Harborough Magna, 14th-century with Victorian additions and interesting stained glass.

A branch to Norman's and Walker's Lime Works leaves on the right. It is followed by Newbold Tunnel, 189m long with twin towpaths, unusual for 1834 when it was constructed as part of the canal straightening. The new alignment is NW–SE whereas the old tunnel ran N–S, its mouth remaining next to the 15th-century church.

Newbold Wharf at Newbold on Avon precedes Newbold Quarry Park. The Boat public house has Northamptonshire or table skittles, unusual for the area. A canal loop to Cosford was made redundant by the Cosford Aqueduct, a magnificent structure with a 7.2m cast-iron trough 4.6m wide x 2m deep with four cast-iron segmental ribs underneath. Three of these have had to be replaced by steel ones after being damaged by road traffic. The Brownsover Arm Feeder, the longest remaining stretch of loop, has been saved from being converted into a road for the Swift Valley Industrial Estate and joins by the Harvester Restaurant in front of the A426.

Toilets are available before Brownsover with its wharf and mill. The canal then crosses the River Avon which flows westwards, although it flowed north-eastwards

from the vicinity of Warwick before the Ice Age. A rifle range can make the area noisy but the three railway lines that used to cross in the next kilometre are now silent. One is remembered by a large brick viaduct over the river. Where Clifton New Wharf and the branch to Clifton Mill were once of major importance, Clifton upon Dunsmore now relies on a minor road that crosses the canal.

After 600 years as an agricultural town, **Rugby** became a railway town, Dickens' Mugby Junction, then went over to heavy electrical engineering. The name has

become synonymous with rugby football, first played at Rugby school in 1823, the school featured in *Tom Brown's Schooldays*. Another building of note is the 14th-century St Andrew's church with 1879 nave and tower by Butterfield, probably the only church in England with a double peal of bells.

Rugby disappears behind the railway embankment with nothing more intrusive than a golf course to the fore.

Hillmorton has three locks to begin the 23m thirteen lock rise to Marston Doles. In 1840 these three were doubled with connections between the pairs to save water. Unusually, the pairs remain operational and pleasure boat use is so heavy that they are needed. The excellent Hillmorton Yard was reopened in 1998 after restoration, its drydock and former Newcomen engine

house being among the attractions. The small village has a medieval cross in the centre and a church from about 1300, with additions to the 18th Century. Local residents included 16–17th-century botanist and entomologist James Petiver.

Dominating the area was the Rugby Radio Station, established in 1924 on a 3.6km² site,

31

The twinned Hillmorton locks.

The restored Hillmorton Yard stands off the lock flight.

transmitting around the world. The tallest masts were 250m high. The last mast was demolished in 2007.

Suddenly the canal changes from its south-easterly course to head south-west. The West Coast Main Line passes over for the last time, together with the line to Northampton, which has just broken away. Next to the A428 bridge is the Old Royal Oak with a notice banning mooring because canoes are in use, apparently a rather old notice that is ignored.

There are stables at Kilsby Road Bridge which carries the B4038 over Rains Brook, the Warwickshire–Northamptonshire border.

Moorings on the left are accompanied by fenced-off patches of land, one with a large dovecote similar to one in the garden of a lock cottage at Hillmorton. A loop of the old canal goes off on the opposite side, its profile in

Canalside stables at the B4038 crossing.

the field clearly visible from the present towpath, and returns at the M45 bridge.

The dolls of Barby live over the hill to the left. Less sociably acceptable youngsters are housed in HM Young Offender Institution at Onley on the other side. There are long views over the Rains Brook valley to a water tower on the ridge at the edge of Rugby. The former railway from Lutterworth to Woodford Halse followed the canal closely for 3km. There is cutting as the route moves back briefly into Warwickshire, the canal also being in a cutting frequented by bats and cowslips.

A keen gardener displays the results of his labours by the next bridge and the Rose in Willoughby touts for business from the canal. At Willoughby Wharf the agricultural associations go back further with more medieval ridges and furrows.

Braunston is one of the best known village names on the canal system although most of the activity is to the east on the Grand Union Canal. The approach from the north is quiet enough, herons and swans minding their own businesses below a ridge topped by a disused windmill and a church. This has a tall Victorian spire and is in a churchyard which includes many boatmen's graves. The Oxford Canal originally continued south-west but this section was taken over as part of the Grand Union Canal main line and southbound boats now traverse this section in the opposite direction from when it was first built by the Oxford Canal Company.

Incorporation of the Oxford section and the 800m Braunston Branch saved £50,000 in construction of the Grand Union Canal route from Birmingham to London. An idea of the importance of this new transport link can be gauged from the fact that goods being sent from London to Abingdon were sent via Braunston rather than up the River Thames because of the poor state of the latter. The Oxford Canal arrives under the A45 at a triangular island with two of the Horseley Ironworks semi-elliptical 15.2m span bridges and a brick bridge carrying the towpath across, the high ends of the arches allowing room for passage on foot underneath for users of the other towpath line. The Braunston Turn formerly had a stop lock.

The Oxford Canal reclaims its name at Napton Junction. From here the Grand Union Canal turns north past Napton Reservoirs, which are a source of water for the Grand Union Canal and of dragonflies that frequent the reeds in the summer. A new marina has been built at the junction.

The southern section of the Oxford Canal had horse-drawn boats long after other canals had given them up and there can be few long sections of canal better suited to the pace of the horse. Joseph Skinner's mule-drawn boat was in use until 1959 and is now in the National Waterways Museum, Ellesmere Port. The rural feel is complemented by bollards made from roughly hewn tree trunks. The southern section of the canal has many angling contests.

Napton on the Hill takes its name from the Anglo-Saxon *cnaepp tun*, meaning hilltop homestead, the hill being over 120m high. On top is a windmill, restored in the 1800s, although there has been a mill here since at least 1543. Materials for building the 13th-century St Lawrence's church were left by the green, ready to start work, but overnight they reappeared at the top of the hill so it was built up there. Napton Nickelodeon of Mechanical Music has monthly theatre organ concerts and music machines on display, including a barrel organ, mechanical violin, Wurlitzer Photo Player and Crompton Cinema Organ, said to be the best museum in the country on the subject.

The Olde King's Head is just up from Napton Wharf, the base for Napton Narrowboats. The Napton Bridge Inn is next to the A425. A pile of bricks marks the site of Napton brick and tile yard, which was claimed to

Braunston Turn, facing the Grand Union Canal. The old Oxford line goes right.

The bottom of the Napton flight with the windmill above.

have the longest kiln in Europe. On the other side of the canal, craftsmanship is alive and well with some quality hedging using the hawthorn so prevalent on this canal.

By Napton Bottom Lock is a farmhouse which served beer to boat crews, becoming the Bull & Butcher and now the Folly, noted for its selection of pies. A notice warns that it is the last pub for five hours. There are ten locks to be negotiated in the next 3km up to Marston Doles.

As the canal rises, so the scenery opens up to give extensive views across the open farmland, much of it used for sheep grazing. Towards the top of the flight is the Old Engine House Branch which was 820m long and had a steam pump. The pumping house remains are still there. The canal company were permitted to use any water found within 910m of the canal.

The top lock is by the Welsh Road at Marston Doles. Picnic tables wait by the lock side and the Holly Bush Inn in Priors Marston offers a transport service to and from the lock. Horses used to be stabled here. Farm buildings of 1865 were refurbished for British Waterways' offices but are now used by a commercial company.

The summit pound runs for 18km and is the most extreme example of a contour canal. It is a ruthlessly rural area with very few buildings along its length but the canal seems to go out of its way to visit all of them and search for more as well. It suffers from water shortages in the summer but a back-pumping scheme is being initiated to get round the problems and flows have been augmented with mine water from Wolverhampton. Consideration is also being given to the enlargement of Boddington Reservoir but, not unreasonably, the locals insist that they don't want this to result in a country park. Where dredging work has been done on the canal, plastic pipes have been used to extend water vole burrows.

As the canal twists and turns two aerials rotate to and fro around the traveller, a TV tower south of Hellidon and a lattice radio mast towards Bishop's Itchington. A

shed of battery hens is left at Marston Doles and skylarks lift the mood. Fields of EU-approved oilseed rape are occasionally replaced with richly scented fields of broad beans. There are occasional pillboxes, flimsy prefabricated structures, not the usual solid blocks more able to withstand bombardment, as found at the southern end of the canal.

Parts of the stone church in Priors Hardwick date from the 13th Century. The village was badly hit by the black death and many of the houses were pulled down by Cistercian monks in the 14th Century. Along the road, Stoneton Village has disappeared almost without trace, Stoneton Moat Farm being one link with the past.

Dandelions, cow parsley, cowslips, white dead nettles and ground ivy add colour to the canalside in the spring. There is plenty of time to look as the meandering is at its best here, the canal progressing just 870m in 3.2km. Rooks watch from their rookery in a wood at Wormleighton Grange Farm as the canal completes its most extreme loop.

Wormleighton Village, with its extensive ridge and furrow system and earthworks, was depopulated in 1498. Its more recent namesake has a 13th-century brownstone

Former stables by the top lock at Marston Doles.

The graceful cast-iron roving bridge in the Tunnel.

The first of the distinctive lifting bridges near Claydon.

Poetry at Cropredy Lock.

church with Perpendicular screen and Jacobean wood-work, a 16th-century manor house with a huge stone gatehouse of 1613 and Victorian mock Tudor and thatched cottages. This is where in 1984 Canada geese first bred on the canal.

The Wharf Inn sits by Sherne Hill Bridge where the A423 crosses at the foot of Shirne Hill. On the next corner is a large marina at Cowroast. At this point the Coventry to Didcot railway comes alongside the canal, following it closely all the rest of the way. Up the hill is Fenny Compton, a brownstone village with a 14th-century and Victorian church, which has an offset tower.

The canal uses a gap cut by overflow from an Ice Age glacial lake between Leicester and Stratford-upon-Avon. The next kilometre is called the Tunnel because it was one until opened out in 1868. It is now a steep-sided cutting. The A423 crosses on a concrete structure, dwarfing a graceful cast-iron roving bridge.

Mussel shells lie on the bank adjacent to Wormleighton Reservoir, a canal feeder. The canal then passes between the blue brick portals of what was once a bridge carrying a railway line to the former railway junction of Woodford Halse, now without railway lines going in any direction.

The boundary between Warwickshire and Oxfordshire comes directly after the first of the lifting bridges that are so distinctive of this canal, built with heavy balance beams but no superstructures. They are particularly to be found south of Banbury and are usually left open. Moving south, the red-brick bridges gradually give way to those of Cotswold stone.

Five locks at Claydon start the 30-lock fall through 59m to the River Thames. Old stables at the top lock now sell traditional canal ware.

St James the Great's church in Claydon is partly Norman in brownstone. Its tower has a saddleback roof and clock with chimes but no face. The Granary Museum of Bygones has been opened to support the church's finances and has a respectable collection with extensive local relics, domestic crafts, steam engines, stationary engines, shops and a 19th-century cottage kitchen.

To the right are the remains of the Gilbertine Clattercote Priory, founded in 1209, above which is Clattercote Reservoir. Unlike its better known namesake, the village of Farnborough has nothing to do with flying. Chipping Warden airfield, facing it across the valley, is primarily a road transport base these days.

Three more locks follow, Varney's with a ridge and furrow system right to the side of the towpath. Prescote Manor is now a craft centre at the approach to Cropredy

Lock, the first of several places where poetry has been presented along the canal, here on what look like corroded metal gravestones.

A bridge across the canal carries the Oxfordshire Cycle Way past the Bridge Stores, which are over two centuries old and were a resting place for men and horses until the 1940s. The River Cherwell is now alongside and to be followed closely to Thrupp. There was a bridge across the river by 1314 but it was the bridge in place in June 1644 that was the focal point for the Civil War Battle of Cropredy. Cromwell's forces, under Waller, attacked a smaller Royalist group in an attempt to force a passage towards Oxford. They were beaten back and lost their artillery. Some of the soldiers were buried in the churchyard of the sandstone church of St Mary the Virgin, which has fine woodwork. It also has a mark on the vestry floor that may be the bloodstain from a young messenger killed there after the battle. The brass eagle lectern spent 50 years in the river after being hidden there from the troops by the villagers. The footsteps of a man, woman and girl killed by a Roundhead are sometimes heard at the 15th-century Red Lion. The Curfew Bell is still rung three evenings a week. Perhaps the Reunion Folk Festival features material from that era.

The Old Canal Wharf used to be a toll collection point. These days people pay for the afternoon teas it provides.

The canal bends round towards Williamscot House, where Charles I was supposed to have spent the night before the battle. It runs peacefully past Great Bourton and Little Bourton until the arrival of the M40, which passes over and is to remain in the vicinity until Clifton.

Below Salmon's Lock the railway passes over by a junction with another line that used to run from Woodford Halse. Rudd and perch are more likely fish today.

The bridge by Bourton House.

Looking back to Cropredy.

Tooley's Boatyard, now tarted up and surrounded by a modern shopping centre.

buildings and restriction of access to suit shopping hours mean it is now a working museum rather than business premises.

Banbury has a poor record of civic vandalism. The cross of nursery rhyme fame was pulled down by the Puritans in 1602. The current cross is a replica of 1859. The Perpendicular Cathedral of North Oxfordshire needed renovation; instead it was blown up (a cheaper option), the 'unsafe' portion surviving the explosion. Its replacement of 1793 by S. P. Cockrell is described as looking like a gaol. Banbury has been famous for its currant-filled cakes made to a Tudor recipe since 1608 but the original cakeshop was demolished in 1968 to make

A Banbury balcony with a difference.

Banbury begins with a foundry which takes its cooling water from the canal. The A423 brings significant traffic noise along the bank by the industrial estate, residents of which include General Foods. A westerly wind brings the aromas of Maxwell House coffee, cocoa, Bird's custard and baking bread from the Fine Lady ovens.

After the A422 crosses there is a spur on the right, formerly the canal's line before roadworks caused a diversion. This is Spiceball Country Park although there are few clues from the water other than a firebeacon by the canal.

On the right is Tooley's Boatyard, established in 1790 and the only surviving one on a narrow canal. It used the principle of building and launching narrowboats sideways into the canal, this being one of the first narrow canals. The drydock was the first on the canal. Its most important commission was adapting the narrowboat *Cressy* for residential use for L. T. C. Rolt. He then published *Narrow Boat* in 1944, leading to leisure use of the canals. It is now at the centre of a shopping development and is part of Banbury Museum, across the bridge on the other bank of the canal. Demolition of non-listed

The misty peace at Grant's Lock belies the fact that it is a stone's throw from the busy M40.

Looking down the canal with the spire of Kings Sutton church rising out of the mist.

way for offices. The canal basin was filled in to create the bus station. The Germans added their contribution in September 1940 with a direct bomb hit on the lock. Clarkes Mill burned down in 1992. At least the lockside has been well restored, complete with more poetry gravestones for those who enjoy them. Arts lovers are advised to look next door to the Mill, which now operates as a first-rate arts centre.

Banbury, from the Old English Banna's stronghold, was a wool town with the largest cattle market in Europe. It was used for filming *3 Men & a Little Lady*.

After the A4260 Cherwell Street crosses, there is a selection of industrial premises on the right, from a builders' merchant to a hire crane depot. Tucked between them is an unobtrusive but attractive piece of art in the form of a balcony railing, made from old pieces of metalwork, a chain, hook, horseshoes, billhook head and the like.

Residential housing pulls away from the canal at Calthorpe, initially leaving a grass recreational area with picnic tables and a barbecue, backed by a garish blue-caged basketball court.

The first of the lifting bridges south of Banbury comes as the canal heads out into farmland. The M40 passes over again, a distant roar for the next 5km.

On the River Cherwell, Twyford Mill is an animal feed factory.

Twyford Wharf boasts a pair of telephone kiosks. It used to be a narrowboat hire centre but customers wrecked too many engines to make it profitable. Earlier, it had a brickworks which supplied bricks for construction of the canal.

From Twyford Wharf the canal continues southwards past gorse, dandelions, cow parsley, white dead nettles and reeds to Kings Sutton Lock, also known as Tarver's from a former lock keeper. One building was the company blacksmith's forge. Most conspicuous, though, is the 15th-century church's magnificent 60m spire which forms a line with Newbottle and Adderbury's churches. It is surprising how many spires and towers do line up: Bodicote–Milcombe–Wigginton, Wigginton–South Newington–Barford St John–Aynho, South Newington–Barford St Michael–Deddington, Barford St Michael–Barford St John–Overthorpe and longer lines if more tolerance is permitted.

Brick abutments mark the crossing point of a former railway branch to Chipping Norton. The M40 crosses for one last time before climbing away to the south-east. The canal makes a rather sharper turn, passing the Old Wharf, which stands at the foot of the hill. This leads up past Bo-Peep Farm to Adderbury, a village which,

like Deddington to the south, is well known in the field of Cotswold Morris dancing.

At Nell Bridge the Waterways Art Gallery has been established in a farm, with canal artists and canal ware. The original narrow bridge of 1787 across the bottom of the lock has been incorporated into a new structure carrying the B4100. Not only does the road have to be crossed but the towpath changes banks. On the left, new angling ponds have been excavated. This is an area for the brandy bottle, a name given to the yellow water lily because of the shape of its seed pods.

Above Aynho Weir the River Cherwell crosses at canal level and passes over a long weir on the right, on top of which a large viaduct carries the towpath. The lock has only a 300mm drop to resist flashes but is hexagonal, wide enough for four narrowboats and able to pass a normal volume of water without requiring larger gates. For 2km the canal scrapes the heel of Northamptonshire.

A traditional Great Western Railway signal box up the hill marks the start of a complex railway junction with flyover and impressive brick viaducts as the Marylebone line leaves.

Narrowboat moorings surround Aynho Wharf.

Geese at Souldern's Wharf Farm make a lot of fuss on the canal. The scenery now opens up and farmland stretches in all directions by Somerton Deep Lock, one of the deepest inland locks in the country with a fall of 3.7m.

Approaching Somerton, four sections of balance beam have been carved, inscribed with poetry and set up by the towpath. Although still quite new, the lettering has already become barely legible.

Decoration is more durable in the crucifixion carved on the tower of the battlemented 14th-century St James' church in Somerton, also noted for its 16th-century tombs and stone carving of the Last Supper inside. There are excellent views to the south-west, towards the Astons, across water meadows which flood in winter, suitable for waterfowl including swans and for growing interesting flora.

It has not always been as idyllic as it seems, however. Above the ridge on the left is the Upper Heyford Airfield, one of the two bases from which the U. S. Air Force bombed Tripoli. Now closed, it was one of the three finalists for the national sporting academy venue and the British Olympic Association's preferred choice but lost out to Sheffield.

Hanging judge Sir Francis Page, who lived from 1661 to 1741, sold Middle Aston to Sir Clement Cottrell Dormer and concentrated on Steeple Aston. Here he had graves in St Peter's church moved to make way for a large one for himself and his wife. The 1730 Page monument shows the 70-year-old judge seated in his robes with his 40-year-old wife lying rather less robed at his feet. Meanwhile, Sir Clement pulled down some of the houses in Middle Aston to reduce the poor rate and this had an adverse effect on Sir Francis' new home. Middle Aston was later bought by Spillers. Although all seems peaceful today, both villages and North Aston overlie coal so they could yet become a colliery.

The railway crosses and the canal wanders to Lower Heyford or Allen's Lock, at Upper Heyford, where a magnificent stone barn stands at the top of the canal bank near the church.

Trees line the canal, approaching Lower Heyford, one of them bearing a crafted tree house with veranda over the canal. Also over the canal is an aluminium version of a drawbridge. It may be lighter to operate but it lacks the attraction of wood. The 13th-century St Mary's church has fine stained glass and a sundial in the porch. The B4030 crosses the canal next to a clapboarded wharf building that is a boat hire base with a shop. The village also has an ancient watermill.

Unusual towpath viaduct passes over as the canal crosses the River Cherwell on the level above Aynho Weir.

Somerton Deep Lock with the Oxfordshire farmland at its best.

The railway runs alongside. In 1992 a Reading to Banbury train overheated and the driver had to borrow a bucket from a passing boat to fill the train's radiator with canal water. It was also near here that canoeist William Bliss discovered white violets and springtime, having his best meal ever, as he described in *The Heart of England by Waterway*.

A 12th-century church on the right faces Rousham Park. The castellated house of 1635, in early Tudor style, was a Royalist garrison in the Civil War and has bullet holes in the front door to prove it. It is set in 8ha of hanging woods and is the only surviving example of landscaping by William Kent from 1730. It includes Rousham Eyecatcher, a sham castle gate on the hilltop.

Dashwood's Lock is surrounded by primroses. North-brook Bridge joins a packhorse bridge over the River

Near Middle Aston a kingfisher sits on a hawthorn branch as a pair of boats pass.

The church and massive stone barn at Upper Heyford.

Tree house over the canal at Lower Heyford.

Moorings of Oxfordshire Narrowboats in a wide section of canal opposite Heyford station.

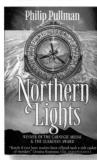

Cherwell. A bridge of red and blue bricks at Nethercott is chamfered to prevent damage to towropes and heads.

Fields of thoroughbred horses surround Manor Farm. In Tackley, John Harborne's manor house has a two-storey stable block of 1616 and pigeon house of similar size, both magnificent buildings. The Roman Akeman Street crosses at a point that is also on the line of the Aves Ditch earthwork. Along the line of the Roman road to the east is Kirtlington Park, with its horse polo pitches. The road then passes the airfield at Weston-on-the-Green, from which aircraft will frequently be seen dropping parachutists.

Washford Hill Stone Quarries and cement works have been disused since the 1920s, when the Oxford Portland Cement Company had a fleet of 10 narrowboats. The quarry is now laid out for public use and leads into a wooded cutting, the steep banks covered with ivy and bluebells in the spring.

The Oxfordshire Way footpath crosses at Pigeon or Enser's Mill Lock, where the Three Pigeons is now a private house.

A bulrush-lined pound passes a golf course and arrives at the Rock of Gibraltar, the Oxford Canal Company's Brindley Head in Enslow until 1807. The railway crosses over the next loop twice, hiding the satellite Earth station with its three large dishes. The canal is lined with sedges and reedmace, the haunts of sedge and reed warblers. Flowering rush is also to be found.

Below Gibraltar or Baker's Lock, the canal joins the River Cherwell for over 1km, relatively wide and faster flowing with some significant twists through water meadows. The large Bunkers Hill cement works makes its presence felt, the chimney being a local landmark.

Vertical wooden rollers assisted towlines on sharp corners. The canal leaves the river at Cherwell or Shipton Weir Lock, another hexagonal structure. The Oxford Canal Company proposed this as a major canal interchange with the Hampton Gay Canal or London &

Shipton Weir Lock, one of the hexagonal ones.

Western Canal of about 1792, forming part of a broad route from London to Exeter, avoiding the problems of the Thames, but it was rejected by Parliament and the Grand Union Canal as it would have added 48km to the journey. The burnt-out ruins of a Jacobean manor after a fire of 1887 are another failed project. In the 1590s there was a peasant uprising when the village was cleared to make way for sheep pastures.

The railway crosses again at a point where nine carriages fell before and onto the frozen canal on Christmas Eve 1874 in the GWR's worst accident, killing 34 passengers and injuring 65 more. The 15-coach train was carrying 500 passengers behind a pair of single driving wheel engines when a wheel on the leading coach broke, with disastrous consequences. Amazingly, the Oxford stationmaster had a rescue train underway just half an hour after receiving a telegram about the accident.

The abutments of another bridge immediately after it used to carry a branch line to Woodstock, passing round Shipton on Cherwell, which has recording studios and a picturesque church, rebuilt in 1831.

At Thrupp Wide the canal again uses the course of the river but this time the river has been moved to a new channel on the east side. The pound has white water lily, water crowfoot, submerged water plants in the clearer water and nesting swans among the reeds. There are moorings here – local boats include on called *Thrupp'ny Peace*. Thrupp has been described as the perfect canal village; cottages stand back from the grassed canal embankment. The Boat Inn provides refreshment, as does the Jolly Boatman, formerly the Britannia.

Light aircraft circle round from Oxford Air Training School at Oxford Airport, one of only a score of airline pilot training schools throughout the world.

The canal turns sharply under the A4260 at Langford Lane Wharf, the lane leading to the airport and Campsfield House immigrant detention centre. Alderman Frank Wise was a signalman, the adjacent Wise Alderman being previously the Railway Hotel and, before that, the Anchor. **Kidlington** begins here, a dormitory for Oxford, once claimed to be the largest village in England and a producer of apricots and other fruit. Residents have included pill magnate Thomas Beecham. Entry to **Oxford** seems much more extended than departure from Coventry. Industrial units line the canal, including that of Tom Walkinshaw motor racing, and a recent bridge over the canal serves more industrial estate. The Coventry to Didcot railway line crosses back over for the last time after Round Ham Lock and heads past Yarnton sewage treatment works, which discharged its effluent into the canal until 1998. St Bartholomew's church, originally Norman but rebuilt in the 1230s and enlarged in the 17th Century, has good monuments in the Spencer chapel, wooden screens, medieval glass and alabaster carvings.

Transport links come in quick succession, beginning with crossings by the A44 and a bridge which used to carry a railway line from Bicester to Worcester. Below Shuttleworth's or Duke's Lock, the Duke's Cut goes right as the first of two connections to the River Thames, built by the 4th Duke of Marlborough across his land in 1789 as a source of canal income. Duke's Cut Lock, on the branch, opened either way as the river could be above or below the canal level. The river level is now maintained above canal level. This reach was used for the climax of Colin Dexter's *The Wench is Dead*. The A40 and A34 complete the road arteries.

Wolvercote Weir has an open bypass channel.

The Plough is passed at Upper Wolvercote and the Bicester to Oxford line is the final railway crossing at Summertown, where the playing fields of St Edward's School occupy the left bank. Beyond the railway to the right is Port Meadow, the largest open space in Oxford and the best biologically documented plot in England. Round Hill is in the middle and it occupies 2km^2 of the bank of the Thames.

Rosamund the Fair cruising restaurant moored in Jericho.

An electric lift bridge across the canal connects sections of the Unipart works, now somewhat dilapidated. The Anchor is near the canal. Houseboats are moored and a large rudder is carefully decorated and installed on the bank. Along this pound are Edwardian houses. Jericho may have been the model for Beersheba in *Jude the Obscure*. It was a location in *Northern Lights*, the author of which, local resident Philip Pullman, was among those demonstrating against a housing development on the Castle Mill boatyard site.

The *Rosamund the Fair* restaurant boat recalls Rosamund Clifford, mistress of Henry II. She was educated and later buried at Godstow Nunnery, the ruins of which are near Wolvercote Lock.

The Hythe Bridge Arm used to terminate at Worcester Street Wharf, infilled in 1949 for Nuffield College and a carpark. The arm now ends abruptly at Hythe Bridge Street with a sculpture of lock balance beams behind the gardens of Worcester College. There are plans to re-excavate the terminal basin.

The canal drops down through Louse or Isis Lock, below an ornate iron roving bridge of 1796, a favourite lock for suicides. Crossing Castle Mill Stream, it is possible to proceed up the Sheepwash Channel to the River Thames or Isis opposite the Tumbling Bay bathing place, which could be restored. This route formed part of the improved Liverpool to London link as the Thames was in poor condition upstream to the Duke's Cut. A local improvement was the decision to fix a swing railway bridge in the open position in 1984. Until then, railway crews had to unbolt the track every time a boat wanted to pass. In 1995 the station, which features on the opening page of *Zuleika Dobson*, had the proud boast that improved security had reduced the number of thefts from its carpark over the year from 46 to a mere 16.

Oxford is usually taken to be the ford where oxen crossed but it may be significant that the Old English for salmon was *ehoc*. Confusingly, the River Ock is downstream at Abingdon. For Hardy, this was Christminster.

Over 2,000 years ago, Lud had the realm of southern England measured and discovered that Oxford was at the exact centre. St Frideswide built a monastery in 727 and began the walled town. In 912 Oxford was used by Edward the Elder as a buffer between Wessex and the invading Danes. It was the sixth largest town in England in 1066 and Oxford Castle was built in 1071 for William the Conqueror by Robert d'Oilly. It was Charles I's headquarters in the Civil War and Adolf Hitler planned to make it his capital after the Second World War. The university was in existence by the 11th Century, the oldest in the English speaking world. The oldest of the 35 colleges dates from 1249. The colleges were heated by Moira coal, brought via the Ashby-de-la-Zouch Canal.

Oxford buildings of note include Blackwell's bookshop, the Bodleian Library and the Ashmolean Museum, dating from the 1840s although the collection was begun in 1683. There are 900 buildings of architectural or historical interest in the city centre and the best concentration of urban gardens outside Japan. The city has been used for filming *Inspector Morse*, *Brideshead Revisited*, *A Fish Called Wanda*, *The Madness of King George*, *Shadowlands* and *Waiting for God*. To Matthew Arnold it was the City of Lost Causes and it was where Lewis Carroll composed *Alice in Wonderland*. J. R. R. Tolkien

Distance
37km from Hawkesbury Junction to Braunston Turn and 81km from Napton to Oxford

Navigation Authority
British Waterways

Canal Society
Friends of Oxford Canal & Basin
www.foxcan.co.uk

OS 1:50,000 Sheets
140 Leicester
151 Stratford-upon-Avon
152 Northampton & Milton Keynes
164 Oxford

10 Grand Union Canal

The Grand Union Canal is the backbone of the canal system, Britain's longest canal, joining Britain's two largest cities, yet it was not named as such until 1929. It began life as at least eight separate canals with William Jessop doing the major building work between 1793 and 1805. He completed the line, which avoided most of the winding Oxford Canal and the Thames in poor condition, shortening the route to London by 100km, although there were many locks. During the Second World War it became an important route for taking arms equipment to **Birmingham** and coal to London, often by teams of three female workers who, from their IW badges, were nicknamed Idle Women. Freight carriage did not end until about 1970, later than on most canals, although there was a marked decline after the severe winter of 1963.

The start was originally at the Warwick Bar, where it met the Digbeth Branch. In order to relieve congestion in the city centre, the Birmingham & Warwick Junction Canal from Bordesley Junction was opened in 1844 to meet the Birmingham & Fazeley Canal at Salford Junction. The Tame Valley Canal was also built to the same point and opened at the same time. These days the sky above the junction is dominated by Junction 6 of the M6, Gravelly Hill Interchange or Spaghetti Junction, as spurs loop in all directions. It is a long time since this was the Old English Beormingsham, the village of Beorma's people.

Below all the concrete spans the start of the Grand Union Canal is still crossed by a steep iron towpath bridge and then immediately crosses the River Tame on an aqueduct. The Saltley Cut took the line away. Nechells Shallow Lock is usually left open. A former loop supplied Nechells power station, the site being scrubland until being taken over for the Star City venue. Gorse and birch trees are among the vegetation to have got a foothold, reedmace grows in wet areas close to the canal and Canada geese and mallards are resident, despite a slightly oily scum on the water in places, together with a small amount of flotsam.

An aqueduct takes the canal across the River Rea. A pair of large gas holders stand by the Aston to Stechford railway crossing. Bales of paper wait above the canal to serve a papermill. The A47, built in 1992 and running along the line of Saltley Reservoir, filled in for the purpose, crosses with the Nuneaton to New Street railway line and an extremely skewed bridge. Near Garrison Bottom Lock, brick factories stand derelict, blocks filling in behind windows which have not been removed except by stone throwers. A car scrap recycling yard faces across to Birmingham City football stadium. Brimstone, peacock and orange tip butterflies add flashes of colour.

The Spaghetti Junction of an earlier age is now met above the canal with the West Coast Main Line and Derby and New Street to Cheltenham Spa railway lines passing over in quick succession. A wharf has been refurbished but every one of the decorative lights hangs broken by the canalside.

At Bordesley Junction, the Digbeth Branch of the Birmingham Canal Navigations runs right to the original

Bordesley Junction at the foot of the Camphill Locks.

A Pullman train passes Garrison Top Lock.

terminus of what was the Warwick & Birmingham Canal, opened in 1799. When the first National Waterway Walk was opened in 1993, to commemorate the 200th anniversary of the start of the construction of the canal, it was along that original line from Gas Street Basin to Little Venice. A cast-iron arched bridge crosses the canal before the Camp Hill lock flight. The canal has been moved to accommodate the A45; a wall containing a narrowboat in red brick has been built in recognition.

Between the Birmingham to Didcot Parkway and Derby to Cheltenham Spa railway crossings is a grand church with no spire or tower but located on a hillside overlooking the canal, as if lesser buildings have been cleared in this generally industrial area of Sparkbrook.

The top lock raises the canal to the 16km long 117m level, the first of the Grand Union Canal's summit levels. The top lock is also the last of the narrow beam ones. Once the smaller canals were amalgamated it was decided to widen the whole route to broad beam in the 1930s as a work creation scheme during the Depression. However, the final locks into the centre of the city were never converted.

The canal is more open at first as it heads south-east. It is crossed by the B4145 over Golden Hillock Bridge, near which is a mosque, then by the Birmingham to Didcot Parkway railway. Beyond this is 32ha occupied by the Ackers Trust, an outdoor centre with its own dock, formerly serving the British Small Arms factory. Beyond the River Cole crossing is the large Tyseley incinerator. Smoke from the opposite side of the canal may be from the Birmingham Railway Museum which is based in a former Great Western Railway shed and has live steam with engines, wagons, coaches, the royal carriage

The imposing church overlooking Camphill Second Lock.

of Edward VII and Queen Alexandra and a travelling post office attacked in the Great Train Robbery. The former BSA factory displays motorbikes and guns.

The next reach sees the canal undergo a personality change. At first it is derelict industrial, enough to suggest spray paint may be a profitable business in which to invest, and one building has electrified wires all round its roof although the missing window now leaves easy entry. By the other end it is in cutting lined with trees, bluebells and red campion, magpies and more. For most of the rest of the built-up part of the city it remains in deep treelined cutting and the view from the water bears little resemblance to what is shown on the map. An indication of the rate of growth is given by the fact that Edith Holden wrote *The Country Diary of an Edwardian Lady* about **Olton** in 1906, the lady herself subsequently drowning at the far end of the canal in the Thames at Kew. Indeed, at the time of the *Domesday Book* it was one of the most sparsely populated areas in England.

The banks have squirrels, foxes, nettles, brambles and holly among other trees and the south side looks very wet and susceptible to slipping. On the far side of the B425 bridge is the factory that builds a British icon, the Land Rover. This familiar workhorse has been manufactured for over half a century yet one of the original half dozen has been discovered still at work somewhere in Africa.

Swans nest in the rushes and rather possessive herons also frequent the area.

Railings show where the towpath has been improved at Elmdon Heath and then, suddenly, the canal is into open country. Although this was the main arterial route of the 19th Century it is almost totally rural from here to Berkhamstead, with the exceptions of Warwick, Leamington and Milton Keynes. It is not always quieter, though, initially passing a kilometre from the flightpath of Birmingham Airport, which lies not far to the north.

The first hamlet is Catherine-de-Barnes with the cricket pitch near Bogay Hall.

Hawthorns have predominated among the canalside vegetation but there are oaks, willows and ferns.

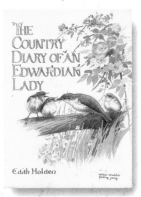

Buzzards hunt the area. The canal picks its way between various important old buildings, Walford Hall Farm, Berry Hall and Ravenshaw Hall. It approaches both sides of the River Blythe on embankment, beside which British Waterways have a dredging spoil disposal area. At the end of the embankment the M42 passes over; screening shuts out much of the traffic noise. After the motorway there are extensive views to the east over the Blythe valley. In 1998 there was an embankment slip near the footbridge giving access to Copt Heath.

Grimshaw Hall is a 16th-century timbered house with gables and notable brickwork, a precursor for Knowle, which has a battlemented church of 1402, buildings from the Middle Ages, a timber-framed guildhouse and the Chester House library, which shows in one building the development of timber house frames from the 13th to the 15th Centuries.

There are moorings between the B4101 and the top of the Knowle lock flight but it seems that many powered craft do not bother coming up any further than the foot of the flight, resulting in increased traffic below. The locks, as often elsewhere, are conveniently grouped together. Five broad beam locks stand beside the six narrow ones they replaced, covered by concrete slabs. The new ones have distinctive 1930s Ham Baker candlestick paddle gear, tall white cylinders leaning slightly outwards, seen from here to Napton Junction. Around them are an assortment of side ponds and flower beds planted with pansies. Strangely, the new concrete locks are above ground level rather than being recessed where the ground could help resist the pressure of the water on the lock walls.

The views to the east over the Cuttle Brook are, again, extensive. Foxes might be seen in the vicinity.

The A4141 crosses King's Arms Bridge. The adjacent public house is the Heron's Nest, this being Heronfield. Close by is a zoo animal sanctuary although most of the animals have come from circuses. Some of the sheep near the canal have more than their fair share of horns. The Black Boy offers an alternative canalside watering hole. The West Midlands boundary follows the canal for a kilometre before giving way to Warwickshire.

One of the most notable houses in the area is the moated 1300 brick and stone manor of Baddesley Clinton. It was a haven for persecuted Catholics and has three priest holes where nine men stood ankle deep in water for many hours in 1591 to evade capture. Little has changed since 1634, except for the 1920s–30s equipment in the medieval kitchen. It is noted for its fine fireplace in the great hall, stained glass, 19th-century Catholic chapel, 49ha of garden with daffodils, stewponds, lake and nature walks. There is also a 16–17th-century church.

Common vetch, creeping buttercup, hop trefoil, cow parsley and sedges are found alongside the canal near the

Did this moorhen deliberately choose a floating nest platform?

There is little to indicate that this is in the middle of the urban sprawl of Olton.

The Knowle lock flight, below which many powered craft turn.

Shrewley Tunnel with its unique horse tunnel climbing away.

B4439 bridge and it is kestrel hunting territory. Following this is Kingswood Junction, where a 200m spur links to the Stratford-upon-Avon Canal, which runs parallel, the two routes never actually meeting in a common junction. The spur is crossed by the Birmingham to Didcot Parkway railway, which comes alongside the Grand Union Canal briefly. The M40 is also only 300m away for the majority of the next 4km, heard rather than seen.

An immaculately manicured hamlet at Turner's Green includes the Tom o' the Wood public house, named after a former windmill. Kingfishers and pheasants live in what was once part of the Forest

of Arden. Rowington has a 13th-century church with a fine peal of bells, timber framed-houses, 17–18th-century buildings and, near the canal, Shakespeare Hall, which belonged to a branch of the bard's family and which has extensive views to the east, as the canal also has here.

Dandelions are widespread and are joined by a mat of cowslips as the canal turns back towards the motorway. The occasional mussel shell is seen on the bank. The

sides begin to rise into fragmented rock cutting. Ahead is the first tunnel, Shrewley. It is 396m long and is wet inside. The towpath passes ferns and, uniquely, enters 54m of tunnel of its own, climbing steeply with brick ribs to assist horses to grip, curving upwards with concrete steps at the top. Ivy proliferates and there can be an odour of fox at the far end.

The motorway is kept away from the canal by a triangular railway junction that used to lead to Cheltenham but now runs out of steam at Stratford. For reasons which soon become apparent, there are fine views eastwards on both sides of the canal. Celandines line the bank and the cuckoo may be heard towards Hatton Country World, located in 19th-century farm buildings.

It has a nature trail, rare breeds, vintage farm machinery, an antiques centre and what is claimed to be England's largest craft village. Hatton stands back on the other side of the canal, even distanced from its Perpendicular and Victorian church.

The Stairway to Heaven is the popular name for the Hatton lock flight, the best locks on the Grand Union Canal: 21 locks drop the canal 45m over 3.1km. Former stables near the top lock house a canal shop, while the Waterman offers stronger liquid refreshment with a fine view from above the fourth lock. A play area includes a boat on dry land to show the basis of piling. The next 11 locks form a line straight down the hillside, or the narrow locks did, the broad ones having been placed on whichever side was most convenient. Directly in line, 4km away, is St Mary's cathedral. To its right is **Warwick** Castle, founded in 1068 by William the Conqueror, some from the 12th Century but mostly 14th Century. Set in 24ha of grounds, it is said to be the finest medieval castle in England. The name comes from the Old English *wering wic* meaning workplace or trading centre at the weir.

The narrow locks act as side ponds and are covered with concrete slabs when on the towpath side. Herons, terns, swallows, goldfinches, yellowhammers, mink and mice might be seen. Below the fourth lock is a Victorian maintenance yard, formerly with blacksmiths, sawpit, hand crane and drydock. It now looks extremely smart and is used as a British Waterways Heritage Skills Training Centre. A large stainless steel dragonfly sculpture has been added opposite for visitors. Also on the flight, Asylum Wharf served Hatton mental hospital,

Kingswood Junction and the spur to the Stratford-upon-Avon.

Immaculate Turner's Green and the Tom o' the Wood pub.

patients unloading the boiler coal themselves, doubtless useful for working off excess energy and stress.

Budbrooke supplied timber for Henry VIII and also for Warwick Castle. St Michael's church was the garrison church of the Royal Warwickshire Regiment and their regimental colours hang on the church wall. A third bell was borrowed from another church for an important wedding and never went back. Neighbouring Hampton Magna is on the site of 1877 barracks, which later housed the Royal Warwickshire Fusiliers, but they were closed in 1960.

After the bottom lock, the canal is crossed by the A46 Warwick bypass and the built-up area begins as suddenly as it finished in Birmingham. To the right lies Warwick horse race course. Ahead is an arched bridge with a surprisingly welcoming arched sign. The welcome is not to Warwick but to the Salterford Branch at Budbrooke Junction, the original terminus of the Birmingham & Warwick Canal. Also opened in 1799 was the Warwick & Napton Canal, called the Warwick & Braunston Canal at that time. Its beginning is well hidden to the left of the arch, looking more like a minor connection or a winding hole.

There is a wild flower garden and a picnic area. There are also two Cape locks, the first by the Cape of Good

The smartened-up Victorian maintenance yard.

The Hatton lock flight, the best locks on the Grand Union.

A452 crosses. There are mallards on the water and hawthorns, willows, horse chestnuts and other trees on each side, sheltering the canal from the adjacent Ford factory on a reach which ends at the B4087 with the Grand Union Restaurant and the Lock, Dock & Barrel.

Royal Leamington Spa was noted for its saltwater springs from 1586 but did not get its royal charter from Queen Victoria until 1838. It has Georgian and Victorian housing, is the home of lawn tennis and hosts the women's national bowls championships each summer, a level of elegance not seen from the canal. Near the station is the art gallery and museum in

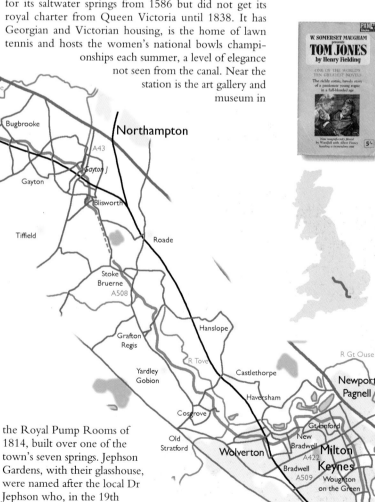

Hope public house and the second taking the canal to its bottom level.

The Lord Leycester Hospital near the canal was founded in 1571 in timber-framed guild buildings of the 12th to 16th Centuries. At various times a council chamber and a grammar school, it is now used for retired or disabled servicemen. At Warwick's Norman gate, it has a chantry chapel, great hall, courtyard with galleries, guildhall, brethren's kitchen, master's gardens and regimental museum of the Queen's Own Hussars. It has been used for filming *Pride & Prejudice*, *Moll Flanders* and *Tom Jones*.

Beyond the A429 crossing the canal is flanked by sycamores, ashes and willows and leads past Emscote Mills, an 18th-century textile factory.

The canal crosses the 70m long Avon Aqueduct on three 13m arches that carry it 9.1m above the River Avon. Powered boaters want to build a deep navigable connection and open up this section of river to powered craft although canoeists, able to portage, are only allowed to use it one day per year.

It is not the Avon that the canal follows, however, but the River Leam. Another noteworthy aqueduct takes the canal over the Birmingham to Didcot Parkway railway on four 6.4m cast-iron spans at 45° to the railway.

New housing has been influenced by canal architecture. It almost surrounds a garden centre, with refreshment facilities, on the A425.

A housing estate leads the canal without ceremony from Warwick into **Royal Leamington Spa**. The Moorings and Tiller Pin stand on opposite banks as the

the Royal Pump Rooms of 1814, built over one of the town's seven springs. Jephson Gardens, with their glasshouse, were named after the local Dr Jephson who, in the 19th Century, developed the spa water concept and was responsible for the town's medical reputation and mini Eden. All Saints church, of 1843, is in Gothic style with a west window by Kempe and a north transept rose window based on that in Rouen cathedral. Canal House of 1820 has wrought-iron balconies.

The canal is crossed for the last time by the Birmingham to Didcot Parkway railway and heads past Sydenham where the noisy Fusilier faces Newbold Comyn Park. Near the old Thornley Brewery the A425 crosses; this area is where it is hoped to make a link to the River Leam to give access to the River Avon.

Rock walls, pines and rhododendrons are met at Radford Semele, the reed ford of Norman landowner Henry de Simely in the days of Henry I. Roman coins have been found and flints and axes show the settlement to be at least 30,000 years old. The hilltop church is now Victorian although its appearance seems to have more in common with its 1100 rebuild. Radford Hall is reconstructed Jacobean with wood carvings, completed by travelling Hugenots in 1622, rebuilt in Victorian times and now divided into several dwellings. There was a popular brewery until 1969. The park grounds of Offchurch Bury run almost to the canal on the north bank as the river is left. The tall grey stone church tower at Offchurch is partly Norman, the name coming from King Offa, who probably had a hunting lodge here.

The *Prince Regent II*, a luxury Edwardian-style cruising restaurant, is based below Radford Bottom Lock. Above this is a viaduct that carried the former Leamington to Rugby railway. More locks follow. There are shallows on the left before Fosse Middle Lock, above which is the Fosse Way Roman road.

Rushes, reedmace, cow parsley, ashes and oaks grow alongside the canal and there are brimstone and white butterflies. Another bridge carries the Welsh Road, a drove road, from north Wales to London.

The toll house at the top of the Bascote flight.

Summer lunchtime at the Blue Lias.

Views over the valley of the River Itchen become extensive, first to the north and then also to the south as the Bascote flight of locks is climbed, a flight including the only staircase pair on the Grand Union Canal main line. At the top is a toll house, the top lock having more the appearance of a private garden, with apple orchard, red hot pokers and yellow roses growing around the lock.

Another bridge used to carry the Leamington to Daventry railway overhead. This area is much more poorly served by railways than in the past. The canal crosses the River Itchen, perhaps with a grass snake swimming harmlessly in it. There are also rare white-clawed crayfish, for which havens have been built in the canal.

The spire of the 13th-century church in Long Itchington was struck by lightning and collapsed during a service in a 1762 storm. The village has many 17–18th-century half-timbered houses and an impressive stand of poplars around the pond. Wulfstan was born here in 1012, going on to become Archbishop of Worcester and St Wulfstan. Elizabeth I stayed at the Tudor House with the Earl of Leicester. The A423 crosses Cuttle Bridge with the Two Boats Inn on one side and the Cuttle Inn on the other and leads south to a model village and a prominent cement works. Quarries produce blue lias stone and lime plus a good supply of large fossils, of which the most notable in 1898 was a 5.8m long near perfect apatosaurus, 9m down and 20 million years old. A reservoir has become a stocked angling lake beside the Kaye's Arm, dug to serve the cement works. Before Stockton Bottom Lock is the popular Blue Lias public house with its apatosaurus sign and flocks of marauding Canada geese, used to being fed by customers.

Half-way up the flight, an adjacent ivy-covered, square, tapered, brick chimney hints at past industry.

Next to a marina and public house, the A426 crosses Birdingbury bridge, dangerous because it is a fast road with a hump hiding the public house car park exit and the minor road junction on the opposite side. This is one of a couple of places on this road where useful signs have been installed, lighting up to warn fast traffic, but their use in inappropriate locations elsewhere has become so widespread that their effect is now devalued.

Elder and birch add to the vegetation. The former Leamington to Daventry railway crosses again just before the site of Caldecott Village. The Calcutt flight is also known as the Wigram's Three.

Above the locks are reservoirs, which were an important source of clay. These feed the canal and are used for sailing and produce dragonflies to frequent the reeds in the summer. Solar panels beside the canal supply power for their sluices.

Originally, it was planned to take a direct line from Warwick to Braunston but it was later decided that money could be saved by using what was already there. From Napton Junction or Wigram's Turn, with its new marina, the Grand Union Canal has taken over a section of the Oxford Canal. Traffic heading for London by the quickest route now travels along in the opposite direction from when it was first dug. It is a wide contour canal with excellent views both southwards to a ridge of high land and northwards over the valley of the returning River Leam. Because it carries traffic from both the Grand Union and Oxford lines it is a busy section of canal.

Napton was one of three large towns in Warwickshire in the Middle Ages. It was given its charter in the 14th century by Edward II. Since then its size has remained unchanged and it is now known for little more than its windmill.

In totally rural fields the marsh harrier might be seen. At Lower Shuckburgh, the 1864 church has contrasting colours of brickwork.

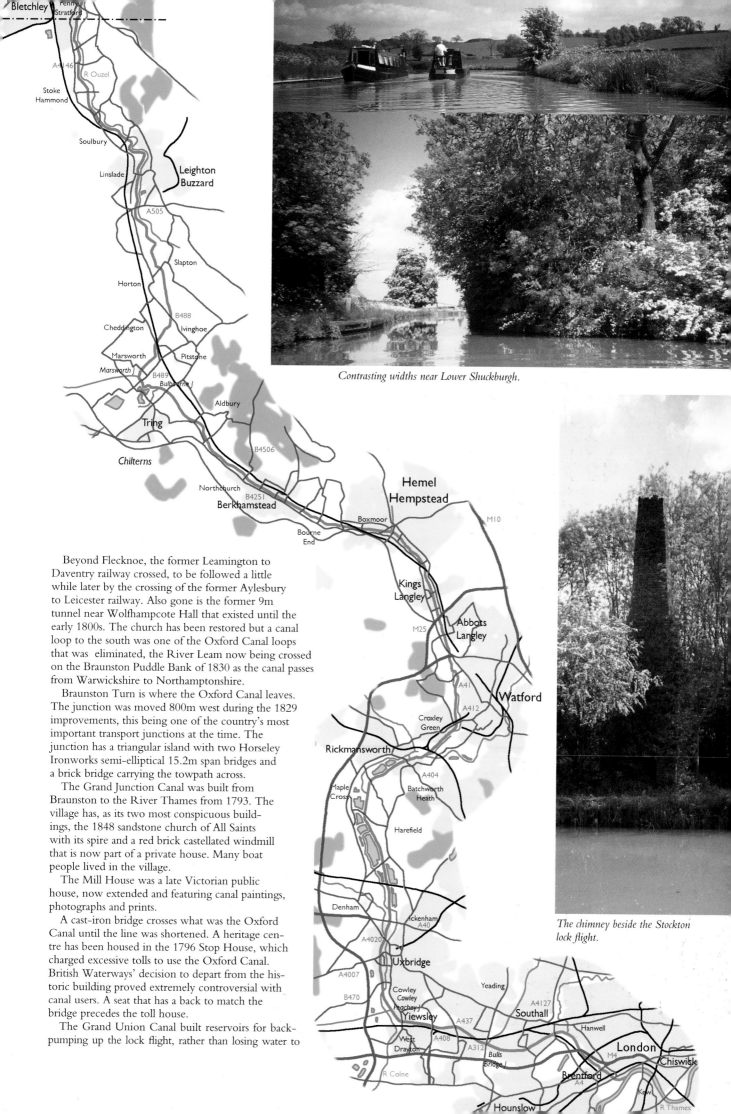

Contrasting widths near Lower Shuckburgh.

Beyond Flecknoe, the former Leamington to Daventry railway crossed, to be followed a little while later by the crossing of the former Aylesbury to Leicester railway. Also gone is the former 9m tunnel near Wolfhampcote Hall that existed until the early 1800s. The church has been restored but a canal loop to the south was one of the Oxford Canal loops that was eliminated, the River Leam now being crossed on the Braunston Puddle Bank of 1830 as the canal passes from Warwickshire to Northamptonshire.

Braunston Turn is where the Oxford Canal leaves. The junction was moved 800m west during the 1829 improvements, this being one of the country's most important transport junctions at the time. The junction has a triangular island with two Horseley Ironworks semi-elliptical 15.2m span bridges and a brick bridge carrying the towpath across.

The Grand Junction Canal was built from Braunston to the River Thames from 1793. The village has, as its two most conspicuous buildings, the 1848 sandstone church of All Saints with its spire and a red brick castellated windmill that is now part of a private house. Many boat people lived in the village.

The Mill House was a late Victorian public house, now extended and featuring canal paintings, photographs and prints.

A cast-iron bridge crosses what was the Oxford Canal until the line was shortened. A heritage centre has been housed in the 1796 Stop House, which charged excessive tolls to use the Oxford Canal. British Waterways' decision to depart from the historic building proved extremely controversial with canal users. A seat that has a back to match the bridge precedes the toll house.

The Grand Union Canal built reservoirs for back-pumping up the lock flight, rather than losing water to

The chimney beside the Stockton lock flight.

the Oxford Canal. The 1897 pumphouse replaced an earlier beam engine, sited in a field on the right. The reservoirs have become one of the largest marinas in the country, with buildings to house blacksmiths, harness makers, ropemakers and carpenters. It was the site of the former Braunston Boat Show, the main inland waterways festival.

Many major canal businesses have been based in Braunston including, until 1970, Blue Line Carriers, the last long distance carriers on the Grand Union Canal. Boats are still moored two deep on both sides of the canal and a workshop with side slips refurbishes narrowboats. Businesses change and an April issue of a waterways magazine suggested it as the site for McDonald's first boat-through restaurant. The Boat Shop is at Braunston Bottom Lock. There is an 18th-century drydock here.

Six locks in the flight take the canal up to another lock shop and the 110m summit level. On the way is the Admiral Nelson, which has Northamptonshire skittles. It was used to film the *Inspector Morse* episode *The Wench is Dead* but has its own story in the form of a black figure who walks through a blocked-up doorway, causing glasses to rattle on the bar and pictures to fall to the ground.

All Saints, Braunston.

Braunston Marina off an abandoned loop of the Oxford Canal.

Double parking at Braunston. In busier times the boats have been moored up to four deep.

The elegant towpath bridges at Braunston Turn.

Stop House at Braunston with seat back to match the bridge.

The summit pound brings moorhens and, more importantly, the 1.9km Braunston Tunnel. It was cut through an ironstone outcrop although 300m of it had quicksand. Setting out errors resulted in an S-bend in the middle so that it is not possible to see through it. This added to the friction in the endless wire loop installed in 1870 and driven by a steam engine at the

Welton end. This was shortlived. From 1871 to 1934 the tunnel was operated by steam tugs that started at hourly intervals from alternate ends. One of the leggers superseded by steam tugs was 75 years old and had worked in the tunnel for 44 years.

The walking route is unmarked. At first it is obvious enough, a track climbing up past rape fields and passing over the English watershed before dropping down near Drayton Reservoir to reach the A361 via a kissing gate badly overgrown with stinging nettles. Opposite is a new concrete farm track. When the track reaches a grass-filled cattle grid and rises to the left it is necessary to take an unmarked footpath to the right.

Pheasants and woodpeckers may be heard. To the north of the canal is Welton, where St Martin's church has a Saxon font that was brought in one piece from East Anglia. Crown jewellers Garrards lived here and were often visited by members of the royal family.

In 1793, permission was obtained to build a canal link to **Daventry**. There are now plans for the work to go ahead. The 5km line will start near the overhead powerlines and follow them towards Daventry Reservoir.

The canal crosses over the stream from the reservoir and there are extensive views to the north over it, enhanced by splashes of yellow in the form of cowslips, buttercups and gorse.

At Norton Junction the Leicester Line of the Grand Union Canal leaves, observed by a single-storey 1914 toll house and by two Victorian cottages, which were occupied by the Salvation Army who helped the boat people.

From here the canal gets very busy for a while as successive major transport links are met. The first in time and position is the A5, the Roman Watling Street, immediately below Buckby Top Lock where, until 1978, a shop used to sell decorated Buckby water cans. It still has the New Inn, the only public house remaining, where there were formerly seven for the six locks. It also has a choice between the two worst portages on the canal. When not obstructed by the lock balance beams there is a very narrow flight of steps leading down between vertical walls right next to the A5, probably not wide enough to stand beside a canoe and no platform at the bottom from which to launch. The much more recent People Pipe under the A5 no longer

Braunston Bottom Lock and the shop.

floods as it did when first installed but the steps down are at right angles to the rest of the subway so it is necessary to lower a canoe in at a steep angle over the handrails at the end, hoping nobody else is using it. Once beyond the A5, launching is not a problem.

Despite the increasingly obvious traffic, a heron still quietly watches the canal.

Next to cross the canal is the West Coast Main Line, here the world's busiest main railway, which follows the canal closely to Blisworth and does not finally leave until Abbots Langley. Guardrails were fitted to stop horses falling in the canal when frightened by steam engines.

Although the M1 does not cross, it follows the canal for 3km, at times just the other side of the hedge, and there is continuous traffic roar.

The Whilton Locks, a source of pottery, are part of the Buckby flight, during the course of which the line of the Roman road from Bannaventa (Whilton Lodge) to Duston is crossed.

Ivy-leaved toadflax grows up lock walls. A waterways magazine carried a report in an April issue that, because of the plant growth on them, the gates of the bottom lock of the Buckby flight had been declared a Site of Special Scientific Interest and use was to be limited, a story horribly plausible with current officialdom.

A marina follows the bottom lock. Despite the noise, the canal is treelined and, in summer, little is seen of the traffic and nothing is seen of the River Nene, which is now being followed.

The motorway leaves the canal to follow the edge of Brockhall Park. Brockhall is a brownstone village and Brockhall Hall is part Tudor, with fine 18th-century interiors. Peace does not return while a peacock has anything to do with it. Passing back under the A5 and beyond a spoil tip area, the canal meanders between the road and the railway but trials bikes find room to make more intrusive noise than the motorway traffic did.

The A45 crosses back as the Saxon Portway at Road Weedon, a village of antique shops. What appears to be an old factory with ornate window panels sets the tone.

A location as far as possible from the coast but with a good new transport system was selected here for the 600m Weedon Military Dock. It had a yellow brick gatehouse with a portcullis, 12 magazine gunpowder stores for 5,000 barrels, separated by blast buildings filled with earth, much small arms ammunition, 800,000 weapons (including 250,000 muskets and 30 pieces of field artillery), accommodation for two regiments, barracks for 500 soldiers, stabling for 200 horses, a riding school and, later, a hospital. There were three large pavilions for the king and cabinet, including a royal pavilion for George III in case of a Napoleonic invasion. Although it was not used after 1920, it was believed to have been a planned staging post for Princesses Elizabeth and Margaret in the event of a Second World War invasion, from where they would have been taken to Prestwick and

flown to Canada. When the railway line was electrified in 1965 the army pulled out and the entry arm was partly filled and built on. It is now private, being used for industrial storage, although it has Grade II listed buildings. In 1991, a boatyard at the end put up a notice saying that they had been there 25 years and warning people not to move in opposite and then complain about the noise. Water voles do not have a noise problem here.

Hidden between the canal and railway embankment at Weedon Bec is the Norman tower attached to a Victorian church. The village's manor had been held by the abbey of Bec Hellouin in Normandy in the 12th Century.

As the canal turns away from the open rolling Stowehill, the railway carries on and blasts straight through it; canal-style brick vent shafts are visible on the hillside above the tunnel. The canal bores under the A5 again and passes the Narrow Boat and a small marina to arrive at a bridge near Flore. The garden corner between the road and canal has a group of topiary animals at the waterside, including an elephant and a rabbit.

From Flore the canal continues through completely

47

Flore's fauna flora.

rural countryside as it moves away from the River Nene, the A5 and the M1, disturbed only by the West Coast Main Line at intervals. There are the inevitable moorhens, mallards and swans plus swallows in summer. Ashes, hawthorns and dog roses line the canal with yellow irises and lilies in the water, hardly what would be expected for England's spine canal.

Powerlines cross the canal three times at Nether Heyford where the church has a monument to judge Francis Morgan who pronounced the death sentence on Lady Jane Grey and then committed suicide with

Rural views at Bugbrooke.

The line of the Northampton Arm from Gayton Junction.

The great warehouse in Blisworth.

Horse and tramway wagon sculpture at Stoke Bruerne.

remorse in 1558. The village has a 2ha tree-lined green and a large Roman building to its east.

Bugbrooke has fine 18th-century houses, a pretty Baptist church of 1808, a 12th-century chapel in parkland and a 14th-century parish church with a notable 15th-century wooden screen and a poem in the bell-tower warning against improperly dressed bellringers. The village had the first soap factory in England, as well as a brickyard, bakeries and mills, of which Hetgate Mill is the only one still operating. From the Wharf public house the canal crosses the Bugbrooke valley aqueduct.

Cow parsley and docks grow strongly in the summer and there are also fine colours in autumn.

These Northamptonshire villages have an air of quiet affluence. Gayton is another with large 16–19th-century stone houses and a large church with an ornamented tower and an old manor house site.

Canada geese flock and reedmace begins to appear near Gayton Junction where the **Northampton** Arm leaves to drop down to the River Nene, currently the most practical route for many powered craft to reach the rivers of the east of England. There is a British Waterways depot at the junction and a roving bridge across it. The A43 crosses, as does the railway which then crosses the old A43 line north of Blisworth on a Robert Stephenson bridge that has particularly good attention to detail. A tank in a garden previously supplied canal water to railway water troughs.

Blisworth has thatched houses with brown Blisworth stone and light Northamptonshire sandstone in alternate layers for decorative effect, a benefit of having had tramways bringing stone from quarries to the canal. It begins at Candle Bridge, where boatmen bought their tallow dips for Blisworth Tunnel. St John the Baptist church dates from the 13th Century and stands behind ornate cast-iron graveyard gates. The Sun, Moon & Stars Inn is now a teahouse and art gallery and Blisworth also had a mill. For the canal user the outstanding building is the great warehouse of 1879.

The village is left but it has one more superlative: Blisworth Tunnel. At 2.8km long, it is the third longest canal tunnel in Britain, being surpassed only by Standedge and Dudley. Opened in 1805, it provided the greatest technical difficulty on the whole canal. The first attempt to cut through the ironstone outcrop in 1793–6 had to be abandoned because of floodwater, the current line being a little to the west. In 1977, part of the invert was found to have lifted by as much as 1.2m so the central section was lined with concrete segments over 1982–4, during a five year closure, yet it still showers water and the walls are covered with calcite. Restoration evolved techniques that were later used for the probe and service tunnels for the Channel Tunnel. A kink near the southern end means that daylight cannot be seen all the way through it. Until 1871 boats were legged through, there being a leggers' hut at the southern end; steam tugs were then used until 1934. Pipistrelle and other bats roost in it perfectly happily.

Although the walking route is not marked, it is followed by a road for most of the way but visibility is not good and there is no footway. At intervals there are tall brick vent columns, the tunnel having been built originally from 19 shafts plus the end portals. These construction shafts were reopened after two boatmen suffocated in 1861 and five children had to be revived after a trip through in 1896, from when women and children were required to walk over. Pheasants and cuckoos are heard and orange tip butterflies add summer colour. Towards the southern end the road bends right and a track ahead descends to the canal. On the way it passes through an embankment which once carried a tramway from Tiffield to Roade. Before the tunnel was built there was a tramway connecting the two canal ends, the first iron railway in the south of England. Above the left bank is a footpath with steel rod sculptures of two deer, a fox and a horse with a tramway wagon. Nest boxes proliferate, as do rabbits.

At the southern portal, one of the concrete lining rings has been set into the sloping bank to show the size to the hordes of spectators visible on any summer weekend, this being the start of Stoke Bruerne. Split by the canal, this is probably the country's best example of a canal village. Built in brown Blisworth limestone, the village had a boat children's school, Sister Mary Ward had a boat people's surgery for thirty years until the mid 1950s and it was the site of what was probably England's last handmade ropeworks. The lock cottage has an Italianesque bay window while the bottom gates have round balance beams and an iron gate from Welshpool Town Lock.

Stoke Bruerne Waterways Museum is located in a restored 1840s stone cornmill, used until 1913. Opened in 1963, it was Britain's first canal museum, with boats, engines, costumes, documents, photographs, paintings, models, a cabin, cabinware, china, utensils, brasses and signs. The museum shop is in an old beam engine house. The Old Chapel Gallery and Crafts Gallery is a museum of rural life and has the work of up to 60 artists in wood, jewellery, ceramics and textiles. There are tea rooms in old stables. The Boat Inn has been run by the same family since 1877 and is one of the few with Northamptonshire skittles. Over the village stands the Norman tower of the Perpendicular church. The village was used for filming some of *True Tilda*.

The Navigation is a newer public house but equally convenient for the canal.

The pond of a former brickworks is now a nature reserve with green woodpeckers, dragonflies, butterflies, shrews, mice, grass snakes, anthills and varied plantlife.

Stoke Park House was built in 1629–36 as the first Palladian house in England. Designed by Inigo Jones, it had flanking pavilions, colonnade and fine gardens but was badly damaged by fire in 1886.

The Stoke Bruerne lock flight is crossed by the A508, beneath which it is like an art gallery. Framed mosaics by local schoolchildren on canal and other themes are mounted on both abutments of the bridge, a striking discovery. Ivy-clad cottages face the bottom lock.

The canal crosses two arms of the River Tove on the level and then follows its valley down to the River Great Ouse. The river passes over a long weir, above which the towpath is supported on brick arches. The design is seen again but not as extensive as here.

On a hill overlooking the canal, Grafton Regis has a 13–14th-century church with a Norman font, fine rood screen and many monuments to the Fitzroy family. Regis was added to the name following the visit of Henry VIII and Anne Boleyn to the manor house, home of Elizabeth Woodville, who had married Edward IV here in 1464 and become the mother of the Princes in the Tower. It was destroyed in the Civil War and is now a hospital. Charles II made his illegitimate son, Henry, the Duke of Grafton. The White Hart also has Northamptonshire skittles.

There are extensive views over the Tove valley. In particular, on the skyline 4km away is the Norman church of St James in Hanslope with the finest spire in Northamptonshire.

A sign warning of elderly ducks crossing is the introduction to Yardley Gobion although it could equally have made reference to herons. The church dates from 1864 but there are older buildings of note. A 14–15th-century pottery has become a public house, confusingly called the Coffee Pot. The Royal Oak is where Edward IV met Elizabeth Woodville.

North-east of the Navigation Inn at Thrupp Wharf is Castlethorpe, its motte and bailey to be found behind

One of the tunnel vent shafts in a field.

Former British Waterways chairman Dr George Greener celebrates Blisworth Tunnel's bicentenary.

Boat Inn in Stoke Bruerne – the quintessential canal village.

Children's mosaics beneath the A508 bridge.

The long arched weir feeding the River Tove.

Beware, Yardley Gobion ahead.

Wire boatman sculpture near Cosgrove Lock.

the Norman church of SS Simon & Jude. The church tower collapsed in 1700, to be replaced with a Georgian one. Again, there are fine views over the Tove valley.

The daughter of the house at Cosgrove Priory fell in love with a shepherd. Her parents did not approve and had him deported on a false charge of sheep stealing. The daughter drowned herself in the millrace but her ghost may still be seen at the full moon. The mill was demolished in the 1920s and the priory is now offices.

Entry to Cosgrove, another village split by the canal, is under Solomon's Ornamental Bridge. Dating from 1800, in Gothic style, the bridge is the most elaborate on the canal. St Vincent's Well has a high iron content, good for eye problems, and is protected by Act of Parliament. The canal passes the Victorian rebuilt church with its 14th-century weathercock and loops round the Barley Mow and the bay-fronted 18th-century Cosgrove Hall, in front of which a Roman bath house has been excavated.

From Cosgrove Junction, the Old Stratford & Buckingham Arm heads south-west but it only gets 200m. Last used commercially in 1932, it was closed in 1961 and has mostly disappeared although the line is largely unobstructed.

A large 2003 wire sculpture of a boatman leaning on a pole faces Cosgrove Lock, which drops the canal to its area bottom level of 70m. Narrow gauge railway lines used to serve local gravel pits. On the north-east side of the canal is Cosgrove Leisure Park with eight lakes resulting from sand and gravel workings and now with funfair, water skiing, jet skiing, windsurfing, swimming, angling and a picnic area.

The canal runs straight across the valley of the River Great Ouse on a 1.4km long embankment, originally crossing the river on a stone aqueduct of 1805, although this collapsed three years later. While reconstruction

Solomon's Ornamental Bridge at the entrance to Cosgrove.

The remains of the Old Stratford & Buckingham Arm.

work took place a flight of locks were built down to the river and up again on the west side although these were problematical when the river was in flood. The remains of the locks and the old aqueduct remain in the bushes. In 1811 they were superseded by the Iron Trunk aqueduct, which has less than 200mm of upstand on the west side and which locals called the Tank or Pig Trough. A scheduled ancient monument, it was one of the first of its kind in the world. Beneath the canal there is a narrow oval pedestrian tunnel. A 1981 plan was to place locks on the Great Ouse to make it navigable to powered craft to Bedford. If the Bedford & Milton Keynes Waterway fails, the plan may be revived although small craft are not permitted to use most of it currently. Crossing the aqueduct takes the canal from Northamptonshire to Buckinghamshire; the canal then follows the river to Haversham.

The embankment ends at some houses, one of which has a dovecote, and the Galleon public house. Behind this are an 1815 church in Norman style with a 1729 rectory, the remains of a Norman motte and bailey and the site of the medieval Wolverton Village.

This is the start of the new town of **Milton Keynes**, built on 89km^2 of farmland from 1967 with the planting of millions of trees and sweeping American-style roads. Bill Bryson said 'I didn't hate Milton Keynes immediately.' It took a couple of hours of walking, trying to find the town centre, before he decided it was even worse than Gateshead's MetroCentre. Other views are divided. Clearly, many people like the little boxes hidden away in tree-screened culs-de-sac and few places of this size have such easy rush hour driving. The town is built in a loop of the canal and, where the canal does enter the built-up area, it is largely in parkland.

Initially it is almost in a tunnel of trees with jays and magpies. To the north is Old Wolverton with Europe's largest recycling factory. To the south is **Wolverton** with the abandoned 1838 railway carriage workshops, responsible for building the royal trains, among others. It has been one of the worst areas on the canal for graffiti and litter, with Second World War camouflage, but seems to have shaken off the image, now with bracken and an improved demeanour.

The West Coast Main Line passes over and a 150m long black and white railway mural of 1980 by Bill Billings and the Inland Waterways Association has been painted on an extensive curved wall.

At the start of New Bradwell, a Victorian railway workers' town, is the Grafton Street aqueduct of 1991, the first new aqueduct in England for 50 years and one of the most complicated pieces of civil engineering in Milton Keynes. Near it is an 1803 sandstone smock windmill, sited by the canal for transport purposes, used until 1871 and now restored. An 1858 church by Street contains Victorian stained glass but a Norman chancel arch comes from a ruined church in Stantonbury. There are also the remains of Bradwell Abbey, founded in 1154 by the Benedictines. Waterside trees include birch, cherry and alder.

The pedestrian tunnel and the Iron Trunk over the River Great Ouse – the first in the world of its kind.

After **Newport Pagnell** Road bridge, its abutments spray painted with bright butterfly murals, the canal is back in farmland with Highland cattle among the black and white Fresian cows. It passes the wildfowl centre in Great Linford gravel pits which was greylag geese, a winter gull roost and sailing. There are roach, bream and perch in the canal and British Waterways have stocked it with mirror carp for anglers to catch.

Garden decorations near the Proud Perch include a rowing boat end built into the bank and a skeleton draped over a tree branch.

The canal turns sharply under a former railway bridge and passes the 14th-century St Andrew's church in yellow stone with a 12th-century tower, Georgian box pews and fine 19th-century stained glass. The late 17th-century limestone Linford Manor, with Greek temple, almshouses with Flemish gables, two-storey school section and farm buildings, has become the Courtyard arts centre with pottery, jewellery and silversmithing. It was built by Sir William Pritchard, Lord Mayor of London, who wanted this as his main country seat. Away from the more formal areas there are cowslips, woolly thistles, oxeye daisies, knapweed, butterflies, other insects and considerable birdlife.

The Newport Pagnell Canal served marble quarries from 1817 to 1864, became the Newport Nobby Railway until 1967 and is now the Railway Walk with nature reserve. An old bridge was retained after residents protested against its demolition. The old Wharf Inn at Linford Wharf has become a private house.

Giffard Park draws attention to the parkland environment, which seems to be the sort of English river where narrowboats are less likely to be met than bailiffs.

The Dell is restored brick kilns, used from the 1890s to 1911, the willow and hawthorn screened ponds being the claypits.

Pennyland Basin has large houses with their own moorings, the result of a 1980 design competition which won an architectural award. Damselflies frequent the flowering rush and water dock.

By Gulliver's Land and Willen Lake are the first Buddhist peace pagoda in the West and St Mary Magdalen, the only Wren church outside London. This reach is the starting point for the proposed Bedford & Milton Keynes Waterway. At 32km long, a broad canal, it will connect the canal network with the East Anglian rivers and will be the longest new canal in Britain for over a century. However, previous attempts in 1811 and 1892 failed and success is not assured this time.

One of the surprising things about Milton Keynes is its name. Rather than being called after Bletchley, Wolverton or one of the towns it incorporated, Milton Keynes uses the name of a village that manages to retain clear space between itself and the new town, standing on the far side of the River Ouzel or Lovat, the valley of which the canal is now to ascend.

Poplars line the canal down towards the Peartree Bridge Inn where alders and horse chestnuts surround a marina. The Old Swan Inn at Woughton on the Green was used by Dick Turpin and his ghost has often been seen on his horse, Black Bess. One night, while being pursued, he apparently quickly reshod his horse with the shoes on backwards to fool his pursuers. Assuming he had the tools and noise was not an issue, how long does it take to reshoe a horse? Would such an experimental procedure be the best use of that time unless the pursuers were so far away that a more leisurely departure could have been made?

Woughton Park and Walton Hall have been home of the Open University since it was set up in the 1960s to provide correspondence courses for those not able to attend full time study at university.

From Woughton Park the canal passes Simpson, a beautiful thatched village caught up in Milton Keynes,

Part of the long railway mural at Wolverton.

Thatched cottages in Simpson.

epitomised by its mainly 14th-century church with wooden rood screen and a 1789 monument to John Bacon.

The George Amey Outdoor Education Centre operates from a former lengthsman's house. Trekkers serves those in need of liquid refreshment. Hedgerows of hawthorn, ash, willow and elder shelter reedmace, cow parsley, yellow iris and lilies, with mallards, moorhens and magpies adding life.

A windmill stands back from an attractive wharf setting, just before the A5 crosses.

Before Fenny Stratford there is a house with a garden seat which uses a cartwheel as its back, attractive to look at but probably less comfortable to use, especially with its large boss in the centre. The lock is also awkward to use as it has a swing bridge across the centre of the chamber. On the other hand, the rise is only 300mm,

Lift bridge and private marina at Fenny Stratford.

Fenny Stratford Lock.

An avenue of poplars at Water Eaton.

Looking down the Soulbury Three Locks.

the smallest on the canal. The bottom pound leaked but only near the top of the canal. Rather than solve the leakage it was decided to install an extra lock and accept a slightly lower level in the bottom pound. The old northern engine pumphouse was originally steam powered and has a craft centre, gift shop and tea room, run by handicapped youths, or the Red Lion is close.

The **Bletchley** to Bedford railway crosses and the following bridge is on the line of the Roman Watling Street, accompanied by the Bridge@Fenny. On November 11th six cannons will be heard, the Fenny Poppers, celebrating the dedication of the red brick St Martin's church. It was built from 1724–30 in early Gothic revival style. The tradition was begun in 1760 by antiquarian Dr Browne Willis, the lord of the manor.

Now closely following the River Ouzel, the valley of which it climbs to Grove, the canal is again lined by poplars and inhabited by herons and swans. At the end of Water Eaton it is back in open country although Stoke House stands prominently on the right bank.

It is still some way to Stoke Hammond, which begins with a church that has a squat central tower and a decorative 14th-century font. Stoke Hammond Lock is also known as Talbot's and has a Victorian pumphouse with semi-circular windows.

A golf course and deer farm are passed to reach Soulbury Three Locks or the Stoke Hammond Three. The Three Locks public house is located in the former stables. The area is haunted by a woman and child, the noises of a lock are heard operating at night, pram wheels squeak and a woman calls for her drowned family. Crows have learned to catch crayfish when locks are empty by nipping off their claws. Terns and kestrels also hunt along the canal and pochard and harlequin ducks are less usual species that may be found.

As the canal moves from Buckinghamshire to Bedfordshire there are fine views over the River Ouzel valley, interspersed with gorse as the canal begins to follow the river's exaggerated meanders. St Mary the Virgin's church dates from the 12th Century. At the point of one of these meanders, just as the railway goes into a tunnel,

The Globe on the northern edge of Linslade.

its portal in grey neo-Gothic brick, is the thatched Globe. This is one of the most attractive canal public houses to be found anywhere, narrowly missed by a tree falling across the canal in 2005. Herons also take refreshment.

After Leighton Lock is Wyvern Shipping, one of the oldest hire boat companies, using a site earlier used by carrier L. B. Faulkner and canal painter Frank Jones.

Tesco have a canalside site in **Leighton Buzzard**. This town of 17–19th-century and half-timbered houses includes the 17th-century Holly Lodge and a Georgian high street plus a pentagonal market cross of 1400. All Saints' church of 1288 has a 58m central spire, 15th-century wooden angel roof, sanctus bell, 13th-century font, misericords, brasses, medieval lectern and fine stained glass by Kempe and others. It also has 15th-century graffiti carving on a pillar, while guarding a relic of St Hugh's cloak of the fictitious Simon and Nellie who made the first simnel cake, a rather contorted chain of reality. The bounds are beaten on Rogation Day. The 1633 almshouses were founded by Edward Wilkes and a choirboy stands on his head in front on May 23rd while the will is read. The town hall is 19th-century and the Cedars are where Mary Norton grew up, used by her as the basis of Firbank Hall in *The Borrowers*. The town name may be from the Old English *leac-tun*, a vegetable farm, belonging to the Buzzard family.

Behind the greenhouses by the canal on the Linslade side is St Barnabas' church of 1848, also with Kempe stained glass windows.

The scenery opens up to the left as Leighton Buzzard centre is left and a footway crosses on a bridge which once carried a railway from Linslade to Luton. Tiddenfoot Waterside Park, with a former sand quarry, now flooded, offers pony rides, walks, picnics and angling.

Silica sand and gravel used to be supplied to London by canal until the late 1940s, some of the track now being used by the Leighton Buzzard Narrow Gauge Society. The remains of some 1920s tramway track stands quite high above the canal for easy loading.

The canal moves back into Buckinghamshire.

The Grove Lock cottage has been converted to a public house. At the following Church Lock, the 14th-century chapel with bell turret has been converted to a private house. There is a large flooded sand pit to the left. The Chilterns are becoming more prominent ahead. Bury Farm has buffalo and sells buffalo milk, ice cream, cheese and yoghurt.

The locks on the ascent of the Chilterns are mostly in open country. Slapton Lock or Neale's is not too far from Slapton, where the Perpendicular church has notable 15–16th-century brasses.

Horton Lock gives the best view of the scarp slope of the Chilterns before they become lost to the foreground scenery. To the east are hot air balloons and gliders flying from Dunstable. The 150m Whipsnade white lion was cut in 1935 in the chalk of the Dunstable Downs to promote the zoo, the largest chalk carving in the country.

Church Lock with its wide and former narrower bridge arches.

One of the Seabrook locks.

As the climb continues the locks get closer together and the route twists and turns more frequently. The two Ivinghoe Locks are followed by three Seabrook Locks, the middle of which has the best preserved of the back-pumping northern engine houses ('northern' meaning those north of the Chilterns). To the east are Gallows Hill, with an array of aerials, and Ivinghoe Beacon, the highest point on the Dunstable Downs at 230m. The beacon was prepared for Elizabeth I to summon men in the event of a Spanish invasion. Ivinghoe probably gave Sir Walter Scott the title of his novel, *Ivanhoe*. Beyond Pitstone is Britain's oldest post windmill, the Pitstone mill of 1627. At one time it was owned by the canal company but they sold it in 1842 to the Ashridge estate. It was badly damaged in a squall in 1903, was restored in 1962 and now belongs to the National Trust. Birdlife on the canal ranges from cormorants to swallows.

A low swing bridge crosses the canal, then the West Coast Main Line crosses on a skew bridge with concrete arch ribs running parallel to the tracks. Cheddington was the scene of the Great Train Robbery in 1963. A mail

Canalside tramway track at Leighton Buzzard.

Chiltern farmland to the south-east of Cooks Wharf, Pitstone.

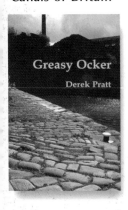

train was carrying £1 million in old bank notes that were on their way to be destroyed. Ronald Biggs and friends stopped the train and attempted to save them the bother by taking the notes to a farm in Linslade. Derek Pratt's novel *Greasy Ocker* adds a narrowboat dimension to the hold-up.

Above the pair of Marsworth Locks, or Norman's, is a line of well heads in a field, one with a small tree growing from its top. Angling is banned on this pound of the main line and the Aylesbury Arm as far as the first lock. As a result, the carp are able to grow to full size and swim about at the surface without fear. Indeed, the locals feed the fish with even more enthusiasm than they expend on feeding the ducks.

The 13th-century All Saints' church is another with a squat tower, rebuilt in the 19th Century. A public house has become a thatched shop and a moat is found near the canal.

At Marsworth Junction there is a British Waterways depot accompanied by a hand-operated crane. The junction is the start of the Aylesbury Arm.

Marsworth, also known as Maffers, is one of the most attractive points on the canal, especially where the B489 crosses by the White Lion at Startop's End. Bluebell's Tearooms are in a lock building. The adjacent car park is notorious for theft. **Tring** Reservoirs were built in 1802–39, with water pumped up to the 120m summit level where the channel is deeper than usual for storage. There are white campion and field pansies by the reservoirs, which have been a 20ha national nature reserve since 1955; hides are located around this migration rest point. There are breeding and wintering birds plus a heronry, edible dormice and orange foxtail, broadleaved ragwort and round fruited rush. The British Trust for Ornithology have their headquarters close by so nothing unusual is going to escape notice.

Seven more locks, the upper one used in filming *The Bargee*, take the canal to the 5km summit level from Bulbourne Junction where there are a Georgian drydock, a large lock keeper's house and the Wendover Feeder Branch, currently being restored.

North of the canal is the College Lake Wildlife Centre in chalk pits, begun in 1937 and cement works using excavated chalk and marl.

An 1848 British Waterways depot with fanlights and a water tower with spire has been the main lock gate construction site for the south of England and built the model for the Foxton inclined plane in 1897. It was closed in 2004, despite outcry from canal enthusiasts, but will be retained for operational purposes. Nearby is the Grand Junction Arms. There is an annual Tring Canal Festival here.

Moving from Buckinghamshire into Hertfordshire the canal is now in Tring Cutting, 2.4km long and up to 9m deep. The site of a wartime bomb strike and a controversial and expensive improvement scheme, these days it is lined with sycamores and other trees, ivy and clematis, and attracts kingfishers. There are horse ramps to help with recovery from the deep water. The railway is hidden in its own Tring Cutting, 4km long, one of the first major railway cuttings.

Marsworth Junction with the main line in the foreground and the Aylesbury Arm going right beyond the British Waterways depot. The Chilterns lie beyond.

The White Lion at Startops End.

The line twists and turns towards the summit at Startops End.

Beyond Aldbury, with its village green, stocks and pond, is a 33m Doric column and urn monument of 1832 to the 3rd Earl of Bridgewater, the canal pioneer who owned the 16km^2 Ashridge estate on which it is sited, whigh boasts good oak and beech woodland. The Chilterns are of chalk, some of the youngest geological formations in Britain, and are noted for their beech-woods and chalk turf.

Toll keeper's house at Bulbourne Junction, the start of the Wendover Feeder Branch.

Bulbourne Dry Dock.

The deep Tring Cutting.

British Waterways' workshops at Bulbourne.

Magnificent copper beech on the Dudswell flight.

The totem pole at Berkhamstead.

Cow Roast is a corruption of cow rest, where there were pens for the use of drovers bringing cattle from the Midlands to London. The lock begins the continuous descent to the River Thames. Crossing the bridge is the Icknield Way Path. While there are various routes suggested for the Icknield Way, the Ridgeway and the current footpaths which take their names, it is clear that the route generally ran along the Chilterns in this vicinity, a Neolithic route founded around 4000–2000 BC, the oldest road in Europe.

The lock also has a pumping station and a pond of large terrapins and is a frequent prizewinner for its flowers.

Watercress beds, using mineral water from the Chilterns, are found often until Springwell Lock, helped by following the River Bulbourne from Dudswell Top Lock.

To the north-east of the canal, Northchurch Common has such birds as nightjars, woodcocks, grasshopper warblers, redpolls and bramblings. Northchurch Top Lock is also known as Bush or Barker's. The church itself, St Mary's, in Saxon flint, is one of the world's oldest, with a handsome 15th-century Flemish chest, fine 19th-century stained glass and the 1785 grave of Peter the Wild Boy, who was found near Hanover moving on all fours and living on grass. There are timber-framed houses in the high street and a section of Grim's Ditch to the south.

The proliferation of lock names continues. The next one, below which is a magnificent copper beech, is called Bushes, Old Ned's, Awkward Billy or Crooked Billet. The two Northchurch Locks, by a park and playground, are also called the Gas Two from a former gasworks. **Berkhamstead** Top Lock is also called Broadwater above a striking bridge painted black, red and gold.

The former market town of Berkhamstead drew its wealth from wool in the Middle Ages but also became the port of Berkhamstead. It has four 17–18th-century coaching inns and a 17th-century grammar school. Graham Greene was born here and went to the school, where his father was headmaster, as a result of which he was bullied and was sent away for six months. Sir James Barrie visited friends in Berkhamstead with five sons, who inspired *Peter Pan* in 1904. Canada geese, kingfishers and even otters are found here.

Berkhamstead Castle has a double moat, a tall 11th-century motte and bailey, curtain walls with D-plan towers and a square tower-like keep. It was built by Robert of Mortain, the half brother of William of Normandy, who received the oaths of allegiance of the Saxon kings here in 1066. In the 12th Century the timber was replaced with stone and flint walls, breached by the stone-throwing machines of Louis of France and the barons opposed to King John in 1216, who took the castle after a fortnight's siege. Projecting platforms of the outer earthworks may be siegework from this campaign. Brick buildings were added later. Thomas à Becket spent time here as Chancellor. The Black Prince lived here from 1336 and it was a favourite royal residence until Elizabeth I. It was visited by Geoffrey Chaucer and Henry Tudor and was used to imprison King John of France but had fallen into ruin by 1495, the materials being used in townhouse building. It still belongs to the Prince of Wales.

The Crystal Palace public house was designed by Parton, who also designed its larger namesake. Opposite is a totem pole that belonged to a Canadian timber company; it was carved in 1967 by a Kwakiutl Indian from Vancouver. However, when they moved to new premises they were not allowed to take their property with them. Bridgewater Boats occupy what was a boat-building yard, boats being launched sideways into the canal. Berkhamstead Bottom Lock, by the Boat, and the following lock, by the Rising Sun which had stables, were both referred to as Sweep's because chimney sweep Eli Oliffe kept a boatman's store. Dalton Wharf, which follows, had the first factory for producing sheep dip.

The 13th-century St Peter's church, one of the largest in Hertfordshire, was restored in 1871 by Butterfield and has several brasses and an east window to William Cowper, whose father was rector when he was born in 1731. The 16th-century Incent House has been restored but the 17th-century Sayer almshouses are intact because of a bequest by Charles II's chief cook. The A4251 follows the line of the Roman Akeman Street.

There are four Bourne End Locks, Lissamer's or Topside, Bottomside (next to watercress beds which were harvested by a narrow gauge railway), Sewerage (adjacent to a sewage works) and Irishman's (after a drowned lock keeper). Little grebes are found near Bourne End Mill, which was damaged by fire in 1970 and rebuilt as a motel. St John's church of 1854 is by George Gilbert Scott. There has been a public house on the site of the Three Horseshoes since 1535. It now suffers from flooding if one of the Winkwell Locks is operated incorrectly.

The West Coast Main Line crosses back over another of the concrete arch viaducts. Above Boxmoor Top Lock are houses with enormously long lawns down to the canal.

Watercress beds are being abandoned for use as a nature reserve.

Boxmoor House has been built at the site of a Roman villa. Boxmoor also has an 1874 Norman Shaw church and two white stones marking the grave of highwayman Robert Snooks. The Fishery Inn draws attention to the Fishery area of Boxmoor.

Hemel Hempstead is known for Kodak and the name is prominently displayed facing across to Hemel Hempstead Cricket Club's ground, which is screened by poplars as the canal approaches Boxmoor Bottom Lock. The A414 crosses Two Waters Road Bridge, adjacent to the confluence with the River Gade, which is now followed. Between the bridge and a timber yard is a B&Q outlet at what was Rose's lime juice works supplied by canal until 1981.

Conifers, willows and copper beeches are among the trees lining the canal, with birdlife including kingfishers, great crested grebes and Canada geese. Mink are also present.

Another important user of canal transport was Dickinson's Apsley papermills, now largely replaced by housing and a marina. The four large Apsley Locks caused a lot of disruption to the mill's water supply so they were replaced in 1819 with five lower volume ones further south. One old mill building does remain beside the canal, just upstream of a new tubular steel footbridge.

This contrasts with an unusual bridge over the canal, with a navigation arch and a flat lower arch which barely clears the water.

Nash Mills were making paper from 1797; they were bought by John Dickinson in 1811.

Ye Olde Red Lion make their own pies. There is a playground close by.

The West Coast Man Line crosses for the last time and the canal wanders down between banks of elder and blackberries and a spring snowstorm of white willow blossom to **Kings Langley**, site of a former papermill and the Ovaltine works and now with a wind turbine.

When the foundations were being dug for the Rudolph Steiner school in 1970 they unearthed the remains of a 13th-century wine cellar. There was a 13th-century friary here and Henry III had a palace. Edmund de Langley, the 1st Duke of York and brother of the Black Prince, was born at the palace in 1341 and is buried in the Norman All Saints church, as is archaeologist Sir John Evans. The Saracen's Head is rather more recent, 17th-century.

Home Park Mill Lock or Five Paddle Lock is by a former Dickinson's site.

A royal hunting lodge site and a moat are overshadowed by a 440m viaduct carrying the M25 over the canal, river, road and railway. The London Orbital is the world's longest city bypass and Britain's busiest motorway and Junction 20 is on the west side of the canal. For practical purposes many people now regard the M25 as London's boundary. Canal piling by the viaduct is tied together with a mixture of standard and narrow gauge railway line.

Home Park Farm Lock is on the outskirts of **Abbots Langley**. In 1931 the Ovaltine Egg Farm was established here, based around a thatched and half-timbered building. It is now a housing estate. In the 11th Century

The M25 viaduct sweeps across the valley at Abbots Langley.

Nicholas Breakspear was born in the town, becoming Pope Adrian IV, the only English Pope.

The first of the Hunton Bridge Locks has an attractive lock keeper's hut that faces across to a bank covered with roses, where a green crocodile peers from the vegetation. In Langleybury, the King's Lodge restaurant was a 1642 hunting lodge for Charles II and its stucco ceilings have arms and insignia of the royal house of Stewart.

Donkeys live by the canal, as do herons. The A41 crosses, to be followed by a spur of the M25, the end of an early section of the motorway. Any biplane aerobatics might be answered by helicopters taking off from the canal bank next to the motorway.

Watford, a market town in the Middle Ages, is the largest town in Hertfordshire.

To the west of Lady Capel's Lock is Grove Park, formerly 77ha of gardens for the Earl of Essex in the 17th Century, landscaped by Stubbs. It is now a golf course with chestnuts, oaks, ashes and a country park feel. The balustraded Grove ornamental bridge was a demand of the Earl of Essex when the canal was built.

Grove Mill was formerly the home of the Earls of Clarendon, now exclusive flats with some interesting pieces of machinery placed like sculptures on the lawn by the canal. The canal has some tight bends and a wooden signal box was built to control canal traffic, the towpath diverted to the east side. The local fauna might include llamas.

The lower of the Cassiobury Park Locks has a Victorian cottage that was wrecked by vandals. British Waterways were demolishing it in 1970 when a buyer arrived, using an Alsatian to stop the demolition and later going on to restore it. A notice on the fence warns about the danger of crocodiles.

Cricket and football pitches lead into 77ha Cassiobury Park, which includes beech and sweet chestnut woods and an avenue of limes planted in 1672. It

Apsley with its old warehouse and new footbridge.

An unusual design of bridge at Apsley.

Helipad next to the M25 spur.

The ornamental Grove Bridge.

Grove Mill with old machinery decorating the lawn.

also has a copper beech planted in 1987 by the Duke of Marlborough to commemorate the canal's bicentenary. Rabbits work to match the various ponds in the woods.

Ironbridge or Watford Lock has seats made from balance beams that look significantly more substantial than the metal ones on the lock gates. The lock is haunted by a man who opens the bottom gate, whatever the balance beam design.

Watford has a carnival at Whitsun, perhaps reached beyond Cassiobridge Lock over the bridge carrying the Metropolitan Line, the canal's first meeting with the London Underground system.

An industrial estate occupies the site of Croxley Mill, demolished in 1982, which used coal brought via the Ashby-de-la-Zouch Canal and esparto grass also delivered by canal. Dickinsons were probably the largest customer for Grand Union Canal transport.

Swans have nested on the east bank directly above Common Moor Lock and have not taken kindly to anyone attempting to use the lock gate.

At the southern end of Croxley Green, Croxley Hall Farm's medieval tithe barn is the second largest in the country. The Metropolitan Line crosses back, without the planned new depot that was refused by the local authority. It is followed by Lot Mead Lock, also known as Walker's, Beasley's or Cherry's after various overseers.

The valley now becomes very complicated as the Gade and Chess join the Colne and the various channels wind to and fro between flooded gravel workings.

A 15th-century house, built for the Archbishop of York, became Cardinal Wolsey's country seat, reconstructed in Baroque style in 1727 by Leoni as the most splendid 18th-century mansion in Hertfordshire, with marvellous interior frescoes. The grounds were landscaped in 1758 by Capability Brown.

Rickmansworth was a market town which got its charter from Henry VIII. When Elizabeth I died it was local landowner Robert Carey who travelled to tell James VI of Scotland that he was also king of England. After getting married, William Penn lived here from 1672 for five years before founding Pennsylvania. There is a brick and timber priory, the half-timbered vicarage is partly medieval and St Mary's church was rebuilt in the 19th Century in a way that retained its early Gothic windows. The Frogmoor Wharf timberyard was used to build and repair boats, many of them canal boats but also some targets for Portsmouth's Whale Island gunnery school, a total of 212 large boats in addition to punts and pontoons. Training of mini submarine crews was undertaken in the canal at Rickmansworth. Batchworth Lock is at the confluence with the Rickmansworth Branch, the dredged part of the River Chess, which served the town wharves, gasworks and gravel workings. It has a rare wide beam lift bridge. In 1973 there was a study for a barge depot to serve road traffic from the M1 after lower canal enlargement.

Batchworth Lock Centre is the normal home of the restored Ovaltine boat-floating restaurant while the

The Rickmansworth Branch joins at Batchworth Lock.

Mural under the A404 bridge.

Stocker's Lock, the rural location for filming Black Beauty.

abutment of the A404 bridge has the silhouette of a mule-drawn narrowboat.

Bury Lake is used for watersports and the Aquadrome canal festival takes place in May.

A restored lock cottage with canal artefacts and an ivy-clad Georgian house are to be found at Stocker's Lock. Otherwise there are 16th-century farm buildings with steep grazing to the south and some very smart horses. It is hardly surprising that *Black Beauty* was filmed here in superb scenery on the fringe of London.

On the other side of the canal Stocker's Lake is flooded gravel workings which produced the gravel for the original Wembley stadium and is now used for angling and birdwatching, including gadwall and water rail; bats can also be seen here.

The canal now moves from Hertfordshire to London. Chalk quarries are cut into the hill and orchids grow in them. Springwell Lock has been used for filming *Blake's Seven* and *Dr Who* episodes. Springwell Lake is also used for angling and birdwatching. It has a nature trail with the Springwell Reedbed Nature Reserve at Maple Cross. Its industrial estate is by the canal and Lynsters Lake. Lilies appear in the canal. A pillbox by the canal is covered in vegetation while derelict factory remains have a large chimpanzee hanging by one hand from a beam high above the water.

There was a breach in 2005 just above Copper Mill Lock at Mount Pleasant. What began as a papermill changed to making copper sheets for boat hulls, the foundry buildings now being flats. There was also an asbestos works here. The name was taken by the Coppermill folk duo in the 1970s, who wrote their own title song.

The private Troy Branch joins at Troy Junction.

Black Jack's Lock takes its name from a black man employed by a mill owner to harass boat people and stop night-time toll evasion. He acquired a stock of windlasses which he stored in a hollow tree. His efforts were not universally appreciated and he was murdered by the boaters, his ghost still haunting the area. The cornmill now houses a restaurant.

Disappointingly, it is not Harefield airfield but Denham Aerodrome which launches small aircraft over the canal but there are Harefield Flashes and Harefield has Elizabethan almshouses and a parish church with notable monuments. An otter holt has been constructed from recycled plastic and hidden near the canal.

Beyond Widewater Lock at South Harefield is the Horse & Barge and a pillbox. Harefield Marina is in a former gravel pit where two dozen working boats were sunk in 1958. Hoveringham Lake is used for sailing and carp angling.

The Chiltern Line crosses on a nine-arch blue brick railway viaduct, 4km from the better known one which takes it over the M25.

The west side of the canal has a golf course, Denham Country Park, a picnic area, a visitor centre, Tudor half-timbered cottages and a church with a 500-year-old Judgement Day mural. An aqueduct takes the canal over the Frays River, a diversion of the River Colne at Ickenham. Below is Denham Lock, the deepest on the canal at 3.4m, used for filming a *Mogul* episode. Denham film studios opened in 1936, including production of *Brief Encounter*, *Robin Hood* and, in 1963, the most famous canal film of them all, *The Bargee*. The area is still surprisingly rural with water voles, grass snakes and woodpeckers among the wildlife.

The Misbourne joins and there is a new gravel wharf with a conveyor for loading barges. The A40 crosses Oxford Road Bridge, immediately before becoming the M40. Gradually the canal becomes more industrial and the public are less inclined to return greetings but it does seem to gather better quality footballs and tennis balls. Sanderson Fabrics have their playing fields at Willowbank. Uxbridge Lock has a roving bridge by a flourmill which was run by Quakers.

Uxbridge is a mixture of half-timbered, Georgian and modern buildings, less placid than it appears. In 1555 three people were burned at the stake during religious unrest. In 1630 there were riots against tolls imposed by the Countess of Derby, the lady of the manor. It was from the RAF headquarters here that Air Marshal Lord Dowding directed the Battle of Britain. The Swan & Bottle is opposite the Crown & Treaty Inn of 1576, which was the home of Sir John Bennet, who hosted the unsuccessful Treaty of Uxbridge in an attempt to end the Civil War. The building's panelling went to the Empire State Building but has since been returned.

Rank Xerox have tower blocks by the canal. At one

time Fellows, Morton & Clayton had one of the largest working fleets on the Grand Union Canal with boats triple-moored here. Denham Yacht Station and Uxbridge Boating Centre have it easy today.

The General Eliott and a Royal Mail depot are by Dolphin Bridge while the B470 crosses above Cowley Lock. In 1896 a barge loaded with chalk drifted onto the cill, broke its back and sank while the crew were in the adjacent Malt Shovel, to which can now be added the Toll House.

At the beginning of the 19th Century there was a four-horse packet boat service to Paddington, with precedence over all craft, operating from what is now the Water's Edge.

Clergyman and forger Dr William Dodd was buried in Cowley in 1777. Cowley Peachey takes its name from the 13th-century Pecche family with the manorial rights.

Cowley Peachey Junction brings in the Slough Branch of 1882, one of Britain's newest canals and very straight but surprisingly narrow at the junction. The wish of canal enthusiasts to link this to the Thames or Jubilee River at Slough faces constraints in the form of buildings, the railway and the M4 so it must remain low on the wish list.

The canal turns east at **Yiewsley** as other traffic begins to intrude. The Reading to Paddington railway comes alongside at West Drayton station and Heathrow airport is 3km ahead, the main runways parallel with the canal.

Parts of St Martin's church are from the 15th Century, including the font, and it has fine monuments, among them the grave of Captain Billingsley, who was lost with the *Royal George* and 800 men in 1782. Opposite an aerial is Stockley Country Park.

The Slimcea bread factory and two large scrapheaps were sited near Starveal Bridge. A cement works is also close but the Broads Dock has been cut off by the railway, now a wildlife area with the Heathrow Express line running alongside.

From the A437 the view is to the railway crossing, usually with a background of a procession of aircraft landing. The coffee smell of Nescafé comes from the Nestlé factory. The canal crosses over Yeading Brook aqueduct, the stream formerly being called Bulls Brook, and is then crossed by the A312 in a former brickfield area. Bull's Bridge Junction takes its name from the old name for the brook and is where the Paddington Arm leaves. Tesco have a supermarket on the site of the headquarters of the former Grand Union Canal Carrying Company depot, a depot where a Bofors gun was manned during the Second World War and where there was a school and social club for boat crews. The British Docks Board research centre for such concerns as tidal flows and erosion was located here with large models. Tickler's Dock was known as the Jam 'Ole and their coal was the last long distance traffic on the canal, occasionally recreated with a Jam 'Ole Run.

The water collects increasing flotsam yet the water is cleaner than might be expected, perhaps an indication of limited boat traffic. A dock has not so much houseboats as floating houses. Opposite the Grand Junction Arms and next to a Porsche repair centre is a new building which complements a wharf crane in excellent condition.

The Old Oak Tree is by the canal at North Hyde and the Lamb by the A3005 Wolf Bridge, named after a different public house. The Maypole Dock of 1913 served the Monstead margarine works, the world's largest, now owned by Quaker Oats. A gold dome to the north is a reminder that

Monkey business at Maple Cross.

Willowbank at the start of the continuously built-up urban sprawl of London.

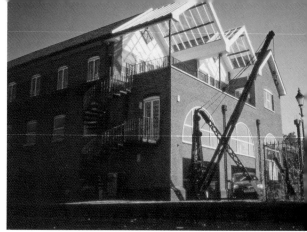

An old derrick crane, new offices and Porsches for unbending at North Hyde.

Below Cowley Lock.

Bull's Bridge at the start of the Paddington Arm.

Southall has long had a large Asian population. There are significant numbers of coconuts in the water, relating to Sikh funerals. There are also some large carp, together with crucian leather and mirror carp, eels, goldfish, various other fish species and Chinese mitten crabs.

Beyond Norwood Locks the canal is crossed by the A4127 on Windmill Bridge, named because it was near the site of Southall Mill, painted in 1810 by Turner. It was also called the Three Bridges, incorrectly as it is a road bridge over a canal aqueduct over a freight railway, the complex designed by Brunel in 1859. Some of his broad gauge track is now used as fencing although, unusually, there is an unguarded drop from the canal to the railway.

The Hanwell Flight of six locks, a scheduled ancient monument, was operated by tow tractors at one time and was used in the filming of *The Bargee*. Alongside is St Bernard's hospital, which was the county lunatic asylum. A conspicuously bricked-up arch was for coal delivery by canal and rectangular holes were for fire hoses. The hospital produced all fruit, vegetable and animal requirements and the surplus was taken away by canal. The flight has horse steps and tubular hitching posts. Less conspicuously, it has black spleenwort, wall rue and hartstongue ferns.

At the bottom of the flight it joins the River Brent from the Welsh Harp feeder reservoir, deepened for barges for the rest of the way to the Thames.

Giant hogweed appears along the banks and a nature reserve has been built on a former refuse tip. Beyond the

The Hanwell Flight drops the canal to the River Brent.

Aviary, Osterley Lock is accompanied by new bird nest boxes at the first of three places where the canal cuts off bends of the river.

The M4 passes over and large houses begin to appear. On the right bank, beyond Wyke Green golf course, is Osterley Park House, a crumbling 1576 mansion which Robert Adam converted from 1763 to 1767 and one of the most complete examples of his work. It has a richly decorated interior, Georgian furniture, wall paintings, sculptures, Gobelins and Beauvais tapestries and fine carpets. A 16th-century stable is still in use. It is all set in extensive park and farmland with Pleasure Grounds, neoclassical garden, lakes, orangery and nature trails.

The Piccadilly Line crosses and the next bridge is Gallows Bridge, a fine cast-iron bridge of 1820, probably the first canal bridge by Horseley Ironworks. The inscription of 'Grand Junction Canal Co. 1820' was altered to 'Grand Union Canal Co. 1820' although the Grand Union Canal was not formed until 1929.

Beyond the M4, which is climbing up onto **Chiswick** Flyover, is Boston Manor House, a fine Tudor building of 1623 with 1670 Jacobean extensions, good plasterwork and one of the best English Renaissance ceilings.

Clitheroe's Lock has high sides.

The A4 Great West Road is next to cross, quickly followed by the Barnes Bridge to Hounslow railway. After these there is a British Waterways depot with warehouse roof over the towpath and the canal, the last remaining overhanging warehouse roof of its kind, although the freight is now mostly carried by lorries rather than lighters to narrowboats. The rest of this reach has been dramatically rebuilt with a dazzling array of new apartments. At the far end of the reach is Brentford Gauging Lock, which is actually two locks side by side, beyond which the water is semi-tidal.

The Three Bridges: looking down from the road onto the canal above the railway.

Gallows Bridge was perhaps Horseley Ironworks' first for a canal.

Large pharmaceutical offices between the canal and the M4.

A new apartment development in Brentford.

Brentford Lock office surrounded by new buildings regenerating this dock area.

Augustus Way with the bridge built by Brunel for his railway serving dock sidings.

Distance
215km from Salford
Junction to the River
Thames

**Navigation
Authority**
British Waterways

OS 1:50,000 Sheets
139 Birmingham
& Wolverhampton
151 Stratford-upon-
Avon
152 Northampton
& Milton Keynes
165 Aylesbury &
Leighton Buzzard
166 Luton &
Hertford
176 West London

Below the lock is Brentford High Street bridge, carrying the A315, first built as the Roman road from London to Silchester. In the time of Edward I the bridge was kept in good repair by a toll levied on every Jew or Jewess passing over. An Act of George IV set a penalty of penal servitude for life for wilfully damaging the bridge. The Wine Bar is one of the few conspicuous buildings that does not seem to be brand new, surprising as **Brentford** was the county town of Middlesex.

The grounds of Syon House have seen plenty of activity over the centuries. In 1016 Edmund Ironsides defeated the Danes. In the 1642 Battle of Brentford, Prince Rupert beat the Parliamentarians and the Civil War might have ended there but the Royalists blew up an ammunition barge and the noise was mistaken for gunfire. Syon House itself began as a monastery for the order of St Bridget and had the dubious honour of being one of the first religious houses to be suppressed by Henry VIII. Perhaps it was just coincidence that dogs got into his coffin and savaged his corpse while waiting here overnight on the way to Windsor for burial in 1547. Katherine Howard was imprisoned here until she was executed, Lady Jane Grey was here when she was offered the crown, Charles I as a prisoner visited his children here and it was loaned to Pocahontas in 1616.

The building was an early Robert Adam restoration in 1762 of a 1547 house and has magnificent interior decoration, furniture and paintings. 81ha of landscaping by Capability Brown includes a 19th-century great conservatory glass dome covering 280m² with free flying birds inside, the London Butterfly House, a heritage motor museum, art centre, needlework centre, aquatic centre, one of the largest garden centres in the London area, a rose garden and 200 rare tree species. It is London's only ducal residence and has been the Duke of Northumberland's London base for over four centuries. The current duke's family were involved with Inland Waterways Association campaigning and running a working fleet on the canal in the 1950s. The grounds include a three-span, very early, wrought-iron bridge of 1790 over an ornamental pond, similar in profile to, but of different materials from, one built in 1812 by the Duke of Northumberland on his estate in Anwick. The venue has been used for filming *The Madness of King George* and *Gosford Park.*

The Dock Road Bridge, a riveted girder construction by Brunel, carries Augustus Way but was built as a railway bridge for a large railway yard on both sides of Brentford Dock (some 30 sidings in all) which is now gone. The dock, which had retaining walls held apart by horizontal arches, was open from 1859 to 1964 but has now been replaced by housing. This was the birthplace of the Bantam push tug.

Downstream there is a large fig tree on the left and, behind that, Fuller's Brewery Tap. There are high walls beyond with not even a ladder up in the vicinity of Thames Locks, not a very satisfactory situation as there is a weir off to the left. Brentford Gut is tidal and the locks can be used by pleasure craft two hours either side of high water or opened through at high water.

It was usual for a man to bring up a lighter on the tide, unpowered and with only a steering oar. One novice failed to make the turning and was carried on up

the Thames, going down and up with the tide and finally getting it on the next pass, very much more tired for the experience.

The Brent Ford got its name from a sand bar produced across the Thames by its tributary and it was here that Julius Caesar crossed the Thames in 54 BC. Excavation in 1859 found stone, bronze and iron weapons as well as Roman coins. There remained a ferry across the Thames until 1939. Ferry Lane is on the north side.

On the opposite bank is the World Heritage Site of Kew Gardens. The 1631 Kew Palace in red brick Dutch style was the smallest royal palace and was subject to flooding while George III spent his final years here. Queen Victoria's parents married in the drawing room, which has portraits by Zoffany, who lived at Strand-on-the-Green on the other side of the Thames. The Royal Botanic Gardens were begun in 1769 by Princess Augusta and cover 1.3km² with 30,000 plant species in Victorian and modern conservatories including the Princess of Wales Conservatory. The Temperate House is the world's largest surviving Victorian glass structure at 4,880m². Kew's highlights include a 1762 50m Chinese pagoda, an orangery, follies, temples, statues, water features, glasshouses, herbaceous bedding, a new treetop walkway, rainforest to desert environments and a herbarium which is used as the international standard for plant identification with an eighth of the world's known species here.

The Ovaltine narrowboat moored at Brentford.

11 Grand Union Canal, Leicester Line

The Loughborough Navigation made the River Soar navigable up to Loughborough in 1778, the most profitable canal in Britain for a while. The Leicester Navigation continued this to Leicester in 1794. The Leicestershire & Northamptonshire Canal extended the line to Debdale Wharf in 1797 and to Market Harborough in 1809. It had originally been planned to go from Debdale Wharf to Northampton but technical difficulties proved too great. Once what was to become the Grand Union Canal main line was built from London with a branch to Northampton, it was decided to change the plan and run from Foxton to Norton Junction instead, this Old Grand Union Canal being completed in 1814. It carried Nottinghamshire and Derbyshire coal, Mountsorrel and Quorn granite and through traffic but was damaged by coal-carrying on the Ashby-de-la-Zouch Canal and by the London & Birmingham Railway from 1838. In 1894, the Grand Junction Company bought the line from Leicester to London, expanding it to form the modern Grand Union Canal in 1926 and extending it to the River Trent in 1932.

It is a contour canal and forms part of the Leicester Ring. It includes three of the dozen longest tunnels still in use and three of the longest half dozen lock staircases.

Norton Junction has extensive views westwards over farmland towards Welton. The banks are lined with alder, ash and hawthorn, with cow parsley, germander speedwell and white dead nettles adding colour to the towpath edges, as does the first marina, near Welton Grange.

For the next couple of kilometres four transport arteries from different eras squeeze together through the Watford Gap. The canal meets them in chronological order starting with the A5 Watling Street, first built by the Romans and improved as the London to Holyhead mail coach road. Next over is the West Coast Main Line on what is nearer a tunnel of concrete beams but retains the facades of the original bridge with their decorative work picked out in colour, a welcome piece of preservation that is only really seen from the canal. Immediately on the right is the M1's Watford Gap Services of 1959, surprisingly unobtrusive but available for provisions or meals even without a vehicle. If something more upmarket is required, the Stag's Head restaurant is on the other side of the B5385. The Ship Bar is boat-shaped and the public bar is in a former blacksmith's sword shop.

After another marina and a field of caravans, seven locks (four in a staircase) raise the canal 16m to the summit level of 126m above sea level, the 20-Mile summit pound being one of the highest and most beautiful with some of the most lonely canal country in Britain. Three side ponds had to be rebuilt in 1994 following damage by vandals. It was planned as a wide canal to carry traffic between the Thames and the Trent but lack of water meant it had to be narrow over the summit, preventing a wide link between the south and north of England. At times the summit has been drawn down as a feeder for the busier main line leaving it short of water along this line; water restrictions are still suffered because the summit is popular with leisure traffic.

Motorway traffic has some of its noise shielded by trees along both sides of the canal as the M1 passes over.

Parkland occupies much of the area over to Watford, where the 17th-century Watford Court with Victorian additions was demolished in the 1970s. Still present is the 13th-century church with interesting monuments. Closer at hand, reedmace grows along the banks as the canal loops round the end of the M45.

The Northampton to Rugby railway crosses over as the canal arrives at Crick Tunnel. The 1.4km long tunnel is wet, having been built on a revised line to avoid quicksand, but has had three roof collapses, 26m being rebuilt in 1987. There is no towpath but there is a benevolent ghost which makes tea in narrowboat galleys. Nobody has thought to give directions for a walking route. The only indication of the whereabouts of the tunnel from above is a line of mounds of spoil brought up during the excavation of the tunnel, left in heaps like molehills, material which rabbits find suitable for excavating burrows. Field edges across the line of the tunnel have electrified or barbed wire fences, hawthorn hedges and ditches of stinging nettles, sometimes three lines of defence at a time. Medieval ridge and furrow systems prevail over the southern end of the tunnel. The correct route is back up a track to the right, along a minor road to Crick, first right down Horse Boat Lane, following the start of a bending footpath for West Haddon, and then cutting left across a field to a white sign above the towpath, a less than obvious route.

Crick, which would not otherwise be seen, has a manor with medieval vaulted cellars and a church with much decorative stonework, the rector of 1622–4 being William Laud, later Archbishop of Canterbury.

Edwards restaurant is opposite Crick Wharf, before the A428 West Haddon road bridge. Crick Marina is the venue for the Crick Boat Show at the end of May.

Running along England's watershed

![Watford flight tranquillity beside the M1.]

Watford flight tranquillity beside the M1.

Spoil heaps above the northern end of Crick tunnel.

Mallards, moorhens, coots and Canada geese are among the birds frequenting the canal. Bulrushes grow along the edge. Approaching Yelvertoft there are good views towards Flint Hill Farm, around which the canal meanders as it follows the contour.

Powerlines cross by Winwick Grange. The semi-deserted village of Winwick is further up the hill with a 16th-century manor house, fronted by a Tudor gateway. Winwick Lodge is a kilometre away and Winwick Manor Farm a further kilometre along the canal.

Beyond Elkington the A14 crosses. The Jurassic Way footpath uses the towpath. Up near Elkington Lodge a dinosaur of a windpump stands on the hillside. The Hemplow Hills lead on to 169m high Downtown Hill, which is used for growing broad beans.

The fields to the left were used by Percy Pilcher, the first Englishman to fly. He was killed in 1899 and a monument to him is in a field north of Stanford on Avon. Flights today take the form of gliding from the airfield between Welford and Husbands Bosworth and also the winter wildfowl on Stanford Reservoir. The site of Downtown Village is on the left bank in front of the reservoir.

A minor road to South Kilworth crosses at Sybole Farm on a bridge with the wing walls held in place by a row of screw props to prevent collapse.

The Rugby to Market Harborough railway followed the canal for about 5km to beyond Husbands Bosworth. A pit, from which gravel was exported by canal, is now a nature reserve.

The canal crosses the River Avon and the Northamptonshire–Leicestershire border on an aqueduct. Views in the North Kilworth and Welford directions are extensive at what is close to the country's watershed. The River Avon flows westwards while the headwaters of the River Welland flow eastwards from just east of the canal. The River Avon is closely followed by the Welford Branch, which was built as a feeder from Welford, Sulby and Naseby reservoirs but has a lock and carried up coal and brought down lime. It was disused from 1938 until restored in 1969 and is a popular destination for narrowboats.

North Kilworth had the first wharf on the canal, receiving coal from the Warwickshire and Derbyshire coalfields. Anglo Welsh have a boat hire base here. The canal heads into a wooded cutting. St Andrews's 12th-century church in North Kilworth is noted for its Jacobean Armada pulpit of Spanish oak.

Husbands Bosworth Tunnel is 1.1km long. This time the horse path is rather more obvious as it climbs over a hill, crosses the A5199 and winds over a footbridge

A balloon festival as the horse path crosses the A5199.

Flat Leicestershire fields to the north of Husbands Bosworth.

Thoroughly agreeable scenery at the foot of the Laughton Hills.

Husbands Bosworth Tunnel with a clear line through.

across the former railway. Perhaps this is as well as there are fewer spoil heaps along the line of the tunnel for guidance. This tunnel also has a ghost, such reports being very useful for keeping away strangers who might have sought work legging the working boats through.

The canal emerges into a cutting below Husbands Bosworth. All Saints church has a 14th-century spire and a Georgian hall, partly from the 16th Century. The village has an annual steam rally.

The rolling Laughton Hills give the canal some of its most attractive scenery, edged with gorse and finishing with a canalside oakwood, which is carpeted with bluebells in the spring.

Foxton is one of the most fascinating points on the canal system. It starts where a tall lock keeper's cottage overlooks the top of the lock flight. There are two staircases of five locks with a passing place in the middle, the best example of staircase locks in Britain. Built 1812, they drop the canal 23m from the summit level. Although they had side ponds to save water, they formed a tremendous bottleneck to traffic, so an inclined plane was added. Based on the one at Blackhill on the

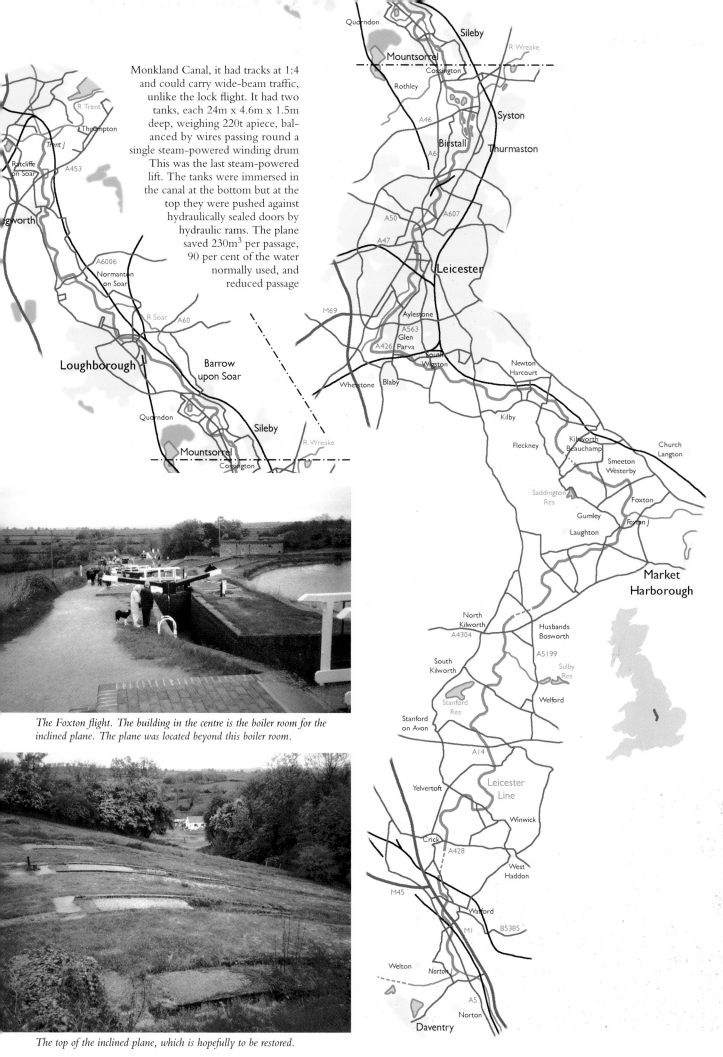

Monkland Canal, it had tracks at 1:4 and could carry wide-beam traffic, unlike the lock flight. It had two tanks, each 24m x 4.6m x 1.5m deep, weighing 220t apiece, balanced by wires passing round a single steam-powered winding drum This was the last steam-powered lift. The tanks were immersed in the canal at the bottom but at the top they were pushed against hydraulically sealed doors by hydraulic rams. The plane saved 230m^3 per passage, 90 per cent of the water normally used, and reduced passage

The Foxton flight. The building in the centre is the boiler room for the inclined plane. The plane was located beyond this boiler room.

The top of the inclined plane, which is hopefully to be restored.

time from 70 minutes to 12 minutes. However, it was uneconomic because there was not enough traffic and it closed in 1910 after a decade of use. There are hopes to restore it or, effectively, build a replica. The plane's boiler room now contains the Foxton Canal Museum, including a working model of the plane. The locks, lock keeper's cottage and junction distance post are all Grade II listed. At the bottom of the flight is Foxton Junction with a horse-drawn trip boat using horse and crew in traditional dress. The Foxton Locks Inn can be very busy with visitors and was the first to be taken into ownership by a British Waterways partnership.

The **Market Harborough** Branch was originally the main line, with traffic including coal and animal bones for a glue factory. It passes Gartree prison and Foxton church, which is built on a pre-Conquest site and has a Norman font and Saxon remains.

One change on the older part of the canal is that bridges have names as well as numbers, useful in this rural area. One of the first carries the Leicestershire Round footpath up towards Gumley, where King Aethelbald of Mercia held council in 749. Gumley Hall has an Italianate tower.

Only the residents of a rookery disturb the tranquility. Red campion appears along the towpath. The canal has been designated a Site of Special Scientific Interest all the way from Debdale Wharf to Kilby Bridge.

Debdale Wharf was to be the point from where the canal headed directly for Northampton but technical difficulties prevented this. Today it has only a marina to show any activity.

The canal clearly runs along the side of 157m high Smeeton Hill to Smeeton Westerby, where Smeeton Aqueduct carries the canal over Langton Brook, a feeder from Saddington Reservoir, built to supply the canal. The embankment collapsed here in 1865 and had to be rebuilt again in 1993 following a leak.

Saddington Tunnel is the shortest of the canal's three, at 800m, short enough for the tunnel profile to be visible all the way through even without lights. It was built crooked although not enough to affect the line of sight. There were slips in the approach cuttings during construction, mounds of soil from these being left near the canal to contrast with the ridge and furrow fields. The horse path over the tunnel is the most direct and obvious of them all. The tunnel acts as a bat roost and has a ghost, Anna, a headless woman.

Once the canal comes out of cutting, there are extensive views back past Fleckney. Powerlines cross over, converging on a substation on the edge of Kibworth Beauchamp. At the time of the Domesday survey, Chiburde was the farmstead of Ciber or Cybba and de

Beauchamp was the name of the family who were to act as the king's chief panteler for centuries, a combination of dresser, butler and banner carrier. The village's main industry has been framework knitting, with factories of up to 400 employees at the beginning of the 19th Century.

The Kibworth locks seem strange for the setting, being the first of the wide beam locks; the second lock is surrounded by spare stone from the front at Blackpool.

From Kibworth Bridge, where swallows swoop over the ashes and hawthorns, the canal drops down through Kibworth Bottom Lock and Crane's Lock to cross the River Sence on a brick aqueduct, a river which the canal is to follow to the River Soar. As it does so, the St Pancras to Sheffield railway line arrives alongside and is to follow closely to Wigston Harcourt. Both keep to the right side of the valley, giving extensive views over the sheep pastures below. On the far side of the valley is Wistow with its Jacobean hall rebuilt in the 19th Century, 18th-century church with Norman work, model railway, model village and garden centre. Wistan was the Saxon heir to the throne of Mercia, who opposed the incestuous marriage of Britfardus, his god-father, to his mother, Elfleda. Britfardus spilt his skull in order to improve his own chances of getting to the throne. A shaft of light appeared above Wistan's grave and human hair is reported to grow from the ground for an hour on 1 June where he was killed. Britfardus went insane and Wistan became a saint.

A brick church of 1834 with a 13th-century stone tower by the canal gives notice of arrival at Newton Harcourt, with its three locks, Wain Bridge, ridge and furrow system and displays of dog roses. This quiet spot was the scene of rioting led by militant navvies during the canal's construction in 1795, resulting in the cavalry being called in to restore order.

Bottom Half Mile, Turnover, Tythorn and Bumble Bee Locks follow, with reeds and cow parsley along the way. Powerlines follow the canal right into the middle of Leicester but they are often far enough away not to be overbearing. The A5199 crosses at Wigston Harcourt. Canada geese add their own noise.

Hardly a building in sight from Tythorn Hill.

The descent continues past Kilby Bridge, Double Rail and Irving's Locks to South Wigston, where the character of the canal changes. After a mostly rural run all the way from the main line near Daventry, the canal now comes within reach of youths from housing estates and the potential for problems increases.

A Rugby to Leicester railway used to cross here and head towards the 14th-century church. A totem pole stands in a picnic area before Bush Lock and some of the gardens are well decorated, notably one with wrought-iron railings and holders with church-sized candles beside the canal. Frogs are oblivious. The B582 crosses Little Glen Bridge and then both road and canal are crossed by the Leicester to Nuneaton railway. Near Little Glen Lock, 6th-century grave ornaments were found in a Saxon cemetery. Across the valley, Blaby's

Looking north-west from the top of Saddington Tunnel.

church is merely 14th-century with a fine 18th-century gallery. Blaby Bridge carries the A426 over before Whetstone Lane Lock and the line of another former Rugby to Leicester railway.

As the Sence meets the Soar, the canal swings through a right angle. Back in a corridor of open country, **Leicester**'s Riverside Park, a heron gazes at the bulrushes and horses reach over fences. Many more people are using the towpath than are using the canal: cyclists and walkers proliferate. The Soar valley was formerly the course of the Avon until the Ice Age.

Beyond Gees and Blue Bank Locks an aerial to the left marks Leicestershire Police headquarters before reaching the A563, Leicester's outer ring road. King's Lock is followed by the canal joining the river with its much more winding course. Just after the confluence the river is crossed by the former railway line which has now become the Great Central Way cycle route where Aylestone Bridge also takes a minor road over and past a church with an interesting 1930 stained glass window.

Leicestershire county cricket ground lies to the east of Aylestone Mill Lock. A gas holder and dyers' premises precede a railway scrapyard, a last chance to try to spot some missing engine numbers. To the east of St Mary's Mill Lock are Aylestone Gas Works, now the first gas museum, including a 1920s all-gas kitchen complete with gas radio.

Leicester – named after Llyr or Lear, king of the sea, meaning darkness and death, or after the Legra, the Celtic name for the Soar – was the Roman town of Ratae Coritanorum by 100. Raw Dykes was a section of aqueduct supplying the town. Indeed, the town began as an Iron Age settlement. Later it was to be where Simon de Montfort called the first Parliament in 1265 and it has developed on the boot, shoe and hosiery trades, also adding its name to the list of prominent cheeses.

At New or Freeman's Meadow Lock the river comes in and crosses the canal to reach a long unguarded weir before the two watercourses combine below the lock. Overlooking the lock are a large electricity substation and Leicester City football stadium.

The river leaves to the left as the canal enters the Mile Straight, a grand water avenue through the city centre, designed in 1890 for flood defence purposes but forming an important part of the city. Old and new bridges across include some ornately decorated Victorian structures. Surrounding buildings feature the same blend of the traditional and those currently being erected. These include the Liberty building of 1921 with a Statue of Liberty on one corner, the De Montfort University and the Newarke Houses Museum that has a Victorian street scene, period costumes, clocks and clockmaker's workshop and history of the hosiery, costume and lace industries. Leicester Castle ruins stand on a motte near St Mary de Castro church, founded in 1107. St Martin's cathedral was a parish church until 1929, originally Norman, with a 13–14th-century interior and mostly 19th-century exterior, a high steeple and a notable south porch. It stands next to the 14–15th-century guildhall with cells and gibbet irons, fine oak panelling and a massive carved chimney piece of 1637. Together with the Holiday Inn these are approached along the A47 from West Bridge, which is also on the line of the Foss Way, the Roman road from Cirencester to Lincoln. The Jewry Wall is the largest surviving Roman building in England. It was erected in 125–130, is 7.3m high and is part of a large public baths, now with a museum and two in situ mosaics. Also of note are the Jain Temple & Museum, the best Indian traditional architecture in the Western world.

The initials UN on a coping stone stand for Union Navigation and mark where the Leicestershire & Northamptonshire Union Canal joined the Leicester Navigation. The river rejoins briefly.

Coots dabble among the lilies in front of a dyers' factory

Wharf buildings and crane at Kilby Bridge.

The lock at Aylestone.

Freeman's Meadow Weir is unguarded.

and swans nest with a controlling view of Evans' Weir, where the river loops away again. At the end of the weir was the terminus of the Leicester & Swannington Railway, one of the first standard gauge lines in the Midlands. Built in 1832 by Robert Stephenson, it allowed Leicestershire coal to compete with Derbyshire and Nottinghamshire and took traffic from the canal downstream. An 1834 bridge for horse-drawn wagons to cross the river is now in Snibston Discovery Park.

The canal returns to the river to allow excess water a second chance to discharge over Hitchcock's Weir before leaving for North Lock. The A50 crosses between the lock and the North Bridge Tavern. A large drain discharges through the bridge abutment, dark but large enough for a small boat to enter.

Canal Carriers Wharves follow to the A5131 St

The start of the Mile Straight running through the heart of Leicester.

The National Space Science Centre rises above Belgrave Lock.

Leicester Outdoor Pursuits Centre on a quiet day.

Margaret's Way, which leads to St Margaret's church, a 13–15th-century building with notable Perpendicular tower and chancel. Abbey Park, with its miniature railway and boating lake, is an important green area in the city. Towards its eastern end is the Belgrave Gate Wharf arm above Limekiln Lock and there is another arm between the lock and the next road. Many coconuts may be found in the canal. The number of alphabets in use on the local college's welcome board are a reflection of Leicester's large Asian population. In the opposite direction the road leads to the National Space Science Centre, rising like a large silver inflatable structure rather than the streamlined tower that might have been expected. The Museum of Technology features canal models and the development of transport and power in the hosiery industry. The Wolsey hosiery works makes its presence known with a prominently labelled factory chimney.

The river returns at Belgrave Lock, over Swan's Nest Weir, and is crossed by the distinctive Holden Street cable stayed footbridge of 1985. Close by is T. S. Tuna.

Belgrave Hall is Queen Anne with 18–19th-century furniture, stables, coaches and an agricultural collection.

Leicester Outdoor Pursuits Centre is dominated by an A-shaped room, fronted by canoe slalom practice gates and completed by a climbing tower.

The A563 crosses back, completing its ring around Leicester, and the river moves out into open meadows and leaves behind the city and Leicestershire's county hall, which stands prominently beside the A607.

Further locks come at **Birstall** and at **Thurmaston**, site of a large Anglo-Saxon pagan cemetery, before the river leaves. The canal enters Watermead Country Park, an area of gravel digging from which gravel was extracted and carried regularly by canal to **Syston** until the 1980s. The park includes a large plastic mammoth.

The Hope & Anchor public house proves popular with everyone, before the A46. The canal joins the River Wreake, at one time the Melton Mowbray Navigation, which it follows back through Junction Lock to the River Soar at Cossington Lock. This detour was required in order that the canal should not be visible from Wanlip Hall. The 16th-century corn and paper mill at Cossington Lock is now a restaurant. Once again, the navigation has to cross the river above a weir to reach the lock cut.

Rothley was the site of a Roman villa and has a granite Norman church with a Saxon cross and a 13th-century chapel with the figure of a Knight Templar. An Elizabethan house was the home of the Babingtons from

The Hope & Anchor wharf – busy on a sunny spring day.

the 15th to 19th Centuries and the birthplace of historian Lord Macaulay. Today it is a hotel. Rothley Brook joins above the sewage works and the river is followed by the A6 to Loughborough. Across the fields, Cossington's church has excellent Victorian stained glass. Perhaps a tern adds interest to the reach.

At Sileby, the vicar hands out oranges to local schoolchildren on Whit Sunday in memory of the victory at Waterloo. A ring of ten bells serves as a reminder of St Mary's large church the rest of the year.

Sileby Lock has two conspicuous weirs feeding the large pool below the lock. The lock island is crossed by the Leicestershire Round footpath.

The canal makes a large loop to **Mountsorrel** rather than following the more direct line of the river. It passes under the A6 facing a granite hill topped by a fire basket and a war memorial.

At the downstream end of Mountsorrel Lock it is necessary for portable craft to relaunch on the far side of the road which crosses but the wall of the Waterside Inn and the humped bridge make the approach of traffic blind from both directions.

A broad arched railway bridge and the Railway remain as reminders that a private railway link of 1860 undermined the viability of Mountsorrel Wharf, which was built to load pink granite from the quarry. Mountsorrel has a covered market cross and the Stonehurst Farm & Vintage Car Museum.

The canal passes back under the A6 and rejoins the river before turning sharp left to follow the St Pancras to Sheffield railway again for the rest of the way.

Approaching **Barrow upon Soar**, there are extensive views over the meadows to the left. This time it is the river that makes the loop while the canal takes the more usual straight line, leaving at the Navigation Inn. Superior housing surrounds lawns sloping down to the canal with the occasional dovecote or tree house.

Barrow upon Soar had a Roman settlement and later limestone quarries. Beyond Barrow-on-Soar Deep Lock, the canal meets the river again below a restaurant and the Soar Bridge Inn, the latter making reference to the venerable red brick bridge across the river. Beyond the inn are more arches, this time supporting the railway as part of a retaining wall built into the side away from the river.

Pilling's Flood Lock is usually left open as the canal leaves the river at the foot of Catsick Hill. The canal crosses Loughborough Moors at higher level than the surrounding land, giving extensive views south-west to Charnwood Forest and northwards over the Soar valley.

An anti-aircraft gun at the front of T. S. Venemous

Sileby Lock with its two weirs and large weirpool.

Railway arches support a retaining wall at Barrow upon Soar.

GCR 8K 2-8-0, Great Central Railway, Leicestershire

gives notice of arrival in **Loughborough**, named after the Old English man Luhdede. Beyond it is the terminus of the Great Central Railway, Britain's only remaining main-line steam railway, running for 13km from Birstall. There is a museum and historic locomotive collection. The line was used for filming *Shadowlands*. The bridge that extended the line to the current freight line to Nottingham still stands across the canal among the derelict factory buildings.

One factory still active is 3M's beauty products mill next to the A60, which crosses on the Duke of York Bridge and leads towards the Bell Foundry Museum. This is sited in the world's largest bell foundry complex, the products of which have included Great Paul for St Paul's in London. A tower of 1922 was Britain's first grand carillon. It is 46m high and now houses an armed forces museum. All Saints church has an exceptional 15th-century roof and clerestory and a magnificent and unusual Somerset tower.

The derelict factories do not bode well for Loughborough but the atmosphere improves and the authorities really make maximum use of the canal as an asset as it curves round the town past the Boat Inn, recent housing and a playground to enter from the north-east. A branch leads back to the south-east to serve Loughborough Basin. Just before the junction is the Chain Bridge, the predecessor of which used to have a chain across at night to prevent boats slipping through in the dark without paying tolls.

A Charnwood Forest branch had also been planned to carry trucks on rafts. Loughborough is a hosiery town and witnessed Luddite riots. Its November fair dates from the 13th Century.

After the junction, the Albion Inn faces the canal although this was not always the case. When built it was a farmhouse facing the other way.

There are two more locks, Loughborough and Bishop Meadow, before the canal rejoins the River Soar with Nottinghamshire on the far bank and a large industrial estate on the near side. From here the navigation uses the River Soar continuously to the River Trent, except for the occasional small loop.

Normanton has a chain ferry. Its buildings include a cruciform church with a 13th-century steeple right next to the river, the Plough Inn, thatched post office and summer house and a number of wooden houses which must surely be weekend haunts, a most interesting little village.

The canal leaves under the A6006, which now has the canal on the north side and a mill stream and weir stream on the south, confusing as powered craft can use parts of various channels. The lock channel cuts a corner at Zouch, returning to the river after a lock. The road runs down the main island, serving the Rose & Crown. A large castle playhouse faces onto the canal.

Loughborough Boat Club's boathouse is passed and the river flows placidly below a steep wooded bank, unusual in this valley, towards the Devil's Elbow and the White House.

The large hill, which has been visible on the left for some while, carries Castle Donington and East Midlands Airport. At **Kegworth** the river is just 3km from touchdown. Kegworth includes a large house with spacious grounds, a conspicuous dovecote and a beautiful rock garden. The river and canal separate round a lock island in front of it.

A caravan site, recent houses and another lock island follow, this one being Kegworth Shallow Flood Lock, which is usually chained open in the summer although the river is susceptible to significant flooding in the winter.

The canal cuts off a loop of river at Ratcliffe on Soar, passing below the A453 to a further lock.

Ratcliff's church dates from the 13th Century and has whitewashed walls but it is totally overwhelmed by the eight cooling towers of the Ratcliffe on Soar power station, plumes of water vapour climbing skywards to provide further conspicuity over a great distance, if not improving the visibility for pilots. Warm cooling water is pumped into the river. This is the source of the various sets of powerlines that have been crossing and following the river since Barrow on Soar.

There is a final island at Ratcliffe Lock with a 1970s lock next to an almost filled-in earlier one. The river and canal meet again by a red gravel cliff streaked with white at Red Hill.

The River Soar finally discharges into the River Trent just above one of its massive weirs. The far bank is part of a large island with two locks so craft need to travel upstream for 300m to Trentlock to go either way on the Trent or up the Erewash Canal.

The church at Ratcliffe on Soar is dwarfed by the cooling towers of the nearby power station.

Distance
106km from Norton Junction to the River Trent

Navigation Authority
British Waterways

Canal Society
Foxton Inclined Plane Trust www.fipt.org.uk

OS 1:50,000 Sheets
129 Nottingham & Loughborough
140 Leicester
141 Kettering & Corby
152 Northampton & Milton Keynes

Ratcliffe Lock picked out in the grass beside its replacement.

12 Cromford Canal

Built to serve the Peak industries

The Cromford Canal was built between 1789 and 1793 by William Jessop and Benjamin Outram, extending the Erewash Canal up as far as Cromford. Cromford Wharf lies just below Willersley Castle, now a hotel. The canal was eventually closed in 1944 by the LMS Railway Act and over its length it now exhibits the full spectrum of states of repair. Derbyshire County Council became the owners in 1974 and the Cromford Canal Society were able to reopen the top section in 1977, working their way down. Restoration methods developed here have been repeated on other canals.

The immediate neighbour of **Cromford** Wharf is Arkwright's mill of 1771, the world's first successful water-powered cotton spinning mill and the first mechanised textile factory, the cotton industry here becoming comparable with that of Lancashire for a while. The cottages Arkwright built for his workers formed the first factory housing development in Derbyshire and, in 1792, he added the small St Mary's church, across the road from the canal basin, where his grave is to be found.

Derwent via Arkwright's mill. The main feeder is at the south-east end of the wharf although an earlier feeder is immediately adjacent to the covered loading bay.

The whole canal is a Site of Special Scientific Interest and liverworts and ferns are particularly noticeable.

The river is the main corridor leading out of the heart of the Peak District. Its tree-covered limestone hills create splendid scenery. The canal is sandwiched between the Derwent, the Matlock–Belper railway and the A6, the latter built in 1820 as the Belper–Matlock Turnpike. The wooded hills rise steeply, hiding the spectacularly eroded gritstone mass of Black Rock to the south-west of the canal popular with climbers.

The immediate surroundings are less dramatic at first as the canal sets off past rugby pitches but with added interest in the form of a woodland water garden on the south-west bank.

The bank is heavily reinforced with many layers of gabions at the site of a stoneworks, where slabs of sliced stone are moved about by large gantry cranes.

The canal leading down to Highpeak Junction.

The canal terminal basin laid out like a small railway station.

Textile yarns and cotton fibres were important parts of the canal's trade, along with building stone, limestone, coal, iron and lead, exceeding 300,000t per year in the early 1840s. A scheduled passenger service also ran to Nottingham.

Warehouses, cottages and office, all of warm pinkish stone, and cranes are grouped around the wharf.

Two horse-drawn passenger boats operate from the wharf during weekends and some other afternoons in summer.

Because of the limited use, the canal water is very clean, feeding in from the River

The canal was to have been the first stage of a route through to Manchester but the engineering requirements were such that it was decided instead to construct the Cromford & High Peak Railway to run from Highpeak Junction to a branch of the Peak Forest Canal at Whaley Bridge. This was no mean feat in itself, a 53km railway with a summit level of 375m above sea level and gradients up to 1:7. There was a transhipment wharf and the railway was initially worked by horses with stationary engines for the steep slopes when it opened in 1831. The horses were replaced by engines in the 1840s with the High Peak Wharf shed being

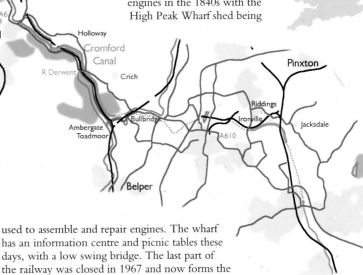

used to assemble and repair engines. The wharf has an information centre and picnic tables these days, with a low swing bridge. The last part of the railway was closed in 1967 and now forms the High Peak Trail.

Covered wharf at the terminus.

The foot of the railway incline at Highpeak Junction.

High Peak Wharf. The loading gauge shows the railway's line.

The quaint Leawood pumping station.

The scenery towards Holloway on the north-east side of the valley is particularly fine, steep limestone hills covered with deciduous trees.

Squeezed between the Matlock sewage treatment works and the canal is one of its prime assets, the Leawood pumping station with its quaint tall chimney. Built in 1849 to raise further water from the Derwent to supplement the normal feed when necessary, its Watt single-acting beam engine was restored in the early 1970s and is in steam once a month and on bank holidays. There is some turbulence when each stroke of 5–6m^3 is delivered.

Next to it is one of the most impressive surviving feats of canal engineering, Jessop's Wigwell Aqueduct over the Derwent, crossing it on a single 24m span. It collapsed the first time he built it, an event he blamed on the use of Crich lime, and he offered to rebuild it at his own expense.

On the far side is the stump of the Nightingale Arm or Lea Wood Branch, built in 1802 to serve Lea Mills.

The canal runs into a heavily wooded section, crossing over the railway on another aqueduct at the south-east end of a tunnel. It is possible to look over the left side and down onto the track as it disappears into the tunnel mouth.

The canal turns sharply into a tunnel of its own, the short Gregory Tunnel, accompanied by the towpath, which is well used by walkers. Exit at the far end is dramatic with the hills rising up to bare limestone cliffs, topped by Beacon Light, directly in line with the tunnel, a memorial to the 11 thousand Sherwood Foresters killed in the First World War.

A temporary end of the canal is reached under the bridge at Leashaw Farm, the latter selling confectionery, ice creams and drinks. Beyond the board dam there is little water and many weeds, becoming progressively more overgrown as the route continues. Cromford Canal Society intend to continue the restoration but are faced by the Derbyshire Naturalists' Trust, who have a nature reserve between Whatstandwell and Ambergate, the kind from which people are excluded, and who have apparently thrown up their hands in horror at the thought of boats on the canal as it continues down between the woods of Crich Chase and Shining Cliff.

The water stops at Ambergate, where Ambergate Gas Works construction has cut the route. Nearby, Bull Bridge Aqueduct across the River Amber was demolished in 1968 to make way for a road widening scheme. Limekilns were established at this point in an area rich in limestone.

The River Amber has been followed successively by the canal, the A610 and the Derby–Chesterfield railway. Buckland Hollow Tunnel was closed in 1969 but it was the collapse of Butterley Tunnel that finally closed the canal in 1944 and killed hopes of through restoration of the line.

The landscape, roads and housing in this area all look new, as if the area was ashamed of its industrial coalfield past and has put a bulldozer through the lot, starting again but economically reusing any earlier routes that could be adapted.

A stream from Riddings flows into the line of the canal 900m north-west of Golden Valley, after which there are varying amounts of water and weed for a while. The canal is emptied over a concrete dam into **Codnor** Park Reservoir, which used to stand beside the canal at Ironville, and then out again over another dam and into a ravine lined with concrete mattresses. The top pound used to finish here with a 3km branch extending up the River Erewash to Pinxton Colliery. Restoration of the arm has begun at Pinxton and is continuing across Smotherfly opencast colliery site towards the B600.

The main line dropped away through 14 locks. Beyond Jacksdale the water runs out again and the line is

Jessop's magnificent Wigwell Aqueduct across the River Derwent.

Gregory Tunnel is quite short and has a towpath on one side.

The canal is overgrown and feeds into Codnor Park Reservoir at Ironville.

Beacon Light is in line with the tunnel.

lost as colliery spoil is reshaped in an area where open-cast mining is still going on.

The final lock, by the towering Langley Mill, has been restored and the last part of the canal is a narrow-boat building yard with a drydock. The final section is needed to link the top of the Erewash Canal with the isolated Great Northern Basin at the head of the Nottingham Canal. The cost of renovating 13 locks and redigging a canal which only leads to a collapsed tunnel is likely to leave the magnificent upper section isolated.

Boatyard at the bottom of the canal at Langley Mill.

Distance
24km from Cromford to the Erewash Canal

Navigation Authorities
Derbyshire County Council, British Waterways

Canal Society
Friends of Cromford Canal www.cromford canal.org.uk

OS 1:50,000 Sheets
119 Buxton & Matlock
120 Mansfield & Worksop
129 Nottingham & Loughborough

73

13 Erewash Canal

The Erewash Canal runs southwards from Eastwood to join the River Trent at Long Eaton. Opened in 1779, it served the Nottinghamshire/Derbyshire coalfield and its prosperity resulted in the building of a number of other canals, the latter now all closed. Bought by the Grand Union in 1932, it was allowed to fall into disrepair, was brought back into use during the Second World War, was again neglected and was finally restored and reopened in 1973 after a combined effort by the Erewash Canal Preservation & Development Association, Derbyshire and Nottinghamshire County Councils and British Waterways.

The canal closely follows the small River Erewash and is in turn followed by the railway as far as Long Eaton. Indeed, it was a meeting in **Eastwood** early in the 19th Century that led to the establishment of the Midland Railway which was to result in the downfall of many of the canals.

The Great Northern Basin at Eastwood has been restored to an attractive terminal with grassed areas, seat-

The Great Northern Basin. Above lies the Cromford Canal.

ing, a swing bridge and the Great Northern public house. Eastwood pottery is made in a modern works and the large Langley Mill stands near the A608.

Two other canals join the Erewash at the basin above the first of its 15 locks. The Cromford Canal has been cut in two by the collapse of a tunnel as a result of mining subsidence. Subsidence has also resulted in the closure of the Nottingham Canal, the line of which closely followed the Erewash Canal all the way to Trowell.

New deep pits on the left are followed by works on both sides before the canal breaks into open country. There are extensive views over the Erewash valley on the right towards Shipley and the scenery is generally at its best as the canal winds through copses. Bridges have flat, cast-iron decks between simple abutments, built in the dark brick of the area. Occasionally, there are collections of old lorries and other debris beside the canal. Graffiti all relates to angling and, although the canal is quite wide, anglers rest the tips of their long rods on the opposite bank to the one on which they are sitting.

Locks often have bypass channels with steep drops at the entry ends. Eastwood Lock has a plaque remembering author D. H. Lawrence on the right side. Lawrence was born in New Eastwood, although the town prefers to forget this prominent writer, who based so many of his books on the area. Part 2 of the first chapter of *The Rainbow* opens 'About 1840, a canal was constructed across the meadows of the Marsh Farm, connecting the newly-opened collieries of the Erewash Valley.' The book goes on to describe the effect of the canal on the community, the railway, the collieries, Ilkeston and Cossethay (a cross between Cossall and Cotmanhay, perhaps) and other aspects of

Contemporary buildings and an interesting roofline on the former stables by the lock at Shipley.

the area, with only minor changes to the detail of what is actually to be found.

His aqueduct over the River Erewash is met after the next bend and, the river being the county boundary, the canal moves from Nottinghamshire to Derbyshire for the rest of its route.

Shipley Lock has the Shipley Boat Inn and a selection of fascinating buildings which were formerly a slaughterhouse and stable block, the latter with a sagging roofline. Not all buildings are so interesting, however, and, after a marshy area, long lines of featureless modern houses with small windows front the canal at Cotmanhay. Having said that, Cotmanhay can boast the old Bridge Inn beside an arched bridge which has settled enough to give problems to some larger canal craft. The cause of the trouble is seen as the large Bennerley coal preparation plant, beside which passes the 430m long Bennerley Viaduct. This lattice trestle viaduct used to carry twin railway tracks right across the valley to Awsworth and is now preserved as a listed structure.

come in between Greens Lock and Gallows Inn Lock, around which are no fewer than three petrol stations and the inn.

An open playing field area with rugby pitch and a track where children might be seen driving miniature replica vintage cars follows and nature pushes back with giant reedmace, bulrushes and perhaps a vole swimming across the canal.

Below Hallam Fields Lock, surrounded by a timberyard, can be seen the end of the Nutbrook Canal, which was closed in 1895. The first of several side weirs is passed, with walkways across for the towpath.

Beyond Stanton Lock is a railway spur into the giant Stanton ironworks at Stanton-by-Dale, the last major commercial users of the canal. Little is seen of the works from here except for a water tower on the canal bank. A much better view is obtained from the M1, which now crosses.

Pasture Lock, at Stanton Gate, still lives up to its name, horses grazing in the meadows. These stretch

The sinking bridge by the Bridge Inn at Shipley.

Bennerley Viaduct and the coal plant.

Stenson's Lock is in open country but by Common Bottom Lock the canal has become hemmed in with houses and small works. An occasional line of poplar trees adds variety and there is another Bridge by the lock. The canal forms the eastern boundary of **Ilkeston**, a market and textile town with a three-day fair each October. St Mary's church, distinguished by its tower, dates from 1150 and has an unusual 14th-century stone screen.

On the approach to Cossall is a most unusual scrapheap as a line of excavators stand next to the canal in various stages of dismantlement. Opposite, a street of terraced houses leads down to the canal and an old fire engine rusts away on the bank nearby.

Modern housing has also been pushed into clear spaces and grassed amenity areas are left to flank the canal. Potters Lock, like some others, has a brick arched bridge across its lower end and forms as attractive a setting as could be hoped for at the edge of a large modern housing estate.

Lorry parks, including a large Marks & Spencer depot,

away to the hill topped by **Sandiacre** church with its original Norman work and 600-year-old font, surrounded by a clump of half-timbered buildings.

A substantial-looking footbridge leads away over the railway to the larger **Stapleford**, which keeps its distance. New factories back onto the canal but some have built canalside terraces with seats and tables and one even has moorings. The pride of the canal's factories is Springfield Mills, built in 1888 for lacemaking, fronted by a prominent chimney with intricate brickwork at the top. It has been converted for residential use. Opposite it is the Plough. The Red Lion and White Lion follow in the centre of Sandiacre, a town that makes a feature of the canal, with lawns to the water, seats and flowerbeds. Other amenities include takeaways and public toilets.

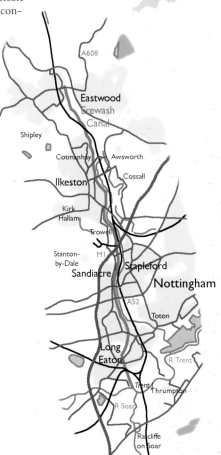

Springfield Mills retain their striking chimney.

The lock cottage at Sandiacre Lock, now the headquarters of the canal society.

The canal flanks an urban boulevard in Long Eaton.

Beyond the busy A52 and tucked in next to the Toton railway yard is the jewel of the canal, the lock cottage at Sandiacre Lock, the base of the canal society. Complete with all the stable doors, it shows rural canal architecture at its best. Just above is the end of the Derby Canal. Abandoned in 1964, it joined the Trent & Mersey Canal at Swarkestone.

Recent housing continues to flank the right side of the canal down past Dock Holme Lock but horses graze on the rough pasture between the canal and railway.

The most depressing reach is that leading down to **Long Eaton** Lock with timber and pipe depots, petrol station, electricity substation and walls with scrawl and broken fittings. Even the lace mill, the basis of the town's prosperity, can't quite match the one at Sandiacre. Below the lock, the canal is flanked by the playing fields near Trent College, by a school and a fire station and by a tree-lined urban boulevard, the playing fields continuing opposite to another school. Facing the canal and its swans is the Barge Inn.

The canal turns sharply under the road by the 15th Long Eaton Sea Scout & Guide Group's premises. It again bends left by a public house with a canalside garden.

Two railway bridges cross with a railway transhipment basin just before the first, the Sheet Stores Basin, which has become a marina.

The canal company's drydock is now part of Mills Dockyard. Moorings. Houseboats line the bank down to Trentlock, overlooked by the garish Steamboat Inn with its funnel and ventilators on the front. The Trent Navigation Inn on the other bank has a landscaped children's area and public toilets in the car park at the back.

Below the lock is a complex junction with the River Trent. A navigation cut comes off at this point and the River Soar, the northern end of the Leicester arm of the Grand Union Canal, enters opposite. Above it all, aircraft descend into East Midlands airport at Castle Donington over the cooling towers of Ratcliffe on Soar power station.

Distance
19km from the Cromford Canal to the River Trent

Navigation Authority
British Waterways

Canal Society
Erewash Canal Preservation & Development Association www.erewashcanal.org

OS 1:50,000 Sheet
129 Nottingham & Loughborough

Trentlock is dominated by the cooling towers of Ratcliffe on Soar Power Station.

14 Grantham Canal

The first canal supplied almost entirely by reservoirs

The canal begins again after being severed by the A1.

Capital of £40,000 for a canal to run west across Lincolnshire, Leicestershire and Nottinghamshire from Grantham to the River Trent at Nottingham was raised in a single meeting in 1791. The contour canal, 54km long, had 18 locks, falling 43m. Constructed between 1793 and 1797, it was the longest of the canals designed by William Jessop in the south Nottinghamshire area and was the first canal to draw its water almost entirely from reservoirs, those at Knipton and Denton. It was built to serve agricultural communities and deliver coal, with plans to extend it to Boston and the Wash. It was sold to the Ambergate, Nottingham, Boston & Eastern Junction Railway & Canal Company in 1854. In 1905 it still carried 19,000t of freight but it was closed in 1936 because of competition from lorries. There is a legal requirement to maintain 600mm depth of water for land drainage, a requirement that is clearly being ignored in most places. It has never carried powered craft The towpath is good throughout and restoration is progressing under the Grantham Canal Partnership as fast as the limited finances will permit.

The **Grantham** terminus with

stretch starts next to a bridge in the Earlesfield area. There is an industrial estate road to the east of the canal while the housing estate on the other side instead provides children, balls and shopping trolleys. The water is clean and has fish although algae collects. Overhanging trees were cut back in 2008. To the north-east is the River Witham and St Wulfram's church, its spire the sixth tallest in England at 86m. Previous local residents included Granta, the man after whom the village was named, and former Prime Minster Margaret Thatcher.

After a piped road crossing, the canal continues to a roundabout by the Ramada Hotel at the end of a spur from the A1. From here it is a 500m walk along the A607 under the A1 and down a footpath to the British Waterways section, which is then largely continuous for much of its length, excluding some four dozen bridges which have been filled in, culverted or replaced at low level since the Second World War.

Although covered with duckweed, the water quickly becomes deeper. The route is mostly rural and the noise of sheep quickly replaces the traffic roar on the A1. There are hawthorns and watercress along the edges. Distance posts – some original, some replica – are found

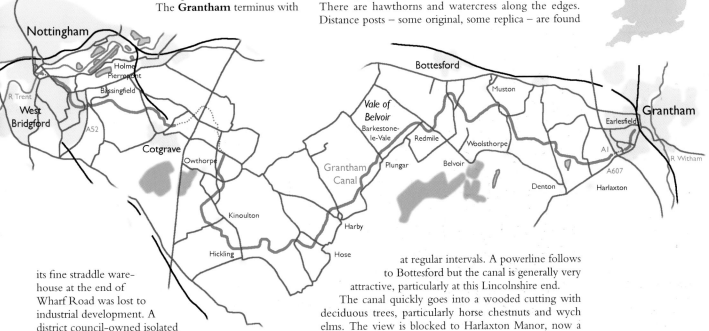

its fine straddle ware-house at the end of Wharf Road was lost to industrial development. A district council-owned isolated

at regular intervals. A powerline follows to Bottesford but the canal is generally very attractive, particularly at this Lincolnshire end.

The canal quickly goes into a wooded cutting with deciduous trees, particularly horse chestnuts and wych elms. The view is blocked to Harlaxton Manor, now a

Horse chestnut trees hang over a duckweed covered surface at Harlaxton.

The Dirty Duck at Woolsthorpe By Belvoir.

college, but it is close enough to the canal for the stone for its construction to have arrived by water to Harlaxton Wharf by bridge 66. Also hidden is Denton Reservoir, surrounded by 1.4km of curved embankment. The 2.6ha manor gardens on seven levels were designed to match any in Europe and have been restored. A slipway and canalside picnic area have been located at Denton Wharf, halfway from here to Casthorpe bridge, which has been replaced to give navigation clearance, the first to be tackled.

The Viking Way footpath crosses but the bridge which used to carry an ironstone railway across, to follow the towpath for 2km, has now gone, along with the tracks. The first three locks were restored and ready for traffic but Woolsthorpe Top Lock has since cracked. By Woolsthorpe Locks is the Dirty Duck public house, which adds a campsite and children's playground to its visitor attractions. On the canal are swans, Canada geese, coots and moorhens, whilst in the canal are carp, perch, pike, roach and tench.

Rising above the oilseed rape fields and beyond Woolsthorpe by Belvoir is Belvoir Castle on a beech-wooded escarpment of Middle Lias ironstone to the south-east of the Vale of Belvoir. Belvoir Castle, visible over a large area of countryside, was first built by Robert de Todeni, standard bearer of William the Conqueror, who called it Belvedere, this having been corrupted subsequently and now pronounced 'beaver'. A later castle was besieged in the Civil War. The present one, the fourth, was completed in 1830, following a fire in 1816, and features many Romantic Gothic additions. There are two Chinese rooms, fine Rococo ceilings in the Elizabethan Saloon, King's Rooms named after a visit by George IV, a vaulted ceiling in the ballroom, a picture gallery including a Holbein of Henry VIII and others by Poussin, Reubens, Van Dyke, Reynolds and Gainsborough plus Gobelin and Mortlake tapestries. There is also a museum of the Queen's Royal Lancers. Spring Gardens date from the 19th Century and there are peacocks and jousting in the summer. The name of the castle, which is the home of the Duke of Rutland, was also taken for a foxhound breed.

Early in the 17th Century Joan Flowers and daughters Margaret and Phillipa were dismissed as castle employees. Together with three local witches, they weaved spells said to have killed the first two sons and daughter of the Earl of Rutland and stopped the family having any more children. Local gossip increased and the three Flowers were arrested in 1619 and taken to Lincoln. The mother denied everything but died on the way while the daughters confessed and were burnt in Lincoln.

From 1815 to 1918 a horse-drawn tramway ran from the castle to the canal between 1815 and 1918 and a wagon from it, in the National Railway Museum in York, is said to be the oldest vehicle with flanged wheels preserved in Britain.

Knipton Reservoir, a canal feeder in the steep-sided valley of the River Devon, had the largest dam in the country when built at 220m long. The 5km feeder channel, often in tunnel, joins the canal near Muston Gorse although it is doubtful if the Knipton water still gets through.

Locks 12 to 15 lower down the Stenwith–Woolsthorpe flight have yet to be restored. Orange tip butterflies fly among cow parsley and cowslips near an old moat and the cuckoo may be heard in the distance in the spring.

The canal moves into Leicestershire where Muston has an ancient cross. Muston Meadows are a National Nature Reserve for their flowers.

A heron fishes quietly in water which is clean enough for freshwater mussels while larks make plenty of noise overhead.

The Lady of the Vale church in **Bottesford** is the biggest village church in Leicestershire and has the tallest spire in the county. Inside are the tombs of eight Earls of Rutland and a three-tier tomb for the three witches. The market place has the stump of an old cross, stocks and a pillory post.

At the foot of Toston Hill white blossom proliferates in the spring.

There is another piped crossing after the Peacock at Redmile, except the bridge is still intact. This time the pipe crosses the canal just above water level and there is a plank to hold back the water.

The canal sweeps round past a pillbox to another blockage and turns back past another moat to continue along the edge of the Vale of Belvoir, passing the remains of a Second World War fuel depot. This is farming countryside with good flora but only because the attempt to turn it into an opencast coal mine was resisted.

The surface of the canal like a mirror near Muston.

The church spire at Redmile reflected in the canal.

The 8km to Harby is a Site of Special Scientific Interest with little vegetation management and more greenery on the unsurfaced towpath from Muston Gorse to Harby. Fallen willow branches have to be fought through near Barkestone-le-Vale but this is far from the whole story about vegetation. Each kind seems to have its own area on the canal: duckweed, bulrushes with the sharp stems of the previous year's dry seed stalks, greater reedmace usually with a clear but narrow channel through, water soldier like pineapple tops and, worst of all, thick algae. Bulrush dried stems can be sharp enough to cause deep cuts. Areas of each species start and stop with no obvious reason but none is impassable. Restoration will solve the problem but at present passage early in the year is easier before weed growth gets going and stinging nettles start growing on the banks at blockages. Most bridges are now piped. Several farm access bridges are low, some just possible to squeeze under.

After Plungar there is a Small Farm Centre and two places where former railways crossed, the second having run down past Langar Airfield, where much parachuting is undertaken.

Notwithstanding the pipes, there are a couple of rowing dinghies at Harby. An old mill building stands beside the canal and both Harby and Hose had wharves. Hose has a duck house, a canalside seat and a white towpath surfacing leading off into Nottinghamshire. A couple of 1990s swing bridges for farm access are the shape of things to come along the canal. The 200-year-old Clarke's Bridge 32 has been fully restored.

Hickling Basin, faced by the Plough Inn, is the only remaining basin on the canal, here at its southernmost point. Hickling Wharf has its original listed warehouse.

The canal winds north past Kinoulton with its cricket ground to the Vimy Ridge Farm piped crossing, where a picnic table has been sited for canal users before the Devil's Elbow. Beyond Mackley's farm the canal has become a young forest. At least the 7km of towpath to Cotgrave is easy walking.

Owthorpe churchyard contains the grave of John Hutchinson, the Puritan governor of Nottingham who was a signatory of Charles I's death warrant.

After Cropwell Bishop, which makes Stilton cheese, Fosse Locks are either side of the A46 Fosse Way Roman road. Lock 11, at the end of the 32km 46m contour pound, had double top gates in case of problems.

Wharf building with loading doors at Harby.

Cotgrave Colliery worked from 1964 to 1993 and a Geordie accent is as likely to be heard as a Nottinghamshire one. Its regeneration included restoration of a kilometre of canal and two locks with a country park alongside. Cotgrave is a village of red cottages spreading towards the canal and a marina planned in the colliery area. Digging in the past has revealed the graves of Roman soldiers.

There is no longer a windmill on Windmill Hill but an attractive golf course lies alongside the canal. Kingfishers and herons may be seen.

It is accepted that the obstructions at the A52, A6011 and Lady Bay area at the Trent end of the canal are now insurmountable and that a better way of connecting to the rest of the waterways network is to build a new line down the course of the Polser Brook to cross the A52 and pass between Holme Pierrepont Hall and the National Water Sports Centre sprint regatta course, with a marina at Holme Pierrepont.

The canal passes round the northern side of Nottingham Airport at Bassingfield, where there is plenty of light aircraft activity.

The most complicated crossing comes where the dual carriageway A52 and two feeder roads have to be crossed, the one consolation being that the central reserve is wide enough for the boater or other user to stand between the two lines of crash barrier while waiting for a break in the traffic.

A supermarket is built on the bank as the canal works its way round Gamston on the edge of **West Bridgford**, past a selection of locks and fixed swing bridges. The westbound carriageway of the A6011 is carried over the old canal bridge but there is a large pipe across at water level in front of it and a low level eastbound carriageway beyond it. The canal heads into a built-up area, popular with dog walkers. On the south side of the A6011 Radcliffe Road are large houses and small hotels. The established houses of Rutland Road, with their mature gardens, back onto the right bank.

After two more obstructions the canal reaches the point where it has been built across and buried under a complicated road junction at the end of Trent Boulevard. The gap in the fencing is on the left before the large willow tree overhanging the end of the canal. This only a stone's throw from **Nottingham** Forest football stadium and Trent Bridge cricket ground.

The final 200m of the canal lie 300m away, a quiet reach under Environment Agency control with a surprising amount of wildlife considering its position alongside the Brian Clough stand of the football stadium. It also contains the bottom lock, fully restored in 1972 but piled across and not used since 1973. Almost opposite on the far side of the River Trent is the Beeston Canal. In 920, Edward the Elder built a bridge across the Trent at West Bridgford. For many years a roving bridge crossed the river. Any goods crossing the river today once again have to go by road.

Nottingham is named after the village of Snota's people, the Snotings, although the city prefers to associate itself with Robin Hood, despite the fact that it was the sheriff of Nottingham who is remembered as the baddie.

The prominent landmark of Belvoir Castle.

The Polser Brook: a possible future line to Holme Pierrepont.

Distance
53km from Earlesfield to the River Trent

Navigation Authority
British Waterways

Canal Society
Grantham Canal Restoration Society www.granthamcanal. com

OS 1:50,000 Sheets
129 Nottingham & Loughborough 130 Grantham

15 Grand Union Canal, Aylesbury Arm

The failed Western Junction Canal

The Aylesbury Arm of the Grand Union Canal is somewhat less significant than when originally planned. It was to have been the Western Junction Canal, running to Abingdon where it would have been continued by the Wilts & Berks Canal to Melksham and then the Kennet & Avon Canal to Bristol. Abingdon already had the River Thames from London and, not far north, the Oxford Canal towards Birmingham. As it happened, the Western Junction never got beyond Aylesbury and Abingdon missed out on becoming a major transport interchange, as it was to do again later when it rejected the Great Western Railway.

A contributory problem was the starting point at Marsworth, just seven locks down from the summit level of the Grand Union Canal. Because of the inevitable water supply problem and the fear that it would drain the main line, all locks on the Aylesbury Arm were built as narrow beam and this would have meant that only narrow beam craft would have been able to work through to Bristol. As it was, Wilstone Reservoir, which supplies the Aylesbury Arm as well as being pumped to the main line summit level by pumps at Tringford, had to be enlarged twice. Like the **Tring** Reservoirs, it sits between the Aylesbury and Wendover Arms, all doubling as nature reserves with breeding and migratory birds and a small heronry.

Nevertheless, the idea had popular support from the people of Aylesbury and the Marquis of Buckingham and the Aylesbury Arm was opened in 1815, remaining commercial until the 1950s, after which it became semi-derelict. In its heyday it took agricultural produce and livestock out of Aylesbury and delivered building materials, timber and coal in return.

Now fully restored, it poses a dilemma for the narrow-boat user. The Aylesbury Arm is one of the quietest and most rural canals in the country, a charming find within easy reach of London, yet it has 16 locks on the descent to Aylesbury, 32 locks for the 20km return trip on this dead-end canal. Many people pass it by but enough are still attracted to keep it busy at times.

The Aylesbury Arm leaves the main line at Marsworth Junction, 300m north-west of the Lower Icknield Way.

Adjacent to the junction is a stone building which serves as a British Waterways depot, accompanied by a large yard of building materials and a hand-operated crane for loading them onto the many British Waterways craft moored alongside. Surprisingly, this depot is only 2km from British Waterways' Bulbourne depot.

The first eight locks form the Marsworth flight. They average 200m apart. The first pair come as a staircase. Many of the locks have houses alongside, not traditional lock keeper's cottage styling but their locations cannot be accidental in this rural landscape. Several locks have gates dating from 1991 and the design is such that water weirs over.

The line is almost due west across Buckinghamshire although sections of Hertfordshire are intersected in the early part of the route.

The fourth lock, which has a wall of periwinkle at its lower end, swings the canal onto the straight line it is to follow for the next 3km. Looking back, the line points to the tower of Marsworth church and the chimney of the cement works at Pitstone, hidden at the start of the arm. Several of the fields thus far have held horses but the land is now predominantly down to crops.

As the last of the eight locks is approached at Wilstone, the only canalside community, the pristine fields of garish oilseed rape give way suddenly to an apparently ploughed field. From beyond the hawthorn hedge come the satisfied grunts and squeals of a herd of pigs wallowing in a pool just out of sight but not out of range of sound or smell.

Wilstone Lock is counted separately although it would seem to be the final step in the Marsworth flight.

From here the gradient eases as the canal distances itself from the foot of the Chilterns. This Vale of Aylesbury countryside is flat and of unspoilt traditional English open-farming character. To the south the scarp slope of the Chilterns rises steeply to the beechwood-clad Coombe Hill, the highest point in the Chilterns at 260m, 170m above the vale here. Interestingly, 5km to the south-west is another Coombe Hill that exactly matches the 260m height of this one, the latter topped by a monument and hiding Chequers, the country residence of the Prime Minister.

The sky is filled with an assortment of forms of flight: noisy biplanes doing aerobatics, model aircraft towards Aston Clinton, gliders from Dunstable, hot air balloons or a heron seeking quieter fishing.

Two Puttenham locks draw attention to the village of Puttenham to the north. One of the church's vicars was Christopher Urswick. As Richard III's Recorder of London, he supplied information to the exiled Henry

On the Marsworth flight near Wilstone.

Happiness is pig-shaped. The lock's neighbours at Wilstone.

A lockside house with a British Waterways logo-shaped weathervane.

Aylesbury Basin and its incongruous neighbours lie beyond the recent footbridge.

Mallard ducklings, not the traditional Aylesbury ducks.

Tudor, later marrying him to Elizabeth, the daughter of Edward IV.

Another rural interlude is interrupted only by Buckland House and Aston Clinton or Redhouse Lock. A house by the latter sports a weathervane which bears more than a passing resemblance to the British Waterways logo.

The village of **Aston Clinton** was developed largely by the Rothschilds who settled here in 1851 and proceeded to spend some of their great wealth in the area.

Hawthorn bushes grow alongside the canal and bulrushes now appear in profusion, together with the other reeds. Powered craft muddy the water and leave mats of chopped reeds to be blown about by the wind.

Lines of communication converge with the canal as it approaches Aylesbury. To the left lies the A41, the current form of the Roman Akeman Street, while to the right is the course of a former railway from Aylesbury to Cheddington. There are three sets of powerlines but these suddenly descend to a substation.

The land becomes very low-lying around Broughton Lock, giving extensive views, and there has been bank restoration work, sheetpiling and fabric membrane laid on top of the embankment to give stability.

The canal probably follows a former line of the Bear Brook which now runs to the south, possibly diverted to serve a mill. The brook's present course is on the side of a slope and in times of flood it overflows down towards the canal. A flood relief scheme for **Aylesbury**, made use of two areas of low ground as flood storage areas. The Stocklake Brook area lies further north but the Bear Brook flood storage area is located between Broughton Manor, the housing of Aylesbury and the canal. Between the brook and the canal is a marshy area, possibly due to seepage from the canal, that is a haven for rare bog plants.

To the north lies Bierton, its church tower extended up among the trees by a slender spire. Above twitters the skylark.

Aylesbury arrives with Victoria Park and a graffiti-covered concrete bridge carrying a minor road past the youth custody centre. New industrial units had rich soil around them which turned out to contain dangerous industrial contaminants. There are older factories around Aylesbury Lock and Aylesbury Bottom Lock. One business

worthy of note is Hazell Watson & Viney, for many years involved with the printing of Penguin books until bought by Robert Maxwell, at which point Penguin withdrew all their custom. Another was the Aylesbury Condensed Milk Company, attracted in 1870 by the canal and now owned by Nestlé. Before Aylesbury Lock the Bear Brook passes under the canal in a brick siphon that has been renovated as part of the flood relief scheme.

An assortment of exotic ducks beyond the A41 come as a reminder that Aylesbury's name is associated with ducks above all else.

The appearance of the town was substantially changed in the 1960s by redevelopment, often unsympathetic with the character of the town. One recent construction, which does deserve praise for being within the spirit of canal architecture while embodying original thinking, is the footbridge across the entrance to the terminal basin, which is striking but well proportioned.

Aylesbury Basin has wharves, moorings and residential boats. Seeing the basin on a summer's day with narrowboats turning to begin the climb back to the main line is a completely different picture from seeing it late on a winter's afternoon, sawn logs piled up on the towpath and on cabin roofs, woodsmoke from various chimney's drifting silently into the dusk.

Being the bottom end of the canal, a weir is needed and takes the form of a neat semi-circular brick overflow which feeds the head of the California Brook on the south-east side, joining the Bear Brook to the west of Aylesbury. Behind the moored boats opposite is an aerial reaching out of some industrial premises. Beyond is the town centre, gathered around a chunky concrete tower block.

The canal basin terminates at an office block that obscures a 13th-century church but has a mirror window that reaches almost to water level so the boater can have the unusual experience of watching his own arrival. On the other hand, he can see all kinds of reflections in the angular glass edifice beyond it, not a swimming pool as might be thought but the offices of the Equitable Life Assurance Co.

Aylesbury Canal Society are best known for their guide to launderettes near canals, used by boaters throughout the country.

Distance
10km from Marsworth Junction to Aylsbury

Navigation Authority
British Waterways

Canal Society
Aylesbury Canal Society www. aylesburycanal.org.uk

OS 1:50,000 Sheet
165 Aylesbury & Leighton Buzzard

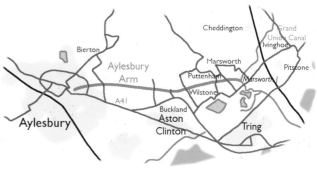

16 Grand Union Canal, Paddington Arm

Linking London's river terrace factories

The Grand Union Canal links Birmingham with London. At the Birmingham end it is well connected but in London the through route leads to the River Thames, convenient for the docks but too low for the industries on the higher river terraces. The Paddington Arm was built as a lock-free pound to link the Grand Union Canal with the north-west of London, terminating at Paddington but subsequently being extended via the Regent's Canal to the docks.

In the days before railways, canals were an efficient form of transport and there is a record of troops taking seven days to travel from Paddington to Liverpool by boat in 1806, much quicker than marching.

At Harlington, the smell of kerosene arrives on the prevailing south-westerly wind, along with the persistent roar of engines as jet airliners touch down incessantly 3km away at Heathrow.

Bull's Bridge Junction is on the main line of the Grand Union Canal, crossed by a Grade II listed arched brick bridge dating from 1801, with stop gates underneath. On the left is a 19th-century toll house, the offices of the Colne Valley Passenger Boat.

The Penzance to Paddington railway crosses and is never too far from the canal for the rest of its journey, Brunel having chosen to locate the terminus of his Great Western Railway next to the canal basin at Paddington.

Each side of the railway are car depots, the one on the far side being surrounded by masses of barbed wire and forming an enormous traffic jam on the site of the **Southall** Gasworks, now represented only by a massive gas holder. The Jam 'Ole was not related to this, though. The Mitre Dock and Ticklers Dock belonged to Ticklers, who made jam and supplied the army during the First World War. The last commercial users of the canal were biscuit makers Kearly & Tonge, who received coal deliveries here.

Among the dumped debris, which the blackberry bushes and other vegetation attempt to disguise, is an overflow weir feeding Yeading Brook. By the A4020 is the Hambrough Tavern.

The extensive **Hayes** brickfields have been landscaped and the canal arms filled in. Spikes arched footbridge crosses, moorhens nest and herons fly over. A pocket of tranquillity has returned. A new development is located around Engineer's Wharf Moorings.

During the Second World War the canal was used for testing lifeboats at Greenford. Glittering murals decorate a bridge.

Recent housing lies within a loop of the canal. Away from the houses, the banks bear yellow irises and elder bushes. Some wooden blocks carved with canal scenes act as seating.

The A40 Western Avenue crosses. After some high tennis nets, High Line Yachting have their moorings and their craft can be anything from narrowboats to Belgian barges, rather than yachts. Factories vary between the new and one that has a wharf but with all the access doors leading to it bricked up. The next railway line to cross over is the Central underground line. One of the sharpest bends on the canal is by the Black Horse, an area where there are also the most interesting nasal attractions. Fresh coffee wafts from the Lyons Tetley works, to be replaced by fresh bread from the Mother's Pride bakeries and, no doubt, the Guinness plant could add to the aromas. Lyon's Dock, marked by its water tower, was built in 1927, the last private dock to be completed in London. The final factory, fronting the A4127, is the Glaxo pharmaceutical works.

After this, the canal breaks into its most rural section. A laminated wooden arch bridge leads towards Horsenden Hill, which rises to 84m on the left bank with rugby pitches. After a large wooden owl, Sudbury Golf Club is located around its lower parts. Opposite is Perivale Wood, established as a nature reserve in 1904, a remainder of the ancient Middlesex Forest. Its 11ha is administered by the Selbourne Society. The rich oak and hazelwood is a riot of bluebells in May and there are 70 nesting bird species. Water voles use the canal where there are not sheetpiled edges.

Horsenden Lane Bridge is also called Ballot Box

Canalside seating with canal motifs at Greenford.

The North Circular Aqueduct at dawn. The Wembley Arch is visible on the left.

Spikes footbridge to the landscaped former Hayes brickfields.

Bridge while Manor Farm Road Bridge, beyond the West London Motor Cruising Club moorings, is better known as Piggery Bridge.

Alperton was a brick and tile making centre after the canal opened in 1801. Sand, gravel, coal and hay were carried to Paddington. A builders' yard is the nearest remaining relative with a set of floodgates nearby. More noticeable these days are a new office block with blue-tinted windows and the Pleasure Boat public house.

The Piccadilly line crosses over and Stonebridge Park offers an industrial estate with canalside terraces.

The North Circular aqueduct and its neighbour might pass unnoticed but for the pillbox next to them and the roar of traffic from the A406. Crossing the River Brent and London's North Circular Road, they survived Second World War bombing, only to burst in 1962.

The *Radio Times* and *Listener* building was established before an 1830 listed canal cottage, out of place in its present surroundings, a set of floodgates being the only other contemporary feature. Nearly a century later, the Heinz factory was built on the opposite bank in 1925 and the beans and other raw materials for their 57 varieties were brought in by canal until the 1960s. The factory has now been demolished. Factories on the Park Royal Industrial Estate were built during the First World War to manufacture munitions.

Acton Lane Bridge is colourful. The Grand Junction Arms next to it includes a three-dimensional mock-up of a working canal boat and a decor of roses, castles and lace-edged plates.

Acton Lane Power Station, built between 1949 and 1954, cannot be ignored. The massive cooling towers on the right bank use canal water, which, consequently, is warmed. Several gloomy walkways and conveyors cross the canal. The cables have been laid in the towpath where the canal water keeps them cool, the towpath being improved in the process. This was an important aspect in establishing the Canal Way walk, which was started in 1978, linking various canalside parks and water-based facilities with the towpath.

The bridges are completed with a complex of railway arches, more following at Old Oak Common. Here, little is seen except a tall aerial and some old factories on the left bank and open space behind a long low wall on the right. Beyond the wall and rather lower than the canal are the massive railway yards which act as a holding area for engines and coaches within easy reach of

Victorian terraced houses back directly onto the canal at West Kilburn. Half Penny Bridge crosses beyond.

Mural from collected scrap at Paddington.

Paddington. Across the railway, Eurostar trains are serviced at the North Pole depot. Soon the Western main line comes right alongside, as is shown by two spans of bridge, the smaller one crossing the canal while the larger one on the right is over the railway. The open space beyond the railway is more open to some than to others. In the 1900s Claude White and others tried to fly from the grounds. Some have tried to fly from the buildings of Wormwood Scrubs since the prison was built in 1874.

Kensal Green cemetery on the left bank was opened in 1832. Its 23ha include three royal mausolea. Railway engineer Isambard Brunel, William Thackeray, Anthony Trollope, Decimus Burton and Emile Blondin are among those interred within its walls. Water gates opening onto the towpath allowed hearse barges to be used, a most dignified way to make a departure.

The Kensal Green gasworks' gas holders of 1882 and 1892 have intricate wrought-iron embellishments surrounding them. Their dock arms date from 1872 and supplied coal until the 1960s. Some hold narrowboats.

Beyond a Sainsbury store, Port-a-Bella Dock was built in 1894 for rubbish barges. These days it has a Sunday

Kensall Green Gasworks.

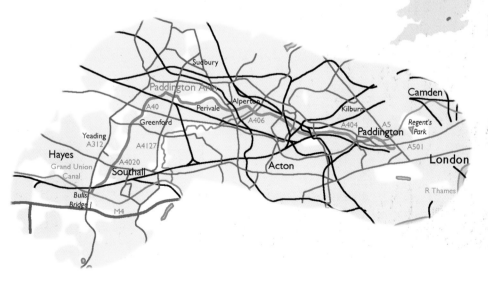

Houseboats are moored alongside the canal as it approaches the Paddington Stop.

market, craft and canalside centre and offers evening canal cruises. The B450 Ladbroke Grove crosses near the site of the 1999 rail crash that killed 31 people and injured some 400, landing Network Rail a £4 million fine.

West Kilburn produces a stark contrast in housing conditions with new tiered developments facing dismal Victorian terraces that back directly onto the canal. Fences and wire keep the canal from a public park and canal users from the windows of the houses.

Wedlake Street Footbridge is also known as Half Penny Bridge, the toll that was charged for using this footbridge.

Housing is increasing and a massive tower block rears up between the canal and the railway, its lift shaft standing clear of the main building, a large room projecting at the top and walkways crossing, precariously it would seem, at intervals.

On a more human scale, Meanwhile Gardens have a kiln and pottery, a buskers' festival in June and a summer festival in August.

Carlton Bridge, with its cast-iron balustrades, was built in 1870. Exactly a century later came the elevated A40 Westway, sweeping out above the canal but not crossing at this point. Opposite it is **Paddington** General Hospital which includes the Victorian Poor Law Union infirmary.

The canal warden's hut is incongruous in front of the Warwick Estate tower blocks. The church of St Mary Magdalene is just incongruous. Built on canal construction spoil by Street between 1868 and 1878, it is in Victorian Gothic and features a prominent red and white striped spire. A wall is brightened up by a colourful mural made from debris collected by Stowe Youth Club with artist Kevin Herlihy.

The Paddington Stop public house is on the straight that leads to the former Paddington Stop toll house, the bridge built at the turn of the last century, and some public toilets. Beyond the bridge is Little Venice or Browning's Pool, one of the best known canal sights in London. Opened in 1801, the basin includes Rat Island or Browning's Island, an island of weeping willow trees, popular with mallards, tufted ducks, Canada geese, moorhens, coots and swans and named after Robert Browning, who lived from 1862 to 1887 in Warwick Crescent in a house that has since been demolished.

Little Venice is one of the most colourful canal scenes in London, an urban canal at its most classy. Jason's Trip operates from here, inaugurated in 1951 for the Festival of Britain, and Jenny Wren cruises operate to Camden along the Regent's Canal, which leaves the northern end of the basin. Waterbus services began from here in 1959.

The canal continues through a park and beneath Westway, the A40 on the top deck and the A404 Harrow Road underneath. The A4206 Bishop's Road Bridge upgrade involved jacking the old bridge up 10m

Bridge complex at Paddington including the A40 Westway.

Cranes on the walls of a warehouse at Paddington Basin.

while its replacement was built underneath, in the process removing on early Brunel bridge for relocation over the canal. Craft are moored in front of the railway terminus. One old building has crane brackets and other ironwork bolted to it to help unload barges while other buildings are recent. A modern cable-stayed footbridge crosses the Paddington Basin.

It all comes to a sudden end, surrounded by walls. The Telecom Tower overlooks all. Marylebone Flyover, carrying the A501 over the A5 Edgware Road, is just a stone's throw away over the buildings. When the basin was opened in 1801 it turned this rural parish into a bustling transport junction. Today the basin is almost forgotten.

Distance
22km from Bull's Bridge to Paddington Basin

Navigation Authority
British Waterways

OS 1:50,000 Sheet
176 West London

Paddington Basin with a helical bridge at the far end.

17 Regent's Canal

A green ribbon around London's heart

The Regent's Canal was completed in 1820, running through open country to the north of London to link the Paddington Arm of the Grand Union Canal with the docks. It was part of John Nash's metropolitan improvements of the Regency, the only coordinated building plan in the capital's history, a scheme that included Regent's Park, St James' Park, Regent Street, Trafalgar Square, the Strand and the Suffolk Street areas. The canal carried much coal, timber, building materials and food produce, serving many factories along its length, but never made the anticipated profit because of competition from the railways. Indeed, the canal's owners attempted to turn it into a railway in 1845. It was the ceasing of munitions traffic after 1945 that really put the canal into decline although working narrowboats and horse- and tractor-drawn dumb lighters continued until the early 1960s and there is still some trade, mostly in timber.

Now, however, its primary function is as one of the most valuable amenity waterways in Britain as it provides a green ribbon around the heart of this built-up area of London. Despite the fact that access gates are locked overnight, the amenity use is appreciated.

The Regent's Canal leaves the Paddington Arm at Little Venice in **Paddington**. Leading east from the basin on each side of the canal are plane tree-lined avenues of fine Regency houses, as well as Junction Cottage, a canalside house.

This reach continues to Maida Hill Tunnel, which is only 249m long but has no towpath. It is difficult to leave the canal because of the continuous spiked railings.

Passing over the straight tunnel is the equally straight A5, originally built as the Roman Watling Street from London to St Albans. Spoil from the tunnel was disposed of on the field of local landowner William Lord. This field, opened in 1809, now forms the home of Marylebone Cricket Club which represents England internationally, Lord's being one of the world's best known cricket grounds.

A house built above the tunnel was the home of Dambusters leader Guy Gibson. Another, where the pedestrian route returns to the canal, has its door at street level and the house is then built downwards, the opposite way to usual.

Canalside Gardens were established in 2006 between this tunnel and the next but have yet to produce horticulture of note at canal level. Pleasure craft are moored near Lisson Grove housing estate, built on the former Marylebone Station Goods Yard. The Aylesbury to Marylebone line, formerly the Great Central Railway line from Manchester, crosses over the canal with the underground Metropolitan line. Many of the main line railway termini are just to the south of the canal while their lines fan out across the canal towards their respective destinations.

A row of extremely imposing red-and-white brick houses gaze down from positions in St John's Wood's Prince Albert Road. Grand in a different way on the other side, at Hanover Gate, is the Central Islamic Mosque.

The canal skirts the northern edge of Regent's Park. Formerly Marylebone Park, which was used as a royal hunting ground by Henry VIII in the 16th Century and until Cromwell's time, the 1.7km^2 Regent's Park was laid out in 1812–27 by John Nash and opened in 1838, being named after the Prince Regent who later became George IV.

There are several interesting bridges along this reach. One carries the Tyburn in conduit. Another is Macclesfield Bridge, named after the Earl of Macclesfield, the Regent's Canal Company's first chairman. It is also known as Blow Up Bridge, from an explosion that took place in 1874. The narrowboat *Tilbury*, loaded with gunpowder and benzol, was one of three being towed by a steam tug when it exploded as it passed under the bridge, reducing the bridge to rubble, damaging nearby houses and killing three crew members. The bridge was subsequently rebuilt, its Doric columns being rotated to put the rope incised sides away from the water. The bridge stands today, its cast-iron capstones proclaiming that they were cast at Coalbrookdale.

The park is best known as the home of London Zoo, one of the world's largest at 15ha and also one of the oldest. The zoo was laid out in 1827 by Decimus Burton and carries out much important research. It houses 8,000 species of animal, bird, reptile and fish, together with a bar, restaurants and cafés. A waterbus landing stage serves the zoo, the canal passing by the antelope terraces from where Arabian oryx gaze at the boater and across at Lord Snowdon's dramatic Northern Aviary of 1965. Not all the wildlife is on the inside,

The Regent's Canal from Little Venice with Browning's Island on the right.

Macclesfield Bridge, also known as Blow Up Bridge.

though, and the canalside vegetation, including pendulous sedge, tansy, feverfew, bur marigold, skullcap, gypsywort and buttercup, ash, elm, hawthorn, sycamore and willow provides habitats for rabbits, hedgehogs, squirrels, willow warblers, chiffchaffs, whitethroats, redstarts and redpolls among commoner species.

Two stones, which mark the old parish boundaries of St Marylebone and St Pancras, stand beside the canal, formerly being positioned under the last of the elegant iron bridges.

The Cumberland Basin now has room for little more than the Feng Shang floating Chinese restaurant. Most of the Cumberland Market Branch was filled with war rubble in 1948 and now forms the zoo car park but initially served a large basin at Cumberland Hay Market, supplying fresh vegetables daily from Middlesex market gardens to this important shopping centre. The original plan to have the canal passing through the park was stopped by objections from those who disliked the idea of a commercial canal running through a fashionable area.

The canal now turns sharply left towards Primrose Hill, with its elegant Victorian houses and canalside gardens and silver birch trees. Before these is St Mark's

Pirate Castle opposite a matching electricity generating station.

church, built beside the canal, a squat stone church with a slender spire. Bridges were designed by Robert Stephenson. One carries the West Coast Main Line. When it was first opened, trains were hauled by cable from Euston up the steep slope over the canal.

The Pirate Water Activity Centre is a club which was formed in 1966 to promote watersports. The building is designed as a pirate castle complete with drawbridge and battlements, club room, canteen, mini gym and workshops; it was opened in 1977. The centre also has a narrowboat, named the *Pirate Princess* in 1982 by Prince Charles, who demonstrated that narrowboat handling was not one of his finer attributes, the tiller being seized from the future monarch's grasp at one point by the centre warden. The club membership peaked at 1,200 and fundraising in the early days included pirate raids on commercial craft such as the *Jenny Wren*, staff trying to get their hands on the booty for development of the centre before it found its way into young pirate pockets.

Across the canal, an electricity generating station has been built in matching style.

The towpath rises over the gloomy entrance to the former Interchange Warehouse, a large railway warehouse with underground vaults, used for storing wine and spirits. It is a listed building.

The former Dingwalls Timber Wharf & Dock has become the **Camden** Lock Centre, opened in 1973. The original stables, with hay lofts above, have become craft workshops, restaurants, weekend market, night club and picture and antique shops. There is a cruising restaurant boat and there are open air jazz concerts, ox roasts and exhibitions, a thriving hive of activity.

Hampstead Road Lock has a two-storey castellated lock cottage which was the first of the buildings in the area to have battlements and is now the canal information centre, reached by a cast-iron bridge over the canal. The lock marks the end of the 43km Long Level pound from the Grand Union Canal main line at Cowley and Norwood Top Lock. From here it is a 26m descent to the Thames.

Lord Snowdon's Northern Aviary at London Zoo.

A heavily laden barge in the remains of the Cumberland Arm.

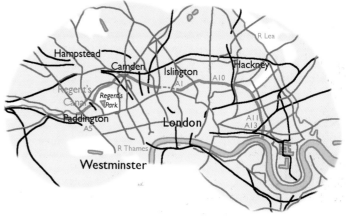

Camden Lock is probably the busiest place on the British canal system.

At the Water's Edge – a futuristic housing development near Kentish Town Lock.

The Constitution's murals brighten up a Kentish Town corner.

All locks were originally built in pairs with paddle-controlled connections so that each could act as a side pond for its partner, saving time and water. This is the only place where the pair of locks remain, the balance beams positioned so that one goes above the other on the centre island. On all other locks one of the pair was converted to a weir between 1973 and 1976 to reduce the risk of flooding.

In fact, this site was used for Colonel Congreve's hydro-pneumatic lock of 1813, which was intended to use caissons to operate without any loss of water but failed to work properly and was converted to a conventional lock.

Beside Hawley Lock is a former brewery.

Modern buildings have been replacing old all along the canal. Beyond Kentish Town Lock, the final one of the flight, is a futuristic housing development known as At the Water's Edge, constructed in rounded metal panels and opened in 1990.

Near Camden Town station the canal passes over the former Fleet, the ground levels having changed considerably over the years. Camden Town was named after the Earl of Camden, who developed the area in 1795.

The Constitution begins a straight reach that is bordered by a large new housing complex and then the Jubilee Waterside Centre. In this urban area naturalists have noted swallows, martins, kingfishers, reed buntings and various warblers while plant life includes ivy, Virginia creeper, buddleia, toadflax, wall rocket, bindweed, ragwort and hemlock water dropwort.

After a hospital the canal is crossed by an oblique bridge which carried the driveway of William Agar, a landowner who delayed construction of the canal with compensation claims. Eventually an Act of Parliament fixed the line across his land in 1816 and he was awarded £15,000.

The Midland Main Line and High Speed 1 Channel Tunnel rail link cross the canal and arrive at St Pancras station with its twin neo-Gothic towers, the station having been extended to act as the European terminal. A complex of connections link to the East and West Coast Main Lines. New Thameslink tunnels approach the station under the canal. The canal passes St Pancras Basin, converted to a pleasure craft area in 1958 from the Midland Railway Company's coal wharf of 1870.

St Pancras Lock has an attractive lock cottage in front of a brick water tower, which has been moved here bodily.

The skyline ranges from the British Telecom Tower to the NatWest Tower but only one of the five St Pancras gas holders remains with its delicate 1860 tracery, a listed structure originally belonging to the Imperial Gas Company. Before this is the Camley Street Natural Park run by the London Wildlife Trust with reedbeds and marginal vegetation, mallard, tufted duck, pigeons and a newt trail.

The extensive King's Cross redevelopment scheme has been taking place on both sides of the canal while the canopy of King's Cross station may be seen on the right, from where the East Coast Main Line passes under the canal. Disused stop gates under a bridge were positioned to protect the railway from flooding in case of a well aimed wartime bomb. The other set of stop gates are under Maiden Lane Bridge, which has fine cast-iron work dating from 1850.

Battlebridge, named after Boudicca's fight with the Romans, became King's Cross. The name is retained in

Waterpoint, the water tower at St Pancras Lock, is a new arrival, acting as a viewpoint and art gallery.

the Battlebridge Basin, also known as the Horsfall Basin from the name of the local landowner at the time of the canal's construction. The 6,000m² basin has the London Canal Museum in a former ice warehouse.

The 878m long **Islington** Tunnel with no towpath is straight with daylight at the far end, albeit rather browny yellow when seen through the exhaust in the tunnel. It was built using a horse-drawn railway along the invert for removing the spoil and was opened in 1816. From 1826 to the turn of the 19th Century it had a steam tug on a platform like a railway engine pulling along a chain laid along the bottom of the canal and passing round two huge drums 2.4m apart. Its successor was said to have been the ugliest boat ever seen on British canals, a coke-fired engine that resembled a corrugated shed enveloped in steam, taking nine loaded barges through in 35 minutes, wide beam lighters being an exact fit in the tunnel. Diesel tugs took over in 1930. An alternative use for the tunnel came in the cold winter of 1855 when skaters were able to use it.

Unpowered craft are not permitted to use it. The walking route is long, follows a network of streets, passes shops and most days takes in a busy market street with a large winged sculpture of tubes and crosses the dual carriageway A1 with central reserve fencing – not a trip to be undertaken by the sensitive. This section of A1 was London's pilot Red Route traffic scheme, yellow parking lines being replaced by red ones, resulting in 30 per cent faster traffic flow and bringing 40 businesses in Islington alone to the brink of collapse because motorists could no longer reach them easily from their illegally parked cars. Crossing the A1 with a boat has not become any safer.

Approach to the far portal is along Colebrook Row, the original line of the New River. Finding the canal again is not easy.

The canal leaves the tunnel mouth between trees and Georgian terraces painted by Walter Sickert as the *Hanging Gardens of Islington*. Islington is from the Old English Gisla's downland, although nothing could be much less like downland these days.

Beyond Frog Lane bridge, built between 1816 and 1820, is City Road Lock and possibly a selection of water lilies and duckweed.

City Road Basin, opened in 1820, was the principal basin on the canal, although only 1ha is left. Neighbours include the Fallen Angel and the London Hydraulic Power Company, who make hydraulic pipes. New residential buildings have been erected opposite.

On the towpath side the stables were used for horses until the 1950s. The Narrow Boat is by Wharf Road Bridge, which dates from 1830.

Wearlock Basin was opened four years earlier, covered 4,000m² and served many industries around its perimeter, now acting as moorings and a refuge for water birds.

Sturt's Lock, named after the local landowner at the time of construction, precedes Whitmore bridge, dating from about 1820. Kingsland Basin was opened ten years later; it is overlooked by warehouses, despite which it attracts mallard, coot, moorhen and tufted duck. It also has a ledge of the tropical grass *Paspalum paspalodes*, which only grows wild at one other site in Britain.

After passing under the A10, the Roman road from Braughing to London, and Britain's first

Eastern portal of the tunnel and Hanging Gardens of Islington.

NatWest Tower and 'Gherkin' hide behind City Road Basin.

The long, the short and the tall at Stepney.

Bench seat with a difference at Stepney Park.

steel bowstring railway bridge, joining up the 170-arch Kingsland Viaduct of the former North London Railway to carry the East London Line extension, the canal reaches Laburnum Basin, which once served the Laburnum Street gasworks.

Beyond the Duke of Sussex are a number of mosaic murals by Free Form Arts and local school children on walls and bridges.

Acton's Lock leads down past the Sir Walter Scott into a reach that seems relatively derelict but has numerous unusual plant species attractive to damselflies, dragonflies and bats.

Beyond the King's Lynn to Liverpool Street railway line is a skew, brick arch in the wall by Mare Street bridge which used to lead to a canal basin but now just serves as a rubbish tip for pushing debris into the canal.

Hackney's Victoria Park was laid out in 1842–5 by James Pennethorne, a protegé of John Nash. It is London's oldest municipal park, the largest in the East End at 88ha, and includes a picnic area. Bonner Hall bridge, with its fine ironwork railings, was part of the original formal entrance to the park.

The engineering workshop and lock cottage at Old Ford Lock.

The park ends at Old Ford Lock with its attractive lock cottage and a British Waterways engineering depot in the building which used to house a steam pump.

Beyond the Royal Cricketers, a cambered cast-iron footbridge of 1830 carries the towpath over the entrance to the Hertford Union Canal or Duckett's Canal. This canal, which provides a link to the River Lea, was bought from Sir George Duckett by the Regent's Canal Company in 1855.

The Mile End Climbing Wall is by the Liverpool Street to Norwich railway line. The canal passes a hospital to reach Mile End Lock, its double-fronted 19th-century lock keeper's cottage sitting uncomfortably among its neighbours of modern brick. The stubby pencil-shape of Canary Wharf, Britain's tallest building, is increasingly visible beyond.

After the A11 the canal passes Tower Hamlets opposite Mile End Park, one of London's newest open spaces with recreational and sporting facilities by the canal on the site of a former busy timber and boatbuilding yard. Canadian, curled and shining pondweeds, azolla, duckweed, hornwort and spiked water milfoil are all to be found in this setting.

Johnson's Lock is left past a gasworks and the Ragged Schools Museum, a charity which benefited from the takings from the lectures John MacGregor gave on his groundbreaking 19th-century travels with a canoe.

A rather ornate brick stack on the left at Stepney is not the redundant factory chimney that it might appear but a sewer vent shaft.

The Fenchurch Street to Southend railway line crosses over the bend leading to Salmon's Lane Lock. Beyond the A13 and beneath the Docklands Light Railway is Commercial Road Lock, the entrance to Limehouse Marina, formerly Regent's Canal Dock. The 4ha basin, in London's tough 18th-century area of Chinatown, has been a collecting point for wood and other debris floating down the canal, as well as for herons, cormorants, tufted ducks, black-headed gulls and great crested grebes. All that has changed with the marina, public houses, shops, housing and offices. Below the basin is the Limehouse Link; at the time of buildings it was the most complicated and expensive British road ever at £220 million for 1.8km. The construction period also involved clearing silt contaminated with arsenic, cadmium, lead and mercury and left a new marina with a link road between the City of London and Canary Wharf, the Isle of Dogs and the Royal Docks passing underneath. The northern half of the basin was filled with North Sea aggregate, pumped from ships on the Thames directly into place on the site. When the road tunnel was finished, the canal was recut through the fill and the remaining material on each side used as a base for residential development up to six storeys high around Limehouse Basin.

The Limehouse Cut connection was opened in 1968, replacing a direct link into the Thames for that canal.

Limehouse Lock was the largest canal lock in London at 100m x 18m, allowing 3,000t ocean going ships through to unload directly into narrowboats and lighters but was only operated for three hours before high tide between 6am and 10pm to avoid draining the basin. The traffic moved downstream in the 1960s. This lock has been replaced by a smaller lock, just 30m x 8m, needing only a tenth of the water of its predecessor. It has computer-controlled sector gates.

Landing is difficult, as is launching onto the River Thames on the far side of the lock. A path leads up the side of the swing bridge, opposite which is the Barley Mow in the former dockmaster's house, from where riverbuses can be seen racing along the river between the City and the Docklands developments and helicopters fly over to the City airport. Dickens' *Our Mutual Friend* included this area.

Distance
14km from the Paddington Arm to the River Thames

Navigation Authority
British Waterways

Ordnance Survey 1:50,000 Sheets
176 West London
177 East London

18 Basingstoke Canal

The canal was completed in 1794 to link the market town of Basingstoke with the capital via the Wey and Thames. With the coming of the railways the canal traffic declined and the canal company went into liquidation in 1866.

In 1966 the Surrey & Hampshire Canal Society was formed to reopen the canal. Surrey County Council supported the scheme with enthusiasm and even Hampshire County Council maintain the towpaths well and have a depot and several maintenance craft on the canal. There are claims that it is the most beautiful canal in the country. Certainly it is one of the most well wooded. However, a rare plant has been found which Natural England believe does not like shade, despite its presence, and the canal society have been told that they may have to cut down many of their trees.

The original canal terminus lay to the south-east of Basingstoke station in an area now overrun with dual carriageways. The first indications of the former line, shallow water overgrown with trees, appear around Old Basing, severed from the remaining canal by the M3. More water is found at Up Nately although this is also shallow. Alongside is a field of alpacas and information about them for the curious, information; boards are commonplace on this canal.

The next bridge is heavily scored by barges. Just before the following bridge, the remains of the Brickworks Arm hides behind a reedbed. The water becomes shallower and it finally dries up before Greywell Tunnel is reached.

The 1.1km tunnel contains a collapse and the western end has been fenced off. It contains an estimated 12,500 bats and is the largest bat roost in Britain, including the world's second largest population of Natterer's bats. For this reason the tunnel is not being allowed to be restored for use by boats, for which it was built. There has been talk of building a second tunnel for boats but there is no

saying that bats would not get into that, too. The walking route is poorly marked.

The other end of the tunnel is amazingly different. A footpath leads to the tunnelmouth, the portal of which was rebuilt in 1975. As with all the bridges on the canal in Hampshire, a location plaque is fixed to the downstream face. Floating weed covers the canal surface but the water is absolutely clear and the canal is wide.

The setting of the first part of the canal is rural, alternating between open meadows and woodland. The wildlife is surprisingly tame. Water voles crouch against the bank as they are passed. Moorhens, coots and little grebes run across the floating weed to wait in the rushes. Pike peer out from under lilies and 500mm long fish come close.

The canal leaves the tunnel on a lefthand curve that brings it next to the River Whitewater, lying hidden on the right. Even when the canal crosses the river it is not obvious. Also hidden is a shallow section at the edge, to allow weed growth, but it ensures wet feet for anyone getting into or out of a boat.

Just after the river, the remains of the three-storey octagonal flint keep of Odiham Castle, built from 1207 to 1210 for King John, stand among the trees on the left, restored in 2008. The castle once had two rectangular baileys, earthworks and a moat with a small outer enclosure, also moated. It was from here that King John set out to sign Magna Carta.

A boom then reaches most of the way across the canal upstream of the first winding hole, used by the S&HCS narrowboat *John Pinkerton*. From here the water becomes murkier as the weeds thin, both results of increased usage by boats hired out from a centre at Colt Hill. After North Warnborough the canal is followed by the Odiham bypass, which can be heard but not seen on the left. On the right, gliders can be seen but not heard as they circle above RAF Odiham. At Colt Hill, a public house has a garden that reaches down to the canal. On the other side of the bridge, the boat hire centre on the same bank sells confectionery.

The most beautiful canal in England at war

The first significant water is met at Up Nately.

Greywell Tunnel eastern portal.

King John's Odiham Castle during restoration.

The John Pinkerton at the top winding hole.

The Barley Mow slipway area at Winchfield.

Trip boat loading point at Dogmersfield Park.

The Tundry Pond is a feeder for the canal.

The canal near Winchfield.

Attractive riparian properties are met quite frequently.

An unusual heath-land canal setting of Scots pines and rhododendrons.

The bypass passes over the canal on a bridge that has red brick arched facades to match the other bridges on the canal, several of which have been rebuilt by the local authorities or the canal society.

Dogmersfield Park has a prominent loading point for a trip boat.

The Barley Mow at Winchfield is the next port of call for the thirsty brigade. On a summer evening the click of leather on willow and sporadic applause break the silence from somewhere beyond the towpath hedgerow. A slipway has been cut to allow small boats to be launched.

Before Dogmersfield a large slip is visible on the left bank but a couple of mature oak trees standing vertically on top of the rotated material show that this was not recent. On the right, the Tundry Pond is visible over the top of the embankment. At 800m long, it stretches two-thirds of the way back to Swan's Farm Bridge, which lies at the other side of the large loop that is just being completed. Pillboxes and anti-tank blocks on the banks act as pointers to the route to come.

A small swing bridge crosses. Between Crookham Village and Fleet is the Fox & Hounds. Houses back onto the canal at **Fleet** and a chip shop and others lie to the north of Reading Road Bridge.

From Fleet, the canal cuts out into wooded heathland. Norris Bridge is a high concrete structure and, on the left, leads up to the National Gas Turbine Establishment but the noise of jets can be deceptive. After another kilometre, on the left bank where the trees clear, fencing blocks a view straight down the main runway at **Farnborough** airfield, the venue for the Farnborough Air Show. A variety of aircraft from the Royal Aircraft Establishment frequently come and go.

All is not noisy, however. From here to Wharf Bridge, several water areas on the south side of the canal are nature reserves and boaters are requested to keep out. Beyond Wharf Bridge

Tank traps and goosegrass obstruct the canal bank at Crookham Village.

Canada geese and goslings at Reading Road Wharf.

landing is not allowed on the south bank, either, but here it is because the land belongs to the army. Ranges and barracks front the canal. Red flags fly on the other side of the towpath when ranges are in use. This is **Aldershot**, Hardy's Quartershot, home of the Army.

Ash Lock marks the end of the top pound and is adjacent to Hampshire County Council's canal depot.

There is an overflow weir into the River Blackwater.

A railway bridge is passed. For the rest of the canal the line is never more than 700m from a railway and is usually much less.

The canal passes from Hampshire to Surrey as the canal crosses an aqueduct over the A331. Alongside are several lakes that have now been set out for anglers.

At Ash Wharf, with its shops, the canal turns north for 6km. Fortifications include a quarter-sized pillbox tucked into the corner of a bridge opposite the Swan.

From Ash Common, the canal links a series of lakes, the end two – Greatbottom Flash and Mychett Lake – being the largest.

The **Frimley** Lodge Miniature Railway, on the left side of the canal, has extensive narrow gauge trackwork and can raise a number of passenger-carrying trains pulled by engines from diesels to narrow gauge steam engines.

A most interesting aqueduct is located at Frimley Green. The four tracks of what was the main Southern Railway between Waterloo and the West Country pass beneath it. Outward facing gates were previously placed on each side to prevent the contents of this section from being drained onto the railway in the event of a breach.

As the canal turns east again, there is a boarded-off bridge on the left, which should give access to a large lake. A boathouse is on the left bank at Deepcut.

The adjacent Deepcut Barracks achieved notoriety after the shootings of a number of young soldiers, all claimed by the army to have been suicides. It is probably not co-incidence that this facility has been earmarked for closure.

There are 14 locks in the Deepcut flight, several with attractive wooded inlets between them. Lock 20 bears plates indicating that its gates were constructed by apprentices from the Royal Aircraft Establishment in 1980.

All around this section are firing ranges, including field gun ranges and the National Rifle Association ranges. The staccato of gunfire echoes across the canal.

The locks of the Brookwood flight follow. Just the other side of the railway from Brookwood Bridge lies the large Brookwood cemetery and its war graves. A heavily buttressed brick wall retains the railway as it comes close alongside on embankment.

The Tuu Peking, Thai and Vietnamese restaurant is next to Kiln Bridge at St Johns. The tower blocks of

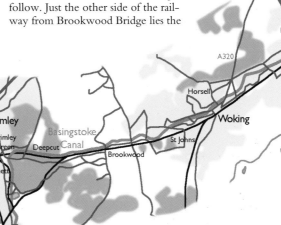

Model railway enthusiasts on the Frimley Lodge Miniature Railway by the canal.

Mychett Lake's natural hollow enlarged for canal water storage.

The heavily butressed brick wall retaining the railway.

Wooded heathland surrounds the Deepcut flight.

Kiln Bridge and the Tuu restaurant at St Johns.

The Basingstoke Canal has a better class of mooring pin.

Distance
56km from Old Basing to Woodham Junction

Navigation Authority
Basingstoke Canal Authority Canal Society Surrey & Hampshire Canal Society www.basingstoke-canal.org.uk

OS 1:50,000 Sheets
186 Aldershot & Guildford 187 Dorking & Reigate

Woking are approached and new buildings beside the canal include a large Debenham's and the New Victoria Theatre.

On the right, just beyond the A320, is a children's adventure playground with a selection of skateboarding half pipes. The trees on the other bank hide a different kind of airfield from Farnborough. It was here on Horsell Common that H. G. Wells would have us believe, in *The War of the Worlds*, that the Martians landed in their invasion of Earth, watched from his home on Maybury hill. These days planes flying in to Gatwick are more likely to be seen crossing the sky.

On the right bank before the next bridge, the remains of a wharf crane remind of the canal's commercial days. The Bridge Barn has plenty of outside seating in the summer.

The canal is now in the stockbroker belt. Houses on the canal banks are well hidden by greenery, however, and do not intrude. There are small boats in gardens backing onto the canal.

The canal's lower locks may only be used during the middle six hours of the day as a water conservation measure. From the six locks of the Woodham flight is an almost continuous line of houseboats, mainly converted narrowboats but some obviously floating houses. After a selection of more recent bridges, the last bridge on the canal is a rounded brick arch, typical of the ones higher up. From the last lock there is a final straight to the River Wey Navigation at Woodham Junction, backed by the M25 viaduct and, beyond that, the remains of the Brooklands race track with its banked bends, still intact in places.

Machinery which H. G. Wells might have seen at Horsell.

Houseboats at West Byfleet.

94

19 Kennet & Avon Canal

Proposals for what turned out to be the only east–west canal across southern England began in 1660, as a broad canal, the Western Canal. It was to be a successful canal but no thanks to the residents of Reading, where the mob, led by the mayor, attacked construction work, which they feared would cost them their road business. The River Kennet was made navigable for barges to Newbury in 1723, one of the earliest and steepest river navigations. John Rennie surveyed the canal route at the age of 29 and it was completed to his design in 1810, after 16 years of construction. Highly profitable for the first 30 years, it carried coal, iron, stone, slate, timber and agricultural products but went into decline in 1841 after the arrival of the Great Western Railway, which follows the line very closely to Pewsey. The GWR bought the canal in 1852 and allowed it to deteriorate. During the First World War it was used for troop training. The closure Act of 1952 was strongly resisted, the annual Devizes to Westminster Race helping to draw attention to the canal and ultimately being a significant factor in its restoration, canoeing's major contribution to our canal network. In 1956 Commander Wray Bliss took a 20,000-signature petition for restoration by canoe and cruiser to Westminster. The Kennet & Avon Canal Trust was formed in 1962 and tackled the country's biggest restoration project successfully, the canal being reopened by the Queen in 1990. Much of the towpath is used as a Sustrans cycle route.

Part of its unique character comes from its chalk downland section as it cuts through the Vale of Pewsey. During the climb to the summit level from Reading, the locks are relatively evenly spaced but on the descent westwards to Bath they tend to be grouped, giving three long pounds.

The navigation uses the course of the River Kennet initially.

Reading was successively an Iron Age then Roman then Saxon settlement on the banks of the River Kennet, rather than on the River Thames, which it joins just downstream. Indeed, the Environment Agency are the navigation authority for Blake's Lock, the last lock on the river.

The Abbey or Forbury Loop takes in several important buildings. The Prudential offices are on the site used by Huntley & Palmer from 1822 to 1970, the world's most successful biscuit company, breaking fewer biscuits by using water transport. Only one of the original factory buildings remains at the east end.

The Scottish baronial-style prison is where Oscar Wilde was imprisoned in 1896–7 for Gross Indecency, writing *The Ballad of Reading Gaol* after his release.

Next to it are the remains of Reading Abbey, founded in 1121 by Henry I, who is buried here. It was built with stone from Caen and from the Roman site at Silchester. King Stephen built a castle inside in 1150 but that was soon destroyed. One of the oldest known songs,

Sumer is Icumen In, was written down here in 1240 and Parliament met here in 1453 and 1464. Only tall sections of wall remain after the Dissolution. Mills were sited alongside, powered by the Holy Brook, which now emerges after being built over for its last kilometre. It has a constitution which requires its full length to be passed by boat every month although the last passage of the underground section was probably in 1984.

The town was made rich by cloth. Henry I was annoyed at having to move aside to allow the wagons of rich cloth merchant Thomas of Reading to pass until he considered the man could be a useful ally, going on to establish the length of his arm (presumably, half his span) as the yard measure for such merchants.

Reading Museum & Art Gallery in the Victorian town hall has the important Roman material from Silchester plus Britain's Victorian replica of the Bayeux Tapestry. The Victorian architecture, which gave Hardy his name of Aldbrickham, is gradually succumbing to Reading's reincarnation as a silicon town.

High Bridge Wharf was in use in the 16th Century and the navigation begins officially at the High Bridge footbridge. In the days of tramlines and cobbles, a policeman without any equipment convicted the driver of an Austin A30 for breaking the speed limit after he saw it get all four wheels off the ground on Duke Street bridge, following.

The first section of the navigation, the Brewery Gut, is the most difficult of all, the flow being fast with blind bends. Use is controlled by users who push traffic light buttons but the buttons are too high to be reached from small craft and are recessed so they cannot be pushed with an object such as a paddle.

What was a mill site became the tram power station and tram and bus depot, Reading being the last British town to stop buying and using trolley buses. Now it has all been replaced by the Oracle shopping centre. Reading is claimed to be one of the top ten retail destinations in the UK and boats pass right through the centre of the complex, fences keeping those on dry land separate from those on or in the water but with ringside seats for viewing any indiscretions.

The river quietens above County Lock. Alongside is a weir with a small drop but it is deep with a long towback and sections of Brunel's broad gauge rail within it. From 1790 Simonds' brewery was sited next to it, where India Pale Ale with its travelling qualities was first brewed for the Empire. The red hop leaf trademark disappeared when it was bought by Courage, who

One of the earliest and steepest river navigations

Girl & the Swan sculpture on the wall of offices near the start of the canal.

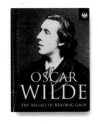

Duke St Bridge in the background is near the navigation start.

Brewery Gut through the Oracle centre, much changed recently.

Fobney waterworks at the start of the first section of cut.

moved out again in 1977. The site is now occupied by housing and a Loch Fyne fish restaurant.

To its north is the distinctive St Mary's church, partly from the 13th Century, built with flint and much stone from the abbey.

The A329 Inner Distribution Road passes over at Katesgrove, which formerly had sailcloth manufacturing, iron foundries, brickworks and underground chalk workings. These are proving more extensive than previously thought and are resulting in collapses under modern housing. Boats of all sizes are moored along the backs of houses. Coots and cormorants are seen among the alders, hawthorns and purple loosestrife.

Vehicles are strongly discouraged, by humps and a very low speed limit, from using the A4, which crosses next. Brunel also wished to bring his GWR across here and up the Pang valley to Pangbourne but his clients insisted that he take it to the north of the town.

A former freight railway has now become part of the A33, which follows the river for 2km before crossing at the confluence with the Foudry Brook, fed by Thames Water's showpiece Reading sewage works. Ahead is a massive wind turbine which distracts users of the adjacent M4.

Leaving the noise behind, the heron fishes from banks of layered silt as the river winds towards Fobney Lock, where water works are now used for fish breeding. Terns might also be seen fishing in the river.

Southcote Mill by Southcote Lock.

into the bank and chunks cut out of the vegetation at intervals.

The canal is crossed by powerlines and then by the Reading to Basingstoke railway line. The line of the canal is not obvious approaching Southcote Lock, except that it is clear that large craft cannot get under the low bridge on the apparent line. Instead, they go under Milkmaid's

British Waterways padlocked a gate to prevent a young motor-cyclist from using the towpath. They were disconcerted to find, on returning, that the padlock had been cut off and a different one fitted, to which they did not have the key.

The canal first leaves the river with a section of cut through water meadows, with elder, yellow iris, lilies, deadly nightshade, dog rose and cow parsley. Swans consider it their territory but reluctantly accept the presence of boats.

Dragonflies are less possessive. Anglers also consider it their territory and steps are dug

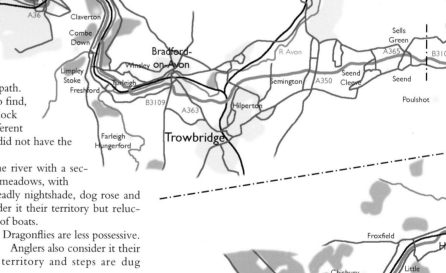

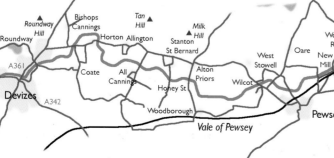

footbridge. Beyond the lock is Southcote Mill, now converted to an attractive house. A pumping station was built here in 1850 to supply water to Reading. Of more recent construction in Southcote are some eight-storey tower blocks, not that impressive these days but relative skyscrapers when built in the 1950s.

The valley has been heavily worked for gravel and sand, pits forming a feature of the line and appearing soon.

At Burghfield the canal picks up the Blue Line which joined from the south and was to follow to Bradford-on-Avon, a 1940 defence line in case of invasion of the south of England. There were 90 infantry and 40 larger pillboxes, forming one of the best preserved defence lines, a conspicuous feature of this canal.

The Cunning Man (or fortune teller) is a popular canalside public house with children's play area. All custom disappeared overnight in the 1980s, however, when the landlord shot an intruding stray dog with a crossbow, resulting in the brewery needing to find a hasty replacement for him. Since then the building has been rebuilt in a totally different traditional style.

A roving bridge was the site of riots in 1720 by 300 people opposed to canalisation of the river.

Burghfield Lock was built in 1968 to replace a turf-sided lock, the remains of which are immediately below it. In the distance are Reading Services on the M4, which is closing on the canal.

Willows and teasels surround the canal, which leads to Burghfield Mill and its weir, a flour mill that became a cocoa mill. From here the route is very obviously river as it winds through water meadows.

The motorway has a driverless vehicle guidance cable laid in the tarmac to form a loop passing round the interchanges each way from the river. It crosses an interesting sequence of bridges. At the time of its construction there were gravel workings on the left before the motorway and a large yachting lake in an area of former workings beyond the M4. It looked likely that there would be large sailing lakes on each side of the motorway so a connecting access bridge was put in to link them. The next bridge north-west of the River Kennet crosses the Holy Brook, some of which may have been dug by monks to power the mills at Reading Abbey. Then there is the Paddington to Penzance railway bridge, followed by Junction 12, where the A4 is crossed.

The gravel pits are good for dragonflies and winter wildfowl. The marsh and reedbeds here and in other former gravel workings west towards Newbury are some of the most important in southern England, with reed, sedge and Cetti's warblers and other rare birds and insects.

Garston Lock, a turf sided lock.

Sheffield Lock, a scallop-sided lock, formerly turf sided.

Scallops are in the original rail rib positions. Its takes its name from Sheffield Bottom, which made its name when the local Stompers jazz group appeared on *Opportunity Knocks*, or might have done if Hughie Green had not announced them as the rather more intriguing Sheffield Bottomstompers. These days Hardy might have chosen a different name than Gaymead for Theale, in the opposite direction. It has Georgian terraces and a church

Garston Lock is the only surviving example of an early turf-sided lock. A restored listed monument, it uses the original stone slabs and a framework of broad gauge rail to protect the sloping sides. Lock balance beams had to be shortened because a pillbox was built too close.

Beyond the next set of powerlines, Sheffield Lock with its scalloped sides is set in an area of mown lawns.

built between 1820 and 1832 by E. W. Garbett in the style of Salisbury Cathedral, although not so obviously as the one at Bishops Cannings, further along the canal.

Cherries and other varied trees plus creepers add interest to the next reach and Canada geese make an appearance.

To the left of Sulhamstead Lock is the 1744 ionic porticoed Sulhamstead House, near the Police Training

College. At Sulhamstead, Folly Farm was built in 1906 by Lutyens in William & Mary style and extended in 1912 in Tudor style.

The swing bridge at Tyle Mill Lock is typical of those along this canal. These included some of the first applications anywhere of ball bearings. This one is now electrically operated. Tyle Mill itself was a flour mill, burnt down in 1914, rebuilt as a sawmill which was used until 1936 and is now a private house. The road leads out to the A4 and the Spring Inn, still known to many as the Three Kings Jacks Booth public house, partly named after Jack of Newbury.

A flint wall in a field on the left before Ufton Bridge is the remains of Ufton Green church. Here, the River Kennet is left below the point where it is joined by the River Enborne.

Between the canal and river is a stand of planted poplar trees, more of which are to follow. The A4 in this part of the country used to be characterised by poplar trees on each side at regular spacings for a considerable distance across the county but they were all cut down as hazards when motorists collided with them as they left the road. These days the thinking would be different and crash barriers would be provided.

A railway disaster killing seven people was caused by someone deliberately parking his car on a level crossing in the path of a high speed train at Ufton Nervet in 2004. On the next corner, at Lower Padworth, the canal, railway and A4 all come together for the first time. Had Brunel started his railway even five years sooner it might be that all our railways would now be broad gauge but standard gauge was already too well established elsewhere and he had to concede defeat. One weekend in 1847 the entire line from Paddington to Penzance was changed from broad to standard gauge with trains running on the new gauge on the Monday morning. It would be interesting to see how long Brunel would have taken to electrify the West Coast Main Line.

Restoration of Towney Lock made Ufton Lock redundant. Here is first met use of rolls of coconut matting along the banks, rather than sheet piling. These contain waterside plants and make it easier for wildlife to get out of the canal.

By Padworth Lock, the canalman's cottage has been converted into a Kennet & Avon Canal Trust visitor centre with a historical and wildlife trail.

Berkshire County Council were held to their promise to replace their fixed link across the canal for the A340 if the canal was ever restored. Aldermaston Wharf was a boatbuilding yard. The spur on the right of the wharf was a transfer point for the railway, which also used canal water to supply water troughs between Aldermaston and Woolhampton. The spur was partly filled and GWR staff were located here on being moved out of Paddington during the Second World War. It has been used as a location for filming *Rosie & Jim* episodes. Aldermaston Lock was formerly called Brewery Lock after the Strange's brewery located there. Hops grow wild in the local hedgerows and make good home brew. An ancient monument, the lock has been rebuilt with scalloped sides and gets more attention than Sheffield Lock as it is easier to find by road.

In 1894, plans were announced to flood the valley between Aldermaston and Newbury as a reservoir for London. The plan was shelved after widespread opposition. Thames Water are now proposing a site across the line of the Wilts & Berks Canal near Abingdon.

Oak, horse chestnut, ash and elm trees are among the broad-leaved trees that border the canal, to where a fixed swing bridge has been replaced by a high level bridge. At the end of this straight the canal rejoins the river. To the left is Aldermaston, on the other side of which is the Atomic Weapons Establishment, scene of

The beautifully restored lock at Aldermaston Wharf.

A high level footbridge with interesting abutment brickwork.

Beside the swing bridge at Woolhampton.

Monkey Marsh Lock: another turf-sided example.

the Ban the Bomb marches in the 1950s. Between the village and research centre, Blue Circle built the new headquarters for their cement company, then had to vacate it as radioactive water seeped beneath it and towards the village. The village still uses the old tradition of a candle auction of church lands every three years, where bids are placed by sticking pins into a candle which then burns down, the rents going to charity.

On the next river section a low swing bridge has been replaced by the high level Wickham Knights footbridge on interesting brick abutments. Near here was an angling notice banning just about every form of pleasure on the navigation, the word 'boats' subsequently being painted out from the banned list.

Woolhampton has a recent swing bridge although its fast current makes it a difficult area for powered craft. The Rowbarge Inn has been much enlarged since the 1970s when the landlord was Larry Naismith, known for his roles in Ealing Comedies although the picture he displayed in the bar was a rather severe portrait of himself in German uniform from his leading role in *Where Eagles Dare*. These days enemy action is likely to be met in the garden in the form of robber robins and chaffinches with the tiniest of shrews hopping through the grass, looking for crumbs. Woolhampton Mill was used until the 1930s, grain being brought by barge and flour being taken away by the same route. It was a coaching village and has a Victorian church and Georgian buildings in Woolhampton Park. The station was renamed as Midgham to avoid confusion for passengers who really wanted to go to Wolverhampton, train announcers having their own version of English that did little for finding the correct platform.

Woolhampton Lock was previously turf-sided but was restored with a brick chamber. Turf-sided locks were cheaper to build but used more water, something in short supply, especially further up the canal. This one is just above the point at which the river is left.

More gravel pits follow to Heales Lock, again with a shortened balance beam as the pillbox was built too close. To the north is Midgham Park.

Reedmace begins to appear beside the canal and then Lock Cottages are passed, a significant distance from any present lock. Midgham Lock is some way ahead.

Briefly, the canal follows the line of the Roman road between Silchester and Cirencester to Colthrop Lock. Colthrop is an industrial estate. One of the largest and earliest occupants has been Colthrop Paper Mills of the Reed Paper Group, major producers of carton and board material. The occasional kestrel might be seen in the vicinity of the factories. Water dropwort is a local plant.

Monkey Marsh Lock is another listed ancient monument, the turf sides replaced with modern materials.

Thatcham probably takes its name from the marsh and reedbeds that provided the material for thatching. It is one of the oldest Berkshire villages, with traces of a Mesolithic settlement. It has had markets from the time of Henry I but suffered badly from the Black Death. There is a 14th-century chapel near the Victorian church.

A straight avenue of reeds, occupied by reed buntings, leads to Widmead Lock. On top of the ridge on the left was Greenham Common airfield, where the Americans had Cruise missiles stationed from the 1970s. Their departure probably followed from the adverse publicity resulting from the women's peace camp at the gate. The airfield itself, which hosted an airshow during that period, has now gone too, an industrial estate spreading across the former site, but the odd dilapidated protest caravan is still occupied at the gate.

As the canal turns to the right in front of gravel workings, the impressive grandstand of **Newbury** horse race course can be seen ahead.

The railway finally crosses on a new bridge. The old pillars remain in place but don't quite meet the new

bridge. Brunel would have loved this, exactly his sense of humour, the line he took when required to shore his bridge over the Thames at Maidenhead. Beyond the bridge there is a nature discovery centre on the right before the River Kennet joins, having picked up the River Lambourn, as Bulls Lock is approached.

Beyond Ham Lock, adjacent to another recent industrial estate, is Whitehouse roving bridge where a breach occurred in 2000 near where contractors were digging footings for a new bridge. It resulted in a narrowboat going aground and being pinned across the breach.

By Greenham Lock is a drydock site and one of three marinas in Newbury. It was also where, in 1811, locals won a £1,000 bet to shear a sheep in the morning and make the wool into a coat by the evening, the sheep that provided the wool joining the subsequent celebrations by being roasted. A local flourmill supplied the Huntley & Palmer biscuit factory in Reading until becoming a power station in 1903. These days the area has much new housing, which seems to be influenced by canal warehouse architecture.

Leaking gates near Thatcham.

The crane and former canal stables at Newbury Wharf.

The A339 was formerly the A34, a traffic light-controlled roundabout creating a notorious bottleneck which typically took an hour to pass on what was the main north–south route in central southern England. Beyond the bridge is Newbury Wharf, or was until it was filled in during the 1930s to become a bus waiting area, then a car park, but may yet be partially restored. A wharf crane stands prominently in front of the Kennet & Avon Canal Trust's sales and display building in former canal stables. These may have been built from stone from the original castle erected in the early 12th Century by the Earl of Perche, stormed in 1152 by King Stephen after a two month siege and later seized by King John. An 18th-century warehouse with an interesting balcony is used to house the Newbury District Museum. This features archaeology, ballooning, old costume, butterflies and moths, prehistoric and Saxon material and the Battles of Newbury. Civil War relics and old cameras are housed in the Jacobean Cloth Hall museum.

The lowest fixed bridge on the canal was the Parkway temporary bridge, built before the Second World War to carry American tanks. It was replaced in 2000, allowing larger craft to use the canal.

Newbury was successively a Roman, Saxon, Norman

KENNET & AVON CANAL
NOTICE
The Captain of every
vessel allowing Horses
to Haul across the Street
will be Fined.
By Order.

and Tudor cloth centre. Most prominent was Jack of Newbury, hero of Deloney's story of the same name. He was apprenticed to a cloth maker, married his widow when he died, inherited the business when she died and employed up to 1,000 local people (150 of whom he took to fight for Henry VIII at Flodden) but declined a knighthood in order to remain on equal terms with his employees. He was subsequently visited by Henry VIII, Catherine of Aragon and Cardinal Wolsey.

In 1643 Cromwell's soldiers caught a witch sailing down the Kennet on a plank. She caught their bullets in her hands but they killed her by the usual method of slashing her forehead and shooting her below the ear. Windsurfers beware. The Waterside Centre is beyond Parkway bridge.

The 1770 bridge over the Kennet with Newbury Lock beyond.

West Mills Yard forms part of Newbury's waterfront.

Hardy was not at his most inspired when calling Newbury Kennetbridge but he did pick the most important feature, the bridge of 1770 which carried the A34 until 1965. This section of river past the back of the Old Wagon & Horses is fast and narrow. Boatmen could be fined for allowing horses to haul across the road, there being no towpath under the bridge so that a long line had to be let down with a float attached. The authorities are still opposed to use of the road, which now has a very low speed limit and humps all along it. The towpath wall and bridge are well gouged with rope grooves. Next to the bridge is a family butcher's shop selling pork pies second to none. Other significant buildings are the 19th-century Italianate Corn Exchange and the Perpendicular church of St Nicholas of about 1500, with its unusual 17th-century pulpit.

The Lock, Stock & Barrel is a weakly named reference to Newbury Lock, in front of which is the channel of the River Kennet. Newbury Lock was the first lock

A field of buttercups west of Newbury.

to be built on the manmade canal. Marked out alongside is the plan of the lock cottage which was occupied until 1958. This was leased by canal restoration pioneer John Gould. Another local canal business is the Kennet Horse Boat Company with motor trip boats here and horse-drawn boats at Kintbury.

There are some very attractive canalside buildings upstream of the lock, including 17th-century weavers' cottages. These were converted for canal labourers in the 1790s. West Mills were bought by Hovis in 1939 but burnt down in the 1960s. Northcroft has had a Michaelmass fair since 1215.

Heading out of Newbury, the canal passes the abutments of a bridge which carried the Lambourn railway line. During its first week of operation a train killed two boys boat spotting on the bridge.

The next two locks, Guyer's and Higgs, are named after the opposing commanders in the First Battle of Newbury, fought on Wash Common in September 1643, the Roundheads beating the Cavaliers in one of the bloodiest battles of the Civil War. The Second Battle of Newbury took place 13 months later at Speen, when the Roundheads drove the Cavaliers out of Donnington Castle but the Cavaliers returned the following week and relieved the castle.

Environmentalists claimed their protest at the building of the Newbury bypass, which crosses between these two locks, was the Third Battle of Newbury. For tens of thousands of motorists each day and a large proportion of the residents of Newbury the removal of the Newbury bottleneck could not come soon enough but hundreds of protesters from as far away as Germany delayed the construction work for the new line of the A34 and set the precedent of major road projects having a significant cost added for security protection.

The railway crosses back over Pickletimber Bridge and the canal approaches Benham Lock with a couple of pillboxes. In the summer, a blanket of buttercups extends in the direction of Enborne, itself 2km from the River Enborne, which featured as an obstacle to the rabbits of *Watership Down*. To the north is Benham Park, laid out by Capability Brown, the house built 1772–5.

Beyond Enborne Copse, Hamstead Park was built by the Earl of Craven, the house resembling Heidelberg Castle to please Elizabeth, Queen of Bohemia, to whom he was romantically attached. The scheme did not succeed and he died a bachelor at 91. He was also a firefighting enthusiast, helped by the fact that his horse could smell fires and take him to them before the alarm was raised. Ironically his own house burnt down in 1718, 21 years after his death. Avenues run through the geometric gardens with the walls and six sets of gate piers remaining. Mottes beside the canal were part of the defence works of the 13th-century Earl Marshall. Up the hill, the church is Norman with 14th- and 18th-century additions. The 19th-century watermill by Hamstead Lock is probably on the site of one recorded in 1086. A pair of pillboxes are more recent. Newest of all, here and further along the canal, are gates where the towpath crosses the road. Crossing was not easy previously for those on foot because of the relatively blind road approaches but at least it was possible to escape onto the towpath. Now the user is trapped on the road while trying to undo gates.

Between Copse Lock and Dreweat's Lock, the ground rises from water meadows to a high ridge, a most attractive area of country, leading to Irish Hill, a corruption of *ebrige* or yew covered ridge. Whiting Mills, closed early in the 20th Century, used quarried chalk to make a fine white powder which was carried by canal to Bristol for use in paint manufacture. The Wilderness on the north side of the canal seems no more deserving of the title than many other sections of the canal.

Kintbury has an air of quiet affluence.

Kintbury Lock is just beyond the Dundas Arms Hotel and the station. Riots in the village in the 1830s against modern tools led to the Grenadier Guards being called, one person being executed, several deported and many imprisoned. The 13th-century church, restored in 1859, has a Saxon burial ground. The large Victorian Gothic vicarage has a couple of large beech trees beside the canal including a particularly fine copper beech. In an area where horseradish is found, the River Kennet makes its highest connection with the canal although the two are to continue beside each other for a further 4km.

To the left of the next bridge is an electricity substation.

Beyond Brunsden Lock are water meadows leading across to Avington with its Norman church with notable font, corbels and chancel arch. The railway crosses and a private road doubles back over the canal before Wire Lock.

From Dunmill Lock the road leads up to **Hungerford** Common past a rare type of pillbox with two anti tank loopholes, used by the Royal Artillery in a BBC *Chronicle* film in 1979. On the other side is a noisy peacock and the 1907 Berkshire Trout Farm which is used to stock most of the tributaries of the Thames for anglers but which lost all the stock in a pollution incident in March 1998, that closed the canal to users for several weeks. The Environment Agency blamed growth of algae but the owners disputed this explanation.

After this, the canal moves away from the River Kennet and follows the River Dun as far as Great Bedwyn.

In 1364 John of Gaunt gave the fishing and grazing rights to the town of Hungerford and donated his hunting horn, which is still in a local bank vault. Two years later he was given the manor, for which the token rent of a red Lancastrian rose is still paid to any monarch passing through the town. At Hocktide, the second Tuesday after Easter, a replica of the horn is blown to call 99 commoners to the Hocktide Court where two Tuttimen are elected. They carry Tutti poles decorated with flowers and ribbons and accompany the Orange Scrambler as he collects a penny a head from each commoner's house. He may kiss every woman in the town

and they are each given an orange in return. Each new commoner is shod by having a nail driven into the shoe and everyone ends up in the Three Swans Hotel at lunchtime to drink Hocktide Punch to the memory of John of Gaunt.

King Charles stopped in the Bear, on the A4, during the Civil War. William of Orange was staying here in 1688 when he received messengers from James II, rejected their proposals and continued his march on London. Georgian coaching inns and 18–19th-century houses are features of this normally quiet market town which shot to prominence in 1987 when a gun enthusiast ran amok and killed 16 people in the street.

Decorative ironwork enhances a house by the A338 bridge. An open grassed area leads to a former granary and Hungerford Lock. Above this is a Gothic church of 1816 that replaces one which collapsed under the weight of a snowfall. Water meadows lead to Hungerford Marsh, an 11ha nature reserve which has recorded 120 species of bird including snipe. Buttercups, daisies, red clover and docks grow along the banks and rabbits scamper about in the evening. Hungerford Marsh Lock has a footbridge which had to be positioned over the chamber as permission to reroute a footpath was denied, suggesting that someone had failed to do the preliminary paperwork.

Cobbler's Lock has another noisy peacock close by.

The three-arched Dun Aqueduct carries the canal over the river near where the U. S. Army were caught replacing a swing bridge by crane during the Second World War after borrowing it to use as a saluting base during a review. There is a pig farm beside the canal and pied and yellow wagtails visit the water. The railway crosses again before Picketfield Lock.

The canal passes from West Berkshire to Wiltshire as the A4 leaves and heads through Froxfield, a village where a Gothic-style gateway leads to a quadrangle of Somerset Hospital almshouses and brick-and-flint cottages, founded in 1694 by the Duchess of Somerset and extended in 1775 and 1813. The Pelican is a family public house that invites canal users except ducks, which have to be ejected from the bar.

On moonlit nights the ghost of Wild Darell may be seen being chased across the fields by the hounds of Hell. He was the owner of Littlecote House in 1575, who murdered a baby by throwing it on a fire. It was thought to be his own by his sister. The cuckoo calls in the spring.

Froxfield Bottom, Froxfield Middle and Oakhill Down Locks come in quick succession with the canal now moving in a southwesterly direction.

Ash and sycamore flank the canal, yellow water lilies add colour and it provides a hunting ground for kingfishers, kestrels and large brown dragonflies.

A metal footbridge over the canal continues over the

Rope grooves on a bridge at Avington.

Brick and wrought-iron at Hungerford.

Footbridge over the canal and railway at Little Bedwyn.

railway with high flights of steps each side at Little Bedwyn. The name has nothing to do with Arabs but comes from the Saxon *bedwine* or bindweed or from the Celtic *bedd gwyn*, white grave. The village had a Roman camp and Saxon mint. To the south is an 18th-century farming village while there are 19th-century patterned brick terraces to the north.

Beyond Little Bedwyn and Potter's Locks, Chisbury lies to the north-west with the 13th-century thatched Chisbury Chapel, rescued from use as a farm building. Chisbury Hill Fort, referred to by Thomas Hardy as Batton Castle, is built on Bedwyn Dyke, part of Wansdyke dug between Inkpen and Bristol in the 6th Century by Saxon King Caelwin, as protection against the Angles from the Midlands.

Beyond Burnt Mill Lock is Great Bedwyn, a Saxon city. Thomas Willis, born here in 1621, was an early member of the Royal Society and the founder of neurology. Although there are thatched cottages, many old buildings were lost in a fire of 1716. It was a rotten borough with two Members of Parliament until 1867. William Cobbett had visited the village in 1821 and talked several times in his *Rural Rides* of the extreme poverty resulting from the system in the village. There is a malthouse with a kiln tower and a local public house sells alcoholic ginger beer. Mill Bridge was one of the first skew bridges to be built in brick, narrower at the crown than at the springing. Because some of the brick courses went wrong, Rennie was accused of not having understood the principles but even modern British Waterways engineers have to use trial and error when setting out skew bridges.

Crofton pumping station, seen over the railway from the canal.

By Bedwyn Church Lock is the 11th-century Bedwyn church with fine pierced embattlements around the top of the tower. Close by is the Bedwyn Stone Museum with stone carvings, tombstones, statues, a dinosaur footprint and the Lloyd's stone masonry which made the original inscription for the Bruce tunnel east portal.

Oak and ivy along the canal add to the feeling of history. Before Crofton Bottom Lock, the start of the Crofton Flight, the canal crosses the line of the Roman road from Cirencester to Winchester, heading south-east past a tumulus and Wilton windmill, the only working mill left in Wiltshire. The other way leads past Tottenham House and through Savernake Forest, perhaps acting as inspiration for the avenues through the forest. The Broad Walk skew bridge of 1796 may have been the first of its kind in England.

The three hectares of Wilton Water are home to mallard, teal, pochard, tufted duck and moorhen. More

The Bruce Tunnel portage.

importantly, it acts as a reservoir to supply the summit level, which Rennie constructed only 4km long and above the springline so it has no feeders. He built Crofton pumping station to raise water 12m. Water difficulties, in what was one of the shortest summit levels in the country when built, mean that back pumping is still required today. The Cornish beam engines are the oldest in the world still in steam and doing their original job, a Boulton & Watt of 1812 with 1.1m bore and 2.4m stroke and one by Harveys of Hale in 1845. Modern pumps do most of the work today. The iron-bound brick chimney, lowered in 1958 for safety reasons, has now been restored to its full 25m height. There is no longer the requirement of the Earl of Ailsbury that there must be no smoke, a sackable offence for any engineman who released any, a stipulation that became rather superfluous with the arrival of the adjacent GWR. The pumping station leat runs on the far side of the railway. Spurs from the railway originally crossed the flight in two places to serve the line towards Andover.

Buzzards circle around Crofton Top Lock where the canal reaches its 140m summit level. Other preying took place to the left at Wolfhall, where Henry VIII courted Jane Seymour. Any visit from a king and his royal entourage was an expensive business. However, the king took over the house here while his prospective inlaws had to move out into a barn.

Tottenham House, set in a deer park to the north of the canal, has a monument to the restoration from insanity of George III, erected by the Marquis of Ailsbury. The current Marquis of Ailsbury is a direct descendent of Richard Sturmy, who is recorded in the *Domesday Book* as holding seven manors in Wiltshire, was then based in Wolf Hall and paid an annual tax of one night's lodging to the royal household.

Bruce Tunnel, 460m long, is named after Thomas Bruce, the Earl of Ailsbury, who wanted a tunnel rather than the proposed cutting through his deer park. It is second only to Netherton for the size of bore of any tunnel remaining open and has chains along the sides for

Leaving the large bore of Bruce Tunnel.

pulling boats through as legging would not have been possible. There is no footpath. The railway crosses on top of the tunnel where Savernake station used to be sited at the junction for the former Marlborough branch. The line of the path over the top is straightforward but the turn under the railway near the far end would give difficulty for anyone trying to get a boat of any length round the corner.

For a while the canal becomes heavily tree-lined, perhaps an effect of the forest. Savernake Forest was a Norman royal hunting forest of 16km², now with ancient oakwood, rare plants, deer, birds and extensive insects, including horseflies which are met on this section of the canal. It includes the 6km Grand Avenue of beeches by Capability Brown, with three other straight avenues crossing at its major intersection.

Timber from Savernake Forest was used for the replica heavy, wood, wharf crane at Burbage Wharf. Adjacent, the A346 crosses on an early example of a skewed bridge.

Cadley and Brimslade Locks are the start of the Wootton Rivers flight. Brimslade Farm has 17th-century tiled buildings with many chimneys. Rivers of any size are not in evidence, however, as the canal cuts through the Vale of Pewsey to Devizes with Lower Greensand and clay, surrounded by chalk hills in a conspicuously dry valley. The canal also runs parallel with the Wansdyke to Devizes.

Wootton Rivers is a village of brick and half-timbered thatched cottages. The church, which was rebuilt in the 19th Century, has a wooden bell tower and a clock made by Jack Spratt to commemorate the coronation of George V, that uses junk such as bedsteads, prams and bicycles and has the letters GLORYBETO-GOD instead of numbers on the face and 24 different quarter-hour chimes.

Wootton Rivers Bottom Lock begins the 24km Long Pound. Gradually the railway pulls away, leaving a quiet reach lined by greater reedmace, rosebay willowherb, thistles and other splashes of plant colour.

To the right of New Mill is Martinsell Hill Fort, to the top of which local villagers used to find themselves pulled by an invisible force on Palm Sundays.

Milkhouse Water has a nature reserve with rare chalkland plants.

Swallows frequent **Pewsey** Wharf. Pewsey is a village with many half-timbered and thatched cottages.

From Stowell Park, with its 19th-century house, the Stowell Park Suspension Bridge crosses the canal, not spectacular in appearance but the last remaining example of a bridge made of jointed iron bars.

A smart village cricket pitch fronts Wilcot. The village also has thatched cottages, the Golden Swan's having the steepest thatch in Wiltshire. A manor and vineyards originated in the 11th Century and there are a school with a prominent bell and the 12th-century Holy Cross church with a Norman chancel arch, mostly rebuilt in 1876 after a fire. A vicar refused to ring the church bell late at night to please a local drunkard so he employed a Devizes wizard to put a curse on the vicarage to the effect that a bell rang loudly in one room every night. It could not be heard from outside although many people went to listen, including James I. The curse stopped when the wizard died.

Wilcot Wide Water and the neoclassical Lady's Bridge across one end were features to appease estate owner Lady Susannah Wroughton. The bridge leads towards Swanborough Tump, where Alfred the Great is believed to have held a Parliament and written his will while being attacked by the Danes. The Bronze Age Ridgeway crosses somewhere in this vicinity, also passing the distinctive Picked Hill and Woodborough Hill with its strip lynchets, relics of Celtic and medieval cultivation. Cylindrical anti-tank blocks on the end of a

Burbage Wharf with its replica wharf crane.

The wharf building at Pewsey.

The now unique footbridge at Stowell Park.

Lady's Bridge at the end of Wilcot Wide Water.

The distinctive Woodborough Hill to the north of the canal.

Milk Hill with its white horse, seen from Honey Street.

bridge near Woodborough are of a kind seen at several canal crossing points.

Alton Priors, at the site of a Stone Age camp, has a Perpendicular church with a large 1590 box tomb, on which is an interesting Dutch brass plate. Neighbouring Alton Barnes, mostly owned by Oxford's New College, has an 18th-century rectory and a heavily restored Anglo-Saxon church with everything in miniature.

Because of the lack of subsequent building, many more ancient sites have survived on the Downs than in most other parts of the country. Knap Hill has a 1.6ha neolithic earthwork enclosure intersected by many causeways. A 2,000m² rectangular earthwork on the east side was a protected Iron Age homestead. There are neolithic tombs and Adam's Grave, a prehistoric barrow. A young lady walking across the hill one summer's day heard an army of horses but the noise stopped as she passed the long barrow. The most obvious horse now is the white one on Milk Hill, cut in 1812 as a copy of the one at Cherhill, one of six in Wiltshire.

Weather-boarded buildings and brick cottages make up Honey Street, where there are also a warehouse and the ruins of a boatbuilding yard that launched canal craft and sailing trows sideways into the canal. Up to 1948 there was still a floating bridge across the canal on the right of way leading to the Tan Hill fair, replacing an earlier ferry. Also important is the popular Barge Inn which had a brewery, bakery and slaughterhouse. After burning down in 1858, during which time the locals attempted to consume the cellar's contents, it was rebuilt in six months. It has been used for filming the *Inspector Morse* episode *The Wench is Dead*. It also has a room devoted to crop circles, with details posted as they become known. Typically, ten on successive days at the beginning of June are revealed and enthusiasts from around the world gather in this building to get the latest news. Activity centres around Avebury with the canal being along the southern edge of the area of sightings.

Stanton St Bernard is marked by its battlemented Victorian church and it also has a 19th-century manor. Beyond is Tan Hill, at 294m, the highest point in Wiltshire. It was near here that shepherds saw a horse-drawn hearse and cortege, the coffin surmounted by a gold crown. Rybury Iron Age camp is on the side of Clifford's Hill.

There was a large 7th-century Iron Age settlement with iron smelter and storage pits at All Cannings Cross. The 14th-century All Saints church at All Cannings has an 1867 High Victorian chancel. The

canal banks are lined with blackberries and convolvulus.

A new low swing footbridge crosses the canal to Allington and its Victorian church, next to the Knoll and in front of the Downs, with continuing large numbers of ancient barrows. Bourton Manor Farm may have been the site of a small monastic cell but became the wealthy manor of the Ernle family after the Dissolution. Much of the house was destroyed in the 19th Century.

The canal loops round Horton, part of a 38km² Crown estate. A high footbridge crosses the canal to Bishops Cannings, named because it belonged to the bishops of Salisbury. St Mary's church is like a miniature Salisbury cathedral and has a spire to 41m, a cruciform plan, rich stone carvings and Early English traces from 1150. At the start of the 17th Century, vicar George Ferebe provided the first peal of eight bells in England since the Reformation and an organ which lasted over 200 years, as well as training an excellent choir. In 1807 the organ was replaced by William Bayly, Captain Cook's astronomer, and is still in use. The church also has a carrel and a single 17th-century penitential seat box pew.

The village has thatched and half-timbered cottages, including the 17th-century thatched farmhouse of the Old Manor and the 17th-century timber-framed Old School House. The tale of Moonrakers, locals claiming to be raking the moon from the surface of a piece of water when surprised by excisemen, comes from various places around the country but Bishops Cannings pond is the most likely venue as it was on the route used to move Dutch gin from the Hampshire coast to Swindon by night. Wiltshiremen are still called Moonrakers. The story of local simplicity was enhanced by a sick drum-maker who took his materials to his room in the Crown Inn, part of the Crown estate, to make a drum for the village friendly society. On completion it was too big to take out so the society were obliged to meet downstairs where they could hear the drum. As with other villages on the edge of the Downs, the location is surrounded by ancient earthworks.

Horton Bridge (which is not the nearest bridge to the centre of Horton) has a Bridge Inn. From here the canal turns west again past Coate, a source of puddle clay for lining the canal. Elder, sloe, ivy and oak all provide berries and fruits along the canal embankment in the autumn. This section of canal is below Roundway Down where, in 1643, during the Civil War, the Royalist cavalry made a surprise attack on the Roundheads, killing or capturing most of them.

The first major buildings in **Devizes** are the Victorian Le Marchant barracks, which are now home to the Wiltshire Regiment Museum, covering from 1756. Almost opposite is the Wiltshire Police headquarters, given an official reprimand in 1999 by the Environment Agency for polluting the canal but stopping short of the fine likely to be imposed on any other such offenders.

Some bridges here are ancient monuments. The A361 makes its first crossing as the canal turns into a straight that runs past the hospital to Devizes Wharf, which is now mostly a car park. This is the site known to canoeists as the start of the Devizes to Westminster Race. The DW, Britain's best known canoeing event, began in 1948 in answer to a £5 bet over whether the local Rover Scouts could travel to Westminster by kayak in four days. They won the bet and the subsequent race has developed into an annual Easter event that has been instrumental in encouraging restoration of the canal, Britain's major canal restoration project. The 1979 race record of 15 hours 34 minutes still stands. Along one side of the wharf is an 1810 granary with a long timbered-balcony which houses the canal centre and canal society, while a theatre is located in the warehouse. It is interesting to contrast the level of activity at the wharf at Easter with the quiet of other times.

Devizes has the largest market place in the west of

England, much influenced by its surroundings, taking its name from the Old French *devises* meaning boundary. The Wiltshire Archaeological and Natural History Museum has finds from prehistoric, Neolithic, Bronze and Iron Ages and Roman material with items from barrows, Avebury and Stonehenge. Many houses are Georgian, which made a suitable venue for filming *Far from the Madding Crowd*. The market cross has the tale of Ruth Pierce, who was involved in a three-way dispute over part of the payment for a sack of wheat in 1753. She asked to fall down dead if she was guilty and duly died, the missing 3d being found in her hand.

The canal curves round behind a former canal forge and gasworks plus Wadworth's brewery of 1885, where the Long Pound comes to a decisive end at Kennet Lock. A low arch takes the towpath under the A342 on the left side of the canal at Town Bridge, where the metal capping is scored by towropes. Sandcliff House was the residence of George Eliot. Northgate House was an 18th-century coaching inn before becoming the lodgings of judges in the 1835 assize courts opposite. Wyndhams was a convent. The Castle was a Victorian folly which is now a private house but there have been three successive real castles in Devizes since the 11th Century, attacked in 1643 during the Battle of Roundway, as the result of which the tower of St James' church on the Green acquired its cannonball hole. Now only a mound and earthworks, the castle was built by Bishop Roger of Sarum, who also built St John's church for the castle in the 12th Century, including the tower. There have been 15th- and 19th-century additions. On the other hand, St Mary's church was built for the town in the 12th Century and extensively rebuilt in the 15th Century.

The Devizes Flight is one of the wonders of the canal system. While Tardebigge has a greater number of locks, here there is a greater fall in a shorter distance. The 29 locks drop the canal 72m in 3.6km. At Dunkirk, beyond the Black Horse, the A361 crosses over Prison Bridge, near the former prison, and the towpath again has a separate arch. The most dramatic section is the Caen Hill Flight, which runs straight down a spur of land at a 1:30 slope. Sixteen locks are closely spaced, each with a 64m x 41m x 1.6m deep side pond; an annual Boto-X inflatable race takes place down these side ponds. The flight was completed in 1810, the last part of the canal to be finished and the last part to be restored. Gas lighting was later added to allow night-time working to reduce the bottleneck. Today's pleasure craft are restricted to a very limited operating day by water shortages. Caen Hill was probably named after Napoleonic prisoners of war billeted here from Normandy.

British Waterways have a depot with a dry dock. Foxhangers House was the canal superintendent engineer's house. Later it was a laundry which discharged soapy water into the canal producing a great deal of foam. Clay pits on the left of the canal supplied brickworks for the canal. After the B3101 crosses the gradient eases but there are several more locks down to where a pumphouse has been built for backpumping.

Powerlines cross below the last lock and a pier of the former railway from Pewsey to Trowbridge stands in the middle of the canal, the canal making a sharp turn while a short spur is located on what appears to be the main line for boats travelling eastwards.

The canal crosses watermeadows and picks up Summerham Brook, which is the only feeder between Devizes and Claverton. The A365 crosses at Martinslade and slate roofs replace thatch. Around Sells Green the canal is on embankment that has proved unstable and needed significant work.

Seend, with its battlemented, Perpendicular church and 18th-century houses, is the current version of a village that housed Flemish weavers in the 17th Century.

While Seend Cleeve is now a quiet agricultural vil-

Devizes Wharf, warehouse and granary.

lage, this was not the case in Victorian times when there was an ironworks halfway down the Seend Flight of locks. Three hundred tonnes a week of iron was mined from higher ground and transported by canal to south Wales, later being processed here in two blast furnaces until 1889. Patrons at the popular Barge Inn on a summer's day would find it hard to believe the change in the scenery.

One hundred and ninety species of plant have been recorded in an 11km length of canal here. A group of apparently new houses have been built in Tudor or older styles. The straight sections of canal are joined by sweeping curves and there are extensive views to the south-west over the Semington Brook. There are four swing bridges before Semington.

Barratt's and Buckley's Locks come in quick succession at Semington, the latter with a line of roses being grown across the route of the shortest portage. An attractive lock keeper's cottage remains, unlike the Wilts & Berks Canal, of which the bricked-off end is located on the right before the bridge carrying the old line of the A350. This canal, which ran to the Thames at Abingdon, is now the country's major restoration project but it would appear that the healthy vegetable garden

Caen Hill Flight: a straight line at a constant 1:30 gradient.

which is the first obstruction on the line is not in imminent danger of being dug up. A line along the River Avon may be used instead.

The Semington Brook passes under Semington Aqueduct and the brook and the canal both pass the village of Semington with the first of the stone houses which are to replace brick from here. Some houses are 18th-century and there is a small stone church with a bellcote and an adjacent village school in similar style. The old world charm is spoiled only by a set of powerlines over the canal.

Lilies survive despite the powered craft as the line continues through open country to the B3105. Staverton has terraces of weavers' cottages. Nestlé had a small factory

The blocked-up end of the Wilts & Berks Canal.

Hilperton Wharf – a mini Docklands redevelopment.

England, which was rediscovered in 1856 by a vicar on a hillside who noticed the cross in the roof patterns. It was largely intact, despite being used as a school and a cottage and having previously been used as a slaughterhouse. More conventional is the 12th-century Holy Trinity church with additions, wall paintings and fine 18th-century monuments. The 13th-century Oratory Bridge over the river, rebuilt in the 17th Century, has a pilgrim's chapel, which was used as a lock-up for a while, John Wesley being an inmate for a night in 1757. It is topped by a copper-coloured gudgeon weathervane. Other significant buildings include the Victorian town hall and a Gothic revival factory.

The pedestrian route past Bradford-on-Avon Lock is long and difficult; it involves crossing the B3109 bridge with blind approaches for vehicles, passing the Canal Tavern, which was used to stable horses, and relaunching past tables of customers who sit outside in the summer taking morning or afternoon tea. From here to Claverton the canal corridor is noticeably well used for recreation. The line from here to Bathampton offers some of the best English canal scenery.

The fourteenth-century Barton Tithe Barn beside the canal is one of the largest in England.

from 1898, formerly producing condensed milk, although the previous users of the site had some of the first powered looms in the area. Now it contains offices. The weaving town of **Trowbridge**, named from its wooden bridge, the Old English *treow* meaning tree, with its 18th-century houses and lock-up, has become the administrative capital of Wiltshire. It was served by Hilperton Wharf, which has now been converted into a mini Docklands housing development, reached under the re-sited Parson's Bridge. A futuristic marina has been built on the other side of the canal. Wyke House, with its Jacobean-style towers, is a replica of 1865. From 1814 to 1832, poet George Crabbe was the local rector.

The canal passes over the Bath to Westbury railway on the Ladydown Aqueduct and then immediately over the River Biss on the fine classical Biss Aqueduct. When a line of poplars, undergrown with dog roses, is passed on the left the canal has for the first time come alongside its target, the River Avon, although the canal remains at a much higher level until Bath.

An impressive fern tree accompanies a line of moorings. These lead to powerlines crossing a marina with the Beef & Barge inside and the Beehive opposite, next to the A363. A huge walnut tree stands over the next reach, which leads to Bradford Upper Wharf with its drydock, previously a gauging dock.

Bradford-on-Avon is a market town in a steep, wooded valley, developed with the help of Dutch weavers, there being 30 weaving factories in 1800. One of the most remarkable buildings is St Laurence's church founded in 705, the most complete Saxon church in

Bradford Clay Farm provided much of the puddle clay for the canal. The 14km pound to Bath has always had leakage problems and there were many ground plugs to allow the canal to be drained quickly to the river below. The section has been subject to blowouts caused by water seeping into the oolitic limestone, forcing out air. The area was also subject to landslips. Eighteen sets of stop gates were installed in 12km and much of it has now been concrete-lined.

On the right, beside the canal, is the Barton Tithe Barn of about 1341, which formerly belonged to Shaftesbury Abbey. One of the largest in England, it is 51m long, has two porches with massive doors and is beamed with a stone-tiled roof covering 930m^2. It now contains farming implements but its uses as a film set have included *Robin of Sherwood* and *True Tilda*.

The dry dock at Bradford Upper Wharf.

Avoncliff Aquedeuct seen from the railway span.

Heavily wooded section on the north bank of the river.

The canal is very obviously on the side of a steep valley, the Barton Farm Country Park and a television mast above.

Below Westwood, with its 15th-century stone manor, noted for its late Gothic and Jacobean windows, is Avoncliff, where there are 18th-century weavers' cottages in a square, the Cross Guns 15th-century coaching inn, the chimney of a former flock mill and one of two mills, retaining its enormous wheel.

The canal now switches to the other less steep side of the valley for a while by crossing the Avoncliff Aqueduct. A neoclassical structure in Bath stone with balustrades at the ends, the 3-arched aqueduct is higher than its near neighbour and longer at 159m. With subsidence in the centre, it has an irregular line to the eye from on top, especially as it kicks sideways over the railway. Leakage meant that an icicle remover had to be employed in winter to clear the section over the railway each morning where the icicles posed a danger to trains. A tramway from quarries in Becky Addy Wood ran across the aqueduct as a feeder for the canal.

From Turleigh the line is mostly through woods. During the First World War, the Red Cross had a narrowboat at Winsley. It was used for taking wounded soldiers on days out from hospital to Bradford. Earlier, Nelson had convalesced in Bath after being wounded in the Battle of the Nile.

Across the valley, the River Frome joins the Avon at Freshford. Limpley Stoke's church, which dates from Norman times, has a collection of carved coffin lids. An impressive viaduct spans the Midford Brook, another left bank tributary of the Avon.

The quarry railway in Conkwell Wood came to fame when it was used for filming the *Titfield Thunderbolt*. The railway has now gone but wooden huts on the

Dundas Aqueduct with its cycle track.

Dundas Wharf looking towards the end of the former Somerset Coal Canal, now a spur used for moorings.

The view north from Dundas Wharf.

right mark its end. A derelict building on the canal's left bank was a lengthsman's cottage and stables.

The courses of feeder quarry railways are visible around Dundas Aqueduct, named after Charles Dundas, the Kennet & Avon Canal Company's first chairman. Completed in 1798, it was Rennie's finest architectural work. It takes the canal over the Avon and the Westbury to Bath railway. In Bath limestone and Doric style, it has three 20m semi-circular arches and flanking 5.8m elliptical arches. There is a 1.2m wide cornice outside the parapet, now broken away in places, along which local schoolboys showed their courage by cycling.

The canal passes from Wiltshire into Bath & NE Somerset, arriving at Dundas Wharf with its toll house, small warehouse and cast-iron crane which worked until the First World War. At the southern end of the wharf is the narrow entrance to the Somerset Coal Canal, now only a 500m spur serving as moorings although it was

noted for its inclined planes when operational. At the northern end of the wharf is a roving bridge and there is also a rusting crank and roller for removing a plug to drain the canal down. Some noise intrudes from the A36, which now runs parallel.

There are animal escape ramps at Claverton, a village where St Mary the Virgin's church has the mausoleum of commercial entrepreneur Ralph Allen, the Man of Bath, who promoted Bath stone and did much to develop the postal system, Bath being where the first stamped letter was posted in 1840. Above the church is Claverton Manor of 1820 in Greek revival style by Sir Jeffrey Wyatville, location for Churchill's first political speech in 1897, when he was 23. It is Europe's only museum of 17–19th-century American domestic life. For the canal, however, the most interesting structure is at the foot of the hill on a channel from the Avon. Claverton pumping station was built by Rennie in 1812 in a former grist

Rennie's unique waterwheel-powered pumping station feeding the canal up the hill to the left.

Bathampton Down overlooks rural English canal scenery as good as it comes yet it is only 3km from the city centre.

mill and is the only one of its kind in Britain. Flow in the river turns an enormous 5.6m diameter x 7.6m wide undershot waterwheel that drives a beam pump to lift 130l/s of water 13m to the canal. The pump is worked on selected days.

Warleigh Manor, across the river, is now a college. The major educational establishment is **Bath** University. It is sited on top of Bathampton Down, around which the canal circles, and is accompanied by a golf course, an enclosure site, the plain square tower of Brown's Folly and a nature reserve with green hairstreak butterflies and great horseshoe bats in old quarry caves. The quarries provided much of the stone from which Bath is built.

The Paddington to Bristol railway emerges between Bathford and Batheaston to cross the Avon on Brunel's Italianate bridge. It is accompanied by the current line of the A4, which has taken a more direct route from Froxfield.

Bathampton has a 19th-century church but the George Inn is 14th-century, formerly part of a priory, and had to have its door moved to the side as the canal passed so close. It was the location of the last duel fought in England. A huge collection of gnomes in a garden will not be to the taste of everyone. However, most visitors have enjoyed using the product of the former Plasticine factory here.

Large gardens slope down to the canal with an air of affluence. This part of the canal, as it heads south-west, gives the best views over the Iron Age hill fort on Solsbury Hill and the dazzling white stone housing of the city. Below, on the river, is the Victorian Bath Boating Station with skiff, punt and canoe hire. Higher up, space is more cramped and part of the canal had to be moved to the left in 1839 to accommodate the GWR. Sydney Gardens were created to hide the canal in a fashionable part of Bath. No. 2 tunnel is 50m long with an Adamesque portal and carries the A36, McAdam's 1830 turnpike, over the top. Two cast-iron footbridges were imported from Coalbrookdale in 1800 and the canal is noted for its Chinese-style bridge. The final tunnel is 54m long and is topped by the Georgian Cleveland House, a stone building of the early 1800s by the Duke of Cleveland. It was the original headquarters of the Kennet & Avon Canal Company and has a trapdoor from the cellar. Nearby is the Holburne Museum of Art in the former Palladian Sydney Hotel, which sets the tone well for Bath.

Cleveland House, the canal company headquarters.

The cottage at Bath's top lock.

109

Bath Deep Lock with its awkward portage.

From Sydney Wharf, the Widcombe Flight continues right down to the Avon, starting with an interesting lock keeper's cottage and cast-iron footbridge of about 1815 by the Bath Top Lock. Halfway down the flight is the ornamental chimney that served a pump engine house and there was another at the bottom lock. Side ponds were dug into the hillside for the locks. After Pulteney and Abbey View Locks, the A3062 crosses and then Wash House Lock has a footbridge of about 1815. Roadworks have resulted in Bridge and Chapel Locks being combined into Bath Deep Lock as one of the deepest locks in Britain at 5.7m. The route on the left goes down a flight of stairs into a tunnel with a piece of railing at the bottom which is unhelpful to anyone portaging. The alternative route is to brave the traffic. Beside Widcombe Lock, the Thimble Mill pumphouse has become a restaurant.

The canal feeds out onto the River Avon, already quite broad, heavily piled and potentially muddy. The River Avon and the south Wales rivers may have originally been dip slope headwaters of the River Kennet, accounting for the canal's route. There were legal powers to make the river navigable for large craft from 1619 although the work was not completed until 1723. There were plans to carry the canal on downstream but concerns that it might interfere with Bath's hot spring water meant the idea was not pursued. In fact, the river remained profitable with traffic between Bristol and Bath as the canal itself declined. The river has a 6.4km/h speed limit and downstream craft have priority.

The first bridge over the river is the lattice Widcombe Footbridge, a footbridge also known as Halfpenny Bridge because of its toll. In 1877, during the Bath & West Agricultural Show, it collapsed resulting in the deaths of 10 people and injury to 40–50 others as they queued on the bridge to pay the toll following the arrival of a train in the station opposite. The station was built by Brunel in a Jacobean-country-house-style, with curved gables and a castellated entrance to the viaduct at the west end of the station. It originally had a hammer-beam roof. His GWR Paddington to Temple Meads railway crosses the river on St James' Bridge and then runs onto the 550m St James' Viaduct which has 73 arches, decorated with castellated octagonal turrets, typical Brunel ostentation.

The iron arched Cleveland Bridge dates from 1827. The major bridge interest is Robert Adam's Pulteney Bridge of 1774, based on Florence's Ponte Vecchio, the only British bridge with buildings on both sides. Below it is Pulteney Weir, with a mill at each end, rebuilt as a boomerang-shaped, stepped structure in 1965 but now dangerous because of erosion under the bottom step and due for a further rebuild.

The bottom lock as the cut joins the Avon.

The Roman Great Bath with Bath Abbey behind.

Pulteney Weir upstream of the canal junction with the Avon.

Bath has been a World Heritage City since 1987. In 863 BC Bladud, the son of King Lud Hudibras, the father of King Lear, caught leprosy and worked as a swineherd but the pigs caught it too. The pigs went into a warm black swamp and were cured so Bladud did likewise. The carved acorns around the city recall the pigs' favourite food. This was the beginning of the taking of the waters for their curative properties. To the Romans it was Aquae Sulis after the local god, Sul. Britain's only hot springs, emerging at 46°C, were used to supply the Roman baths that were used from 65 to 410 and from which the city takes its name. A curse thrown into the King's Spring includes the earliest reference in England to Christians. When rediscovered in 1775, the spa produced the greatest collection of Roman finds in Britain, the temple of Sulis Minerva, mosaics, statues, jewellery, pottery, tombs, saunas, cold plunges and the Great Bath, 22m x 9.1m x 1.5m deep, which still uses the Roman plumbing and its 8.5t of Mendip lead lining. The Gorgon's head from the temple, actually a male, has been copied as the centrepiece of the maze in the Parade Gardens.

The See of Wells moved here in 1088 and Bath Abbey, actually a priory although once a cathedral and wool centre, as recalled by Chaucer's Wife of Bath in *The Canterbury Tales*, was begun in 1499 on the site of a church of circa 680, where Edgar was crowned in 973 as the first king of England. The church was gutted by fire in 1137. The 1499 design was given to Bishop Oliver King in a dream, resulting in the stone carvings of angels climbing Jacob's ladder to and from heaven on the west front. In a piazza, the Perpendicular Gothic building has fine fan vaulting, restored 18th-century cellars with an exhibition of 1,600 years of abbey history, a 49m tower, a heavily carved oak door and a wrought-iron screen. It is called the Lantern of the West because of the 52 windows, especially the brilliant 17th-century 76m^2 window. There are large numbers of wall tablets and other memorials. The National Trust shop is in the home of former MP General Wade, better known for his Scottish military roads.

The oldest building in Bath is Sally Lunn's House of 1482 although it was not until 1680 that she baked her Hugenot brioche buns here. They are still baked to the same recipe. The house has a museum of Roman, Saxon and medieval building and clay pipes in the cellar. Bath buns are sugar-topped, spiced, currant buns, another Bath speciality. Bath Oliver biscuits were invented in the 18th Century by Dr William Oliver as part of an obesity diet.

The city was the first British tourist resort and considered fashionable after a visit by Queen Anne. It became established as a spa by Beau Nash, who was known as the King of Bath for 50 years, banning the wearing of swords and other anti-social behaviour. The pump room was opened in the style of an orangery in 1706 and rebuilt in its current form in 1789–99 by Thomas Baldwin. Exhibits include a 3m longcase clock of 1709 and two sedan chairs, cheaper to operate than the three-wheeled Bath chair, invented locally by Arthur Dawson and able to take the user indoors and up stairs if disabled. The water flows at thirteen litres per second with a high content of calcium sulphate and sodium chloride but low in metals except iron. Many of those treated with the waters were suffering from lead poisoning, especially from exposure to it in various industries and from its use to sweeten and preserve alcohol. Thermae Bath Spa, opened in 2003 after a troublesome building operation, is in a glass and stone rooftop building with indoor and outdoor thermal pools, a whirlpool, neck massage jets, airbeds, steam rooms and massage rooms. It is the only UK bathing site that uses natural thermal waters.

The city offers the best Georgian townscape, much designed by John Wood the Younger, despite having

been heavily bombed in the Second World War. Dolemead Wharf received stone from Combe Down quarries by Ralph Allen's Waggonway of 1730, probably the first British railway to be described in print. The Guildhall of 1766–75, by Thomas Baldwin, with its Reynolds portraits, is claimed to be the finest room in Bath. In the autumn it is one of the hosts of the Mozartfest. The city also hosts a spring Bath International Music Festival.

The market is on the site of a medieval slaughterhouse with a twelve-sided domed interior and, as in Bristol, incorporates the 18th-century concept of the business nail, like a tall small table, so that business was done on the nail.

The Gorgon's head.

There is a book museum while the Victoria Art Gallery features 18–20th-century British art, notably Turner, Gainsborough, Sickert, Whistler, prints, drawings, ceramics, glass and watches.

Residents have included William Wilberforce and Jane Austen in Sydney Place, who set *Persuasion* and *Northanger Abbey* in the city. Sheridan set *The Rivals* and *The School for Scandal* in Bath. Dickens stayed here and commented on the flavour of Bath waters in *The Pickwick Papers*. It has also been used to film *Dr Doolittle*, *Other People's Children*, *The House of Eliott*, *Persuasion*, *The Remains of the Day* and *A Respectable Trade*. Thomas Gainsborough painted *The Blue Boy* here and Tears for Fears were a local duo.

Less decorative are all the shopping trolleys in the shallows of the river. Black-headed gulls seem equally out of place.

The Victoria suspension bridge of 1836 uses James Dredge's patent arrangement that places more tension in the chains at the towers than it does in the centre of the span.

Local points of interest include the Forum, the Impossible Microworld of Willard Wigan and a sculpture of huge nails. Sainsbury's store has a Georgian frontage with Ionic columns and octagonal columns holding wrought-iron arches, this being a leading example of a lesser Victorian station, the Midland Railway's terminus of 1870.

The William Herschel Museum is in the Georgian furnished home of astronomers and musicians William and Caroline Herschel and is where Uranus was discovered in 1781. Stone buildings are all around but none can match the 200m Royal Crescent of 1767–74 by John Wood the Younger. The 1796 home of the Duke of York at the end is now a Georgian museum and has the ghostly footsteps of an unpaid servant who left two of the duke's illegitimate children to starve. It

Dredge's suspension arrangement in the Victoria bridge.

The graceful New Bridge carrying the A4.

is set in the 23ha Royal Victoria Park, a Park of Special Historic Interest with lakes, aviary, botanical garden, bowls, tennis, a children's play area and the Dell, with ghostly sounds from former duels for which it was used. The Georgian Garden has plants and a layout from 1760.

A former railway bridge crosses, as does the Twerton Suspension Bridge. The 583m Twerton Viaduct runs parallel with the river. Twerton produced yarn in the 18th Century.

Dutch Island takes its name from a former miller. Weston Cut goes right past the Dolphin to Weston Lock, also known as Newton St Loe Lock. As with the following locks, smaller craft need to take out up ladders. Unlike the following locks, this one bypasses a weir with large lifting radial gates.

The river cuts through a dark wooded cleft with, on one side, a Morris Minor centre lining up cars high above on the bank.

The former Newbridge railway bridge carries the Avon Walkway and cycleway over although there are plans to restore the Avon Valley Railway along this line to the outskirts of Bath. This is the end of the city with Bath Marina, the Boat House public house and the A4 crossing between them on the graceful 26m Bath stone-arch New Bridge of 1740, which continues as a viaduct of stone on the south-west side of the river.

Oolite hills on the north side become more conspicuous, now clear of their housing. Kelston Park is the one obvious building high above the river, overlooking the site of a Roman villa.

Eerily, trains may be heard clearly as the disused Kelston Park railway bridge is approached. In fact, the noise comes from Brunel's line beyond, about to run along an embankment above the river bank for 2km. Between the two, a meander was straightened in 1977. The embankment hides Corston.

Red kites and jays may be seen and blackthorn is present. Distance markers hung in the trees are for the benefit of the rowing club at Saltford, the members of which train up to Weston lock.

Kelston Lock is at Saltford. The adjacent weir has a 1m drop although this is in two steps on the far right and can be shot by suitable craft when there is adequate water flow.

A dormitory suburb, Saltford has some interesting and attractive buildings near the river. Saltford Brass Mill, formerly a fulling mill, was water-powered, the last of its kind. It closed in 1925, the most complete annealing and brass rolling mill in the country. Brass mills in this valley made their money out of trinkets for the slave trade. Handel is said to have written the *Hallelujah Chorus* after hearing the hammers. By the church, Saltford Manor is one of the oldest inhabited houses in England, its 17th-century facade hiding what is a mostly Norman house.

A wall of limestone strata on the west side provides an unwanted windbreak for Saltford Sailing Club, which uses this reach of the river, although Saltford Marina is likely to be more appreciative of shelter.

Saltford Lock was almost destroyed in 1738 by protesters against the import of better quality Shropshire coal via the river in place of more expensive lower grade Somerset coal, despite the fact that damaging the navigation was a capital offence.

Alongside is the Jolly Sailor, where bargemen who had been promoted to skipper branded the wooden fireplace with a poker and then bought beer for perhaps a score of others. Graduating in debt is not new.

At the top of the lock, the ghost of a barmaid is sometimes heard crying for the father of the illegitimate baby in her arms.

The weir is about 1m high and slopes, more gently to the right of the chute in the centre. Kelston Brass Mill

The moorings at Saltford.

Victorian buildings now house a boatworks.

The ridge rising to the north of Keynsham Hams.

The weir stream past Hanham Lock sees much activity in the summer.

and annealing ovens add interesting architecture and the adjacent workers' cottages are some of the oldest in existence.

The river again approaches the hills to the north-east, now in more open country, and runs below the A431, a road first built by the Romans. Bulrushes appear at intervals.

Swineford Lock bypasses a weir with a 1m drop. Swineford has been suggested as the site of swineherd Bladud's leprosy cure. The village has a brass mill which became a woollen mill and has a corrugated-iron church.

Beyond extensive glasshouses, Bitton's church has a longer history, its nave being Saxon, its detailing Norman and its chancel and decorated tower 14th-century. It is haunted by a lady in grey. There is also a 19th-century Wesleyan chapel.

The Avon Walkway crosses the former Bitton Railway Bridge, alongside which the River Boyd enters, crossed by Ferris' Horsebridge.

The Avon Valley Country Park proves popular, with mini steam train rides on the Strawberry Line, quad bikes, adventure play area, barbecue area, pets corner and a falconry that can be seen from the river.

Keynsham was known to the nation's youth in the post-war years as it was spelled out several times per night on Radio Luxembourg in support of a football pools-beating scheme. It would seem that anyone with a surefire system should be using it for his own bets rather spending money to advertise it to the world in general.

The river meanders across a wide valley of eroded clays and marls. Ammonites are said to be large snakes that were turned to stone in the 6th Century by St Keyne, the daughter of the Welsh Prince Braglan, because converts would not otherwise visit her on land which had been given by the lord of the manor. Abbey Park has the remains of the Augustinian abbey, founded

in 1170. The 13th-century church has a Victorian interior and has an ornamental tower of 1734, which replaces one that blew down in a storm. There has been significant industry in later years, Avon Mill ruins being the site of a large and early brass rolling mill, Albert Mill being the site of a works obtaining dyes from tropical woods, a chemical works being founded in 1881 and the Portavon Marina being not insignificant.

The lock cut passes the Lock Keeper public house to reach Keynsham Lock, built before 1730. The weir is about 2m high with a small step and then a vertical drop. The River Chew joins beside the A4175.

Below the lock is a prominent red-brick Cadbury, chocolate factory of 1918, following a merger with Fry's and a move from Bristol. An earlier building here was the Keynsham Somerdale Roman villa with 50 rooms and a bath suite. The river was to have wharves served by horse-drawn railway lines from a coalfield.

A small tributary arrives conspicuously under a canal-style stone bridge from Cadbury Heath. After this, a wooded ridge rises high on the north side for a kilometre. On the opposite side, Keynsham Hams are used for grazing sheep but the steep banks result in sheep falling in and drowning because they cannot get out again.

The river makes a sharp turn in front of **Hanham** Court and enters a dark Pennant sandstone gorge. Some of the craft moored along this reach are of substantial size. The banks are just as popular because of the Chequers Inn, Boathouse Bar and Old Lock & Weir Ale House. Hanham Toll Office was built at Hanham Lock, the start of the Port of **Bristol**.

The weir can be shot in suitable craft if conditions permit. This was the previous tidal limit but high tides can still reach here or even to Keynsham, despite construction of a weir at Netham.

Distance
139km from Reading to Hanham

Navigation Authority
British Waterways

Canal Society
Kennet & Avon Canal Trust
www.katrust.org

OS 1:50,000 Sheets
172 Bristol & Bath
173 Swindon & Devizes
174 Newbury & Wantage
175 Reading & Windsor

20 Grand Western Canal

Devon's engineering marvel

The Grand Western Canal was part of the scheme to link the Bristol Channel and the English Channel. The Taunton to Topsham section was authorised in 1796 and John Rennie built the Greenham to Tiverton section between 1810 and 1814 as a single-level, wide-beam, barge canal, following the contours. The section through to Taunton was opened in 1838 and was an engineering marvel, having one inclined plane and seven vertical lifts, the only place where these really worked successfully. The lifts required trains of box like tub boats carrying 4–8t each. The lower section was disused by 1867 although signs of the lifts are still visible at Nynehead.

The lower section remained in use until 1924, when there was major leakage, being officially abandoned in 1962. In 1970 it was taken over by Devon County Council (in whose county the whole wide-beam section of the canal lies) and restored as a country park, including returning water to a dry section. Because the canal is not connected to the rest of the canal network, boats have to be brought overland and so it is both a quiet and a clean canal.

The canal is featured strongly in the **Tiverton** Museum.

The missing link in the coast-to-coast canal was the drop from the canal basin in Tiverton (Hardy's Tivworthy) down to the River Exe and then its canalisation to Exeter. The scale of the problem is easily gauged from the Tiverton basin, which looks down on the roofs of the town and commands fine views across to the hills northwards.

From the northern side the complex resembles less of a canal basin and more of a railway terminus from its heavy, arched walls although two massive buttresses are actually old limekilns that were supplied with limestone from Burlescombe quarries by barge, a far cry from the wool and lace on which the town based its early prosperity from Edward III onwards.

These days the canal basin has neatly mown lawns laid out with picnic tables, planted trees, car park, toilets, tea room, canal shop and TS Hermes, all overlooking an immaculate bowling green. The south bank houses the stables, offices and tack room of the Grand Western Horseboat Company which takes horse-drawn barges nearly to the aqueduct.

This is one of those rare canals that has no sign of modern industry; it mostly passes through open and scenic countryside and even has attractive surroundings in the built-up sections, particularly the first kilometre leaving Tiverton where gardens leading down to the canal are obviously kept with pride. Greater reedmace, water lilies and arrowhead grow in the water. Oak trees provide sheltered sections to the canal at intervals. There are many ducks on the canal and swans can be a nuisance in the spring.

The bridges are all named, the first being William Authers, a laminated wood arch.

Views are often good from the water along the canal but almost always only on the north side. As soon as the houses are left behind, it is possible to see across the Lowman valley to the hills beyond.

Crown Hill aqueduct carried the canal over the railway connecting Tiverton with Tiverton Junction. Unusually, the canal has outlived the railway.

After the bridge carrying what is now a minor road, the arch of which stands directly in front of an oak tree, the canal doubles back to the left and then goes round

Former limekilns form the wall of the canal basin at Tiverton.

The landscaped terminus of the Grand Western Canal.

Wooded section to the east of Tiverton.

Water fern in the lined section near Halberton.

Ivy-covered Whipcott Bridge.

three sides of a square to reach Halberton, its contour route taking it past a golf course and then into a lined section and past steep hillsides to the north of the village. The new opening Dudley Weatherley Jubilee Bridge allows foot access across the canal. An embankment crosses a valley that divides three ways, giving marvellous views towards Noble Hindrance.

Weed in the canal is water fern, a brown carpet which collects on the bows and can be a problem despite its beautiful appearance when examined closely.

Either side of the bridge at Rock House are some interesting dwellings. The first has smart canalside gardens around a sundial, the second is a large house and the final one is full of character but derelict in a small quarry.

over from the canal is carried underneath the channel and flows southwards as Spratford Stream to join the River Culm.

A bridge that once carried a railway spur to the quarries at Westleigh is now unused. Aggregate lorries move noisily over the next road bridge. Surprisingly, the railway bridge's girders are supported on a wooden crossbeam which is now showing the passage of time.

Powerlines cross over and back and an electricity substation is located in the meadows.

After a couple of leads off to the left, the water becomes shallow, weed-free and absolutely clean. On the left bank are Wayman limekilns, set in cutting. For the first time Canadian pondweed becomes the plant of the canal.

Without warning, the canal plunges into the short Wayman Tunnel at Beacon Hill.

Warning of the end arrives with a short area of reedmace right across the canal, which comes to an abrupt finish just beyond in Lowdwells Lock, the first on the canal. An overgrown hollow leads towards the nearby River Tone and the Somerset border, leaving a major restoration project for the future although a portage of 300m will enable the canoeist to continue to the Bridgwater & Taunton Canal.

Ayshford Chapel stands close to the bank of the canal.

Wayman limekilns were important canal users.

Weed removal at Sampford Peverell is by a weed-cutting boat. Some of the growth is yellow iris, enhancing the canal in the spring, with a backing of weeping willow trees.

A slipway has been added near the A361 bridge.

The M5 passes within 700m of the canal but is not noticed, the only obvious heavy traffic being over the bridge carrying a minor road. Aircraft noise is another matter with naval jets on low-level flights.

Indicative of quieter days is the distinctive old chapel at Ayshford, which stands at the start of the first of the two long straight reaches on the canal. The Exeter to London railway sweeps past the second of the two bends connecting the straights, overlooked by the village of Burlescombe with its prominent church tower. In turn, it overlooks the meadows on the inside of the bend. Instead of flowing northwards to the Tone, excess water weiring

Distance
18km from Tiverton to Greenham

Navigation Authority
Devon County Council

Canal Society
Grand Western Canal Trust

OS 1:50,000 Sheet
181 Minehead & Brendon Hills

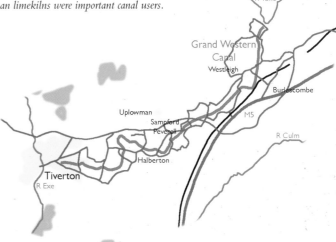

115

21 Bridgwater & Taunton Canal

The Bridgwater & Taunton Canal links the River Tone at Taunton with the tidal River Parrett at Bridgwater. It was once part of the Parrett, Tone, Chard and Grand Western navigations, from which there were plans for a link to Topsham, connecting the north and south coasts. The line from Taunton to Huntworth was opened in 1827. Bridgwater Dock was added after the completion of the canal in 1841 by Maddocks. Most revenue came from through traffic going to the Grand Western and Chard Canals. In 1866 it was sold to the Bristol &

The canal begins behind the bush on the centre. Firepool Lock is beyond.

The A38 crosses the canal at the hamlet of Bathpool.

Creeper shrouds the bridge at Creech St Michael.

Exeter Railway Company whose line closely follows the canal between the two towns.

Commercial traffic ceased in 1907 and the locks fell into disrepair. Somerset County Council purchased the dock for recreational purposes. The Somerset Inland Waterways Society was formed in 1965 to undertake the restoration, just two years after British Waterways took over the canal. The canal is wide, open and clean. Weed collects around the locks and at the lower end of the canal in summer, mostly duckweed although algae can be found at the southern end of Bridgwater.

The canal leaves the River Tone in **Taunton**, Hardy's Toneborough, at Firepool Lock, a lock with a mere 500mm drop. It is tucked in a corner beneath tall trees and the first of the mellow red-brick arch bridges, overlooked by the large and noisy cattle market and the railway goods yard.

Toilets are conveniently next to the river, just by the junction.

The canal winds away past an MFI warehouse and under bridges which range from an old brick arch accompanied by a large riveted pipe to a modern flyover which crosses gravel workings with concrete batching plants and a timberyard with the associated smell of Creosote.

Here the canal breaks out into open country. Until Bridgwater, the canal passes through nothing larger than the occasional agricultural hamlet and meets few main roads. It is an area of attractive rolling Somerset countryside, neither Sedgemoor and the Somerset Levels, which it skirts, nor the Quantock Hills, around which it arcs and which can be seen on the left at this point and on numerous later occasions. Banks are usually low, allowing extensive views.

Passing a large thatched farmhouse, the canal moves gently along to Bathpool, a delightful village with its stone chapel next to the canal. A fixed swing bridge carries a side road over. The A38 crosses on an arched brick bridge. Rowing boats are hired out from near the stone St Quintin Hotel in season. A caravan site is not obtrusive. Mown lawns edge the towpath and ducks potter about aimlessly.

Beyond the M5, a large brick mill on the banks of the River Tone is opposite the first houses of Creech St Michael. On the downstream side of the mill is an aqueduct across the river, one of the few remaining signs of the Chard Canal, which joined the Bridgwater & Taunton here. Operational only from 1842 to 1868, it was something of an engineering marvel with no less than four inclined planes along its length.

There are fine views to the south and the Blackdown Hills at this point. Prominent is wooded Stoke Hill.

This point is also marked by the first of the many pillboxes which line the canal. During the Second World War the War Office requested that the canal be turned into a defence line. Thus, in 1940, the fortifications were added and swing bridges were fixed and strengthened to carry War Department vehicles. These bridges are now all having to be raised to allow canal boats to pass underneath.

Gardens backing onto the canal continue their horticultural efforts at the bank. Red-and-white water lilies upstage the simpler yellow ones that are found frequently along the rest of the canal. Wisteria has not only coated the house on the left of the bridge but has spread across the bridge itself in a blaze of seasonally changing colour.

Perhaps a grass snake might be seen swimming across the canal although eels are much commoner in these waters.

Charlton is a hamlet with a large orchard. The Taunton area is known for its cider. Apple orchards dot the landscape. The land rolls gently, here a hillock with a wind pump on top, there a dip below canal level, making the scenery more interesting, bringing the Quantocks back into view or giving glimpses of the unusual church at North Curry, apparently in Regency style.

The canal now moves away from the River Tone as it becomes tidal. Another divergence comes on the railway, which has been close for most of the journey but only visible when trains are passing. Even then, they are often heard but not seen. A strange asymmetric railway bridge is the site of an oblique junction of major importance. While the Exeter to Bristol line continues to follow the canal, the London route cuts left and then crosses under the more northerly line.

Trees by the canal are scattered but oaks have been planted over a couple of reaches. Weed removal keeps the canal clear and only the gates on the towpath seem to be in bad repair. From the distance they appear to be pairs hanging off their hinges. Closer up they are seen to be of a rectangular shape with a tapered vertical wedge cut out of the centre and a low step to clear at the bottom, easy enough for a walker but keeping animals at bay.

As well as the locks at each end, there are four locks – Higher, Maunsell, King's and Standard – grouped over 3km midway along the canal. These have totally unique features in the form of pear-shaped counterweights on the lock gates; their concrete balance beams are also rare. Wooden benches were added as amenities. The upper gates at Maunsell Lock allow water to weir over and this can be disconcerting as the drop is substantial.

At North Newton the canal turns sharply in front of a meadow with a very ancient and picturesque church behind. This is not the village's only gem. In 1693 the Alfred Jewel was found in North Newton and this is now the oldest surviving Crown Jewel.

From Standard Lock it is possible to look right across the Somerset Levels and just see that most mystical of all the hills in Britain, Glastonbury Tor.

canal, railway and newly arrived River Parrett in one long glide. It is an unreal experience to sit under the viaduct at the canal end and look down that long line of columns as the invisible traffic roars overhead.

The water becomes scruffy as the canal enters the south end of **Bridgwater**, debris floating on the canal as it passes the backs of factory units and caravans. A weir on the right is unprotected although signs warn of reduced headroom on a bridge that is no lower than many others on the canal.

The corner before the A38 has once contained a green surrounded by cottages and it appears that efforts are being made to revive its previous atmosphere.

Another corner, between a water intake and a high fence round what could be a tennis court from the number of balls in the canal, leads into the major engineering feature of the canal, a deep cutting between high walls lined in the New Red Sandstone of the area. The overbearing nature of this cleft, with its solid bridges crossing, is enhanced by a number of heavy timbers shoring between opposite walls, a quotation carved on each one.

The A38 crosses and then there is a bridge high over the top with ornate balustrading.

Bowering's cattle-feed mill overshadows the final lock.

Bridgwater Dock was not closed until 1971 and has now been reopened as a marina. Moorings cover most of the area of its upper basin.

Admiral Sir Robert Blake was born here in 1599. The Admiral's Landing and residential and amenity building are all around and there is a large crane for lifting out cruisers. Nevertheless, old cranes and bases are still in evidence. A conical red-and-white object on the quayside is a Bristol Channel buoy from 1860, standing upside down for stability and exhibited as a feature. A brick bottle kiln on the far side of the river is the last remaining of a forest of them that once made bricks and tiles from the silt in the River Parrett, sending them as far afield as the West Indies. A double-leaf bascule bridge gives access to the smaller lower basin. This is overlooked by Russell Place, a row of Georgian cottages built in 1841 for dock workers, with a double-fronted villa at each end of the row for senior staff. In 1873, at the height of its prosperity,

Maunsell Lock with overspill gates, pear-shaped weights and concrete balance beams.

The large brick farm on the left at Fordgate has a rather neglected air about it, a theme repeated by a pillbox on a constriction in the canal, opposite which a number of pyramidal anti-tank blocks still stand in the long grass.

At Huntworth, present day activity returns with the Boat & Anchor Inn and its colourful children's playground. Neat lawns and trimmed hedges are matched by the orderliness of the pairs of rectangular pillars which support the 900m long M5 viaduct that crosses the

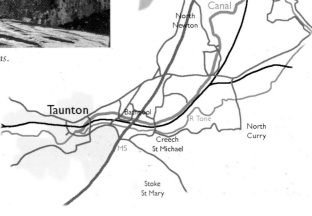

The canal's main engineering work, the imposing cutting in Bridgwater. The timbers have inscriptions.

Bridgwater was Britain's fifth most important coal-importing port.

Over 40 sluices connect the lower basin with the tidal river outside. These were used to flush silt out of the dock after building up a 9m head on the ebb tide. In addition, it was built with a 16m x 4m barge lock and a 13m beam lock, the latter now with a concrete dam across one side and deep deposits of silt on the other. In 1844 Brunel built a steam-powered drag dredger to keep the mud under control and this is the world's oldest working steam boat.

Car access to the dock is easy on the north side from the east end, next to the bascule bridge. This is now electrically operated but, until recently, required pairs of men to turn the handles on each side of the bridge and two more to control road traffic, a team of six in all.

Distance
23km from the River
Tone to the River
Parrett

**Navigation
Authority**
British Waterways

OS 1:50,000 Sheets
182 Weston-
super-Mare
193 Taunton
& Lyme Regis

Bridgwater Dock, now a marina. The buoy is a 19th-century marker from the Bristol Channel, now upside down.

118

22 Gloucester & Sharpness Canal

The Gloucester & Sharpness Canal or Gloucester & Berkeley Canal was built to bypass the uncertain conditions of the upper Severn estuary for shipping intending to travel upriver. Designed by Robert Mylne, its construction began in 1794 but was not completed until 1827, by which time the financial assistance of the Government had been enlisted. When completed, it was the widest and deepest canal in the world and these factors mean that it is still a viable waterway for oceangoing ships. Because of the presence of coasters, it pays to use the canal on Sundays or holidays when commercial traffic is largely replaced with pleasure traffic. Even at the statutory 10km/h, a coaster can depress the canal surface 500mm in the trough following its bow wave with water flowing rapidly from bow to stern behind its bow wave before breaking into turbulent water beyond the ship, waves being reflected to and fro for up to a kilometre behind it. Grain barges travelling above Gloucester still average 280t and the canal was the last canal to run a commercial- as distinct from a pleasure-passenger transport service, the Gloucester & Berkeley Steam Packet Company terminating its operations in 1935.

The canal is all at one level with lock gates feeding from the River Severn into the docks at **Gloucester**. The docks are dominated by a number of seven-storey, 19th-century, brick warehouses which present a lofty atmosphere with an attraction that newer concrete structures lack. This is home to one of the three national waterways museums. Gloucester Docks, with their shipping berths and cranes, were the setting for some of the filming of the BBC's *Onedin Line* television series. Gloucester cathedral is visible. The dock sides are high, often with lips that have been worn to a round profile.

A less conspicuous warehouse, on the left beyond the large ones, is used by British Waterways to store various small craft and other historical canal artefacts.

Use of the canal by ships means that there is always a considerable sediment load as the water is churned up and, inevitably, some oil on the water surface. As the canal leads off down the Vale of Gloucester, following the River Severn in a south-westerly direction, it takes the boater head on into the prevailing wind. This is given free rein by the generally thin spread of trees and the flatness of the vale, which tends to funnel in towards Gloucester, intensifying the wind.

The new lifting Llanthony bascule bridge carries the inner relief road over the canal. Most bridges are swing bridges to allow for high superstructures and masts. Large ships are expected to work straight through without stopping and all the bridge operators are linked by telephone and have the bridges open ready for approaching vessels. All traffic on the canal is controlled by traffic lights and even though small pleasure craft can pass under the first bridges without needing them to be opened, they are required to wait for clearance in case something big is approaching from the opposite direction. Craft normally hoot to draw the bridge operators' attentions.

The City Barge has been converted from a warehouse, which allows people to walk alongside the water but has the first and subsequent floors flush with the water's edge, supported on a row of slender circular columns.

At first the route is commercial with tim beryards on both sides. A swing-bridge base is passed on the right at Hempstead, followed closely by the first swing bridge.

The British Telecom vehicle yard faces a rowing club and is followed by a gas holder and a concrete batching plant – prominent landmarks – with the canal's bucket dredger moored nearby at Podsmead.

The tight Two Mile Bend has been bypassed with a new swing bridge, the second largest in the country, carrying a new road. The canal breaks out into open country. The scarp slope of the Cotswolds moves steadily nearer on the left. The more distant Forest of Dean does the same on the other side of the river. Even the Malverns are visible at times, away to the north.

The banks steepen, resulting in a number of slip failures, some of which have been repaired with old railway line, concrete blocks and other such materials. Overhead, power cables pass on extra high pylons.

At Lower Rea is the first of the lock keepers' cottages. These cottages are completely unique to this canal, built in classical Regency style with pedimented porticoes on Doric columns, all differing slightly in detail and resembling so many Greek temples – an unexpected find on what is still a very commercial waterway.

A footpath crosses a field to the river at Stonebench, a point that is considered to be one of the best positions for viewing the Severn Bore as it rushes up the river at some 16km/h, sweeping through a full 180° bend here.

In the spring, banks are covered with cowslips while teasels take their place later in the year.

The Quedgeley oil depot and tank farm are important as the upper point of navigation for oil tankers up to 1,000t. Nearby is the Pilot Inn.

The bridge keeper's cottage at Hardwicke is boarded up and no longer in use. From here the bridges get lower.

The landscape becomes totally rural, cutting through farmland which is only broken after the first of another grouping of high power lines across the canal.

Wheatenhurst, an attractive village in Cotswold stone with stone roof panels,

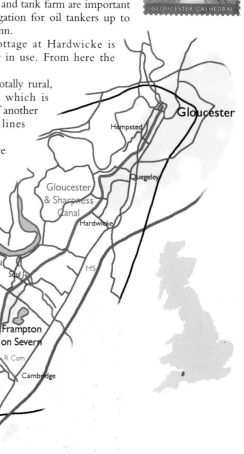

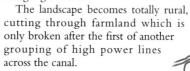

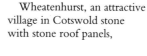

Gloucester Docks with its warehouses, including the one that houses the National Waterways Museum.

The new swing bridge at Two Mile Bend, now completed.

Wooded banks at Hardwicke.

At Quedgeley. The widening is the turning point for the oil depot traffic.

Hand-operated Sellars Bridge and Regency lock keeper's house.

produces the first of several interesting church towers, this one having an upward extension built onto one corner.

The canal crosses the River Frome, followed immediately by Saul Junction, an unusual canal crossroads with the Stroudwater Canal. The older waterway was abandoned in 1954 and the remaining section is used as a marina. In its heyday it connected the River Severn via the Thames & Severn Canal with the River Thames at Lechlade. There is work afoot to reopen it.

Saul Junction boasts a tower crane to service its boatyard and several concrete silos which are not particularly decorative.

Progress along the canal is marked by triangular stone posts which give the respective mileages to 'G' and 'S'.

An orchard reaching to the canal marks **Frampton on Severn**, a village constructed around an 800m x 100m green. At Church End, the 14th-century church has delicate detailing to the turrets on the top of the tower. Inside, it has a Romanesque lead font.

The Regency keeper's cottage at Splatt Bridge serves as an ice cream and confectionery shop for those wishing to look inside one of these buildings.

As the canal heads into the Vale of Berkeley, flanked by reeds and willow trees, a cottage with wooden wall tiles and a large wooden dog kennel marks another feeder, the Cambridge Arm, supplied by the River Cam.

The Tudor Arms stands by the canal at Shepherd's Patch, an area once known for its sheep grazing. Towpath walkers with dogs give way to walkers with field glasses, drawing attention to the more recent attraction on the right, the Slimbridge Wild Fowl Trust, of which only the odd observation point can be seen from the canal. Established by Sir Peter Scott, who also played a central role in setting up the Inland Waterways Association, it has 160 resident species in addition to the migrant population and is a centre for much important bird research. Varieties range from pink flamingos to

trumpeter swans. They generally prefer not to use the canal, which they may find too busy or too dirty.

Sprays play onto the water at Purton and screen rakes stand at the ready where a large waterworks draws $110,000m^3$/day from the canal to supply Bristol.

The village also has a church with an unusual stumpy octagonal spire.

Although the canal follows the River Severn closely, there is only one point where it is clearly visible from the canal, beyond the Berkeley Hunt Inn where overflow weir allows excess water to discharge into the estuary immediately alongside. Here, there are fine views across Waveridge Sand to the Forest of Dean, which is now close on the opposite shore.

Ponds on the left of the canal were used at one time for storing complete tree trunks afloat until they were required by the various timberyards.

An arch beyond these ponds with a round tower opposite marks the end of what was the 22-arch Severn Railway Bridge until October 1960. A pair of tankers coming upriver in thick fog missed the end of the canal at Sharpness and went on to collide with the bridge, resulting in its subsequent demolition.

The canal divides at Sharpness. The arm to the right, now used for moorings, was the original connection with the estuary. The lock is now disused and all traffic is routed through the docks.

The docks, covering 12ha of water, handle five hundred thousand tonnes of freight annually and are one of the

Swing bridges at the entrance to the Sharpness Docks.

A corner of the massive Sharpness Lock.

Junction to the old lock at Sharpness, the Severn on the left.

Sharpness Docks, British Waterways' main commercial workplace.

leading sources of revenue for British Waterways' Freight Services Division. Timber is a major commodity, the first timberyard belonging to British Waterways, but there are also more of the seven-storey brick warehouses with concrete silos reaching up to twice their height. A drydock is available and narrowboats are constructed although these look totally lost among the shipping moored here. Shipping line agents occupy portable cabins and even the Customs & Excise office looks transitory.

North of the lock stand a pair of unusual buildings. One is a geodesic dome with a spiral staircase up the centre, while the other is a wedge-shaped corrugated building some three storeys high that is Scandinavian in appearance.

The final lock down to the River Severn is one of the wonders of the canal network. The tidal range here is the second largest in the world, the river drying out around the dock entrance at low tide so the lock is used only on the upper part of the tide. Not content with this, the lock is some 200m long and 100m wide with mooring fingers projecting out from the sides. It is capable of holding ships up to 5,000t.

From here there is a magnificent view down the estuary past the Berkeley nuclear power station to the Severn Bridges silhouetted on the horizon.

Distance
27km from Gloucester to Sharpness

Navigation Authority
British Waterways

OS 1:50,000 Sheet
162 Gloucester & Forest of Dean

121

23 Monmouthshire & Brecon Canal

The longest pound in Wales, despite the mountains

The Monmouthshire & Brecon Canal runs south-east across Powys from Brecon to Newport, a narrow contour canal followed by one of the best towpaths in the country. Two features that distinguish Welsh canals are their beauty and their proliferation of names, both being applicable in this case, not least because the canal started off as two separate canals. At the lower end was the Monmouthshire Canal, built in 1796 to Pontnewynydd.

The Brecon Canal, Brecknock & Abergavenny Canal or Brecon & Abergavenny Canal was to be built to the River Usk at Caerleon but the two were, instead, connected at Pontypool, a fact which was to save the upper canal in later years as a water feeder for the lower canal. Designed by Thomas Dadford Jr, it was opened throughout in 1812 to export iron ore, coal and limestone but was also used for local coal, lime and cattle transport for farmers in the rural upper section. A plan for an extension to Hay was dropped in 1793 and the proposal for a canal from Abergavenny to Hereford was blocked by landowners. The upper canal was bought by the Monmouthshire Canal Company in 1865 to safeguard water supplies. There was no significant freight after 1870 and the canal was bought by the Great Western Railway in 1880. Tolled trade continued until 1938 although there were boat movements during the war.

The upper canal lies within the Brecon Beacons National Park, the requirements of which have resulted in keeping it beautiful. Restoration of this section was completed by British Waterways and Monmouthshire and Brecon County Councils in 1970 and the change of name recognises their efforts.

Because it is not connected to the national network, the traffic is light but it has extensive views and prolific spring flowers. The many private and hire craft keep the water murky. The canal has roach, perch and dace and is stocked with bream and mirror carp for anglers to catch.

The canal is supplied by a feeder from the River Usk in **Brecon**, a town built at the confluence with the Afon Honddu and Afon Tarell. Mostly 18th-century, the town has Roman origins. Brecon Castle is largely 11th-century, with the remains of a large motte, bailey, walls and two towers. Most of the destruction took place in the Civil War with the assistance of the locals, who were shrewd enough to realise that they would have a quieter life if neither side occupied it. The Priory Church of St John, founded by Bernard Newmarch, is mostly 13th-century but with a 14th-century nave and is noted for its fine glass and side chapels dedicated to medieval trade guilds. It was made a cathedral in 1923. The Brecknock museum of country life includes the dugout canoe from Llangorse Lake crannog, the most famous Welsh dugout, some 1,100 to 1,200 years old, described in *Dead Men's Boats*. It may have belonged to Brychan, the Welsh king who gave his name to the town and whose Irish father, Anlach, would have been associated with canoes and crannogs. The South Wales Borderers museum features a Zulu war room. Local residents included Sarah Siddons and Charles Kemble. The city currently draws devotees to its August jazz festival, one of Britain's best.

Construction of a new canal terminus and a large theatre was undertaken in 1997, 300m short of the original terminus.

The canal slips away quietly past the backs of small gardens, forming the southern edge of Brecon. The road that the left bank then follows was a Roman road and later the A40. Traffic now bypasses the city on the far bank of the River Usk, which will never be too far away until Abergavenny. Hawthorns and occasional low-hanging willow and ash trees give the canal very much the feel of a river; the *Dragonfly* trip boat passes regularly so navigation must be easier than it appears. The trees obscure views of Slwch Tump fort, a hilltop aerial and barracks on the north side. The view is dominated by a wooded ridge to the south. Over the length of the canal, the prevailing wind is either following or is blocked by the mountains which provide imposing scenery to the south. The valley ahead provides extensive views to the east.

The A40 crosses on a concrete culvert-like tunnel at Abercynrig, where the Afon Cynrig joins the River Usk. The beech, sycamore, hazel and alder trees that grow beside the canal are accompanied by yellow archangel, herb robert, stitchwort, cow parsley, broom and ferns.

The single Brynich Lock, with its stone lock cottage, is adjacent to the Usk Bridge, carrying the B4558. A field of long-horned goats between the canal and river pay interest to any activity.

A sharp turn takes the canal over the four-arch masonry Brynich Aqueduct with views down onto the rocky River Usk. A winch is provided to remove a drain plug, allowing the canal to be de-watered into the river.

Brecon terminus with large modern theatre building alongside.

The canal makes its departure from Brecon.

From here to Pontymoel the canal runs along the right side of the valley with the towpath on the left bank. Bridges have been built with the arch higher over the towpath than over the canal. In addition, the right side of any road across is frequently much higher than the left so that the asymmetric arch has a point of weakness above the towpath and many of the bridges have had their arches reinforced to counter this problem.

Sheep and Fresian cattle graze alongside the canal. On the far side of the river, at Llanhamlach, a standing stone is hidden among the trees, just upstream of the 13th-century church.

The Cambrian Marina for hire craft and private craft was built in 1988 at Ty Newydd. An old Brecon Boat Company storehouse was built with a covered loading bay over a winding hole.

Mayflies and other insects swarm over the canal. Cinquefoil, violets, buttercups and red campion join the flora but are eclipsed by the springtime carpet of bluebells beside the towpath.

A low aqueduct crosses the Nant Menasgin with a wooden roller which was a windlass for the chain from the drain plug. The stream flows through Llanfrynach with its 13th-century public house and Roman bath house site at the foot of the Brecon Beacons, the bend giving easy views up to 886m Pen y Fan and other peaks.

A low drawbridge allows access to a farm.

The canal makes commendable use of the moat of the 16th-century former Penkelly Castle on its mound, now occupied by a 16th-century farm with a door from 1542 and a 15th-century refectory table. Canoeing groups from Plas Pencelli outdoor centre are frequently seen on the canal. The Tower is another antiquity down by the river towards Scethrog.

Three more modern drawbridges follow at Cross Oak. Primroses grow along the banks and wrens fly about the bushes. Another standing stone on the river floodplain is the Gilestone.

Flooding was a problem in December 1994 when heavy rain resulted in two breaches of the canal, the first at Talybont-on-Usk. Since then much work has been done on bank repairs. While the vegetation grows down to the water, a boat's wash can be heard running along beneath the edge of the bank. In some places this has been tackled with new clay puddling.

Near the Star Inn the canal reaches another low electrically operated drawbridge of 1970, a notice forbidding it from being opened during rush hours for the local school. The canal passes houses at rooftop level to cross the Caerfanell aqueduct, the embankment having been conspicuously lined with concrete placed around the trunks of all the trees.

The Taff Trail from Brecon to Cardiff has been following the towpath but now joins the former railway line passing over. This Brynoer tramroad supplied the wharf and limekilns until 1865. A pipe bridge on a line from Talybont Reservoir crosses near the railway bridge, a large valve being positioned at the centre of the span.

A barn conversion provides a striking building next to the towpath for those visiting the Travellers Rest.

Up the hill is a fort site, the Roman road from it crossing the line of the canal at Graiglas bridge.

The next reach ends with

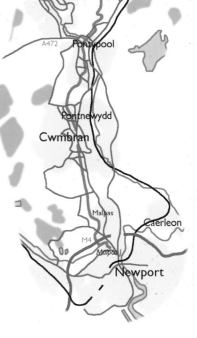

The gauge at the mouth of Ashford Tunnel gives an indication of how far it has dipped in the middle.

the 343m Ashford Tunnel, which has no towpath so boats were pushed through with poles using notches in the walls. The tunnel dips in the middle so there are height marker boards across the tops of the tunnel mouths. A vertical shaft in the centre adds a welcome but initially puzzling pool of light. A chain hangs along the side to give assistance if required. One of the bridges over the canal, giving access to a forestry area, has been reinforced by placing a Bailey bridge over the arch, ugly but it serves its purpose. Workhouse Bridge draws attention to three cottages which became the workhouse, now one residence.

Cwm Crawnon Lock and the four Llangynidr Locks drop the canal to a 40km level pound, the longest in Wales, despite the mountainous terrain. Among the beechwoods, wild cherries, bracken and foxgloves are a stone warehouse used by Country Craft Narrowboats, a British Waterways depot, a picnic area and a wooden roller next to the Afon Crawnon Aqueduct for removing the drain plug. Blue tits have nested under the coping of Cwm Crawnon Lock and the cuckoo can be heard in the woods.

The Coach & Horses Tavern is noted for its French cuisine and dining tables are placed on the bank above the canal, 100m before the last lock. At this point there is another water pipe bridge over the canal with heavy-valve bleed facilities in the centre.

The canal is crossed by the B4560 at Coed-yr-ynys, the road then crossing the river at Llangynidr bridge, a 17th-century Grade II listed structure with six arches divided by triangular cutwaters and pedestrian refuges.

Llangynidr itself has a 19th-century church and the remains of a wagonway.

Although the higher mountains have been to the south, a few of the hills on the opposite side of the valley have been prominent, including Allt yr Esgair, Buckland Hill and now Myarth. Further to the east are 701m Pen Cerrig-calch, with Table Mountain slipping off its edge, and the conically shaped 596m Sugar Loaf beyond Crickhowell, the southern edge of the Black Mountains. Myarth has a standing stone at each end and conspicuously out of place Italianate towers in orange brick at Gliffaes. Glanusk Park owns some of the area with a golf course and duck shooting among the activities for visitors.

Towards Dardy there is a settlement site up in the woods. Concern over settlement of the canal has resulted in this section being lined in concrete and brick. Trefoil and vetch add a little intimate colour. The old workhouse stands on the right. Round the corner at Ffawyddog the Ty Croeso Hotel stands above with forgetmenots and teasels. Rabbits scamper about on the hillside and flocks of mallards paddle along the canal.

Crickhowell, with its 18- and 19th-century houses, is reached via a medieval 13-arched bridge over the river, guarded by the Norman Ailsby's Castle. This was destroyed in the 15th Century except for the motte, bailey, curtain wall and small tower, which partly survive. The 14th-century church has interesting stained glass and other local features include the Porth Mawr gatehouse and a burial chamber. The town was formerly a centre for Welsh flannel production.

Llangattock was formerly a weaving village. Conspicuous features still are the massive limekilns beside the canal with some smaller ones in the field beyond. These were supplied with limestone from the Llangattock escarpment, above on the side of Mynydd Llangatwg, some of the tramroads remaining in place. Caves in the escarpment include Agen Allwedd, which runs for 18km.

Hen Castell's site looks down over Llangattock Park, where larch plantations, California redwoods and other exotic trees border the canal. There are displays of rhododendrons, sometimes so laden with blossom that the leaves cannot be seen, just balls of magenta flame.

Limekilns at Llangattock, much larger than usual for this kind.

Lambs graze at Llangattock beneath Mynydd Llangynidr.

Bluebells coat the banks near Llangattock as the Brecon Beacons rise above the woods beyond.

The Sugar Loaf seen from before Gilwern.

There is another standing stone across the river by Cwrt y Gollen army camp, from where the noise of rapid gunfire might be heard.

Just before an earthwork on the hillside opposite the confluence of the Grwyne Fawr with the River Usk, the canal passes into Gwent.

In 2007 there was a major breach of the canal near Gilwern.

The Bridgend Inn and Navigation Inn stand at opposite sides of the canal beyond the Gilwern Aqueduct, which clears the River Clydach in a single arch. Gilwern had two tramroads, one passing under the canal and along the Clydach valley, to bring down coal from the Valleys and pig-iron from the Clydach gorge ironworks to Gilwern Wharf, now a picnic site. A sign points up the hillside to the Lion, in the direction of the A465 Heads of the Valleys Road, which now comes alongside for a noisy kilometre until it crosses and moves away. An aerial on Gilwern Hill acts as a landmark but from close up it is hidden by the trees.

There is another drawbridge at Llanwenarth and then two lots of powerlines cross before another Bridgend Inn and a canalside general store.

An oblique, disused railway bridge at Govilon has been an opportunity for the bricklayers to demonstrate their skill with heavily skewed courses. It now carries a Sustrans cycleway. Govilon Wharf formerly had an ironworks, limekilns and tramway. Now it is the home of Govilon Boat Club, the first inland cruising club in the country in 1963.

From Govilon, the Monmouthshire & Brecon Canal runs south. Briefly there are gardens backing onto the

The temporary Tunnel of Love trough at Llanfoist.

canal, including one with a prominent pair of cart-wheels, but the woodland surrounding the canal, as it runs along the flank of 559m high Blorenge, quickly becomes undisturbed, a forest of beech trees, gradually being joined by alder, ash, hazel and sycamore plus bracken, ferns, broom and mosses. The canal forms the eastern boundary of the Brecon Beacons National Park as far as Pontymoel.

Beyond the trees lies the gateway to Wales, **Abergavenny** (the Gafenni being the blacksmith's river), with the Ysgyryd Fawr and the Roman site of Gobannium on the Ysgyryd Fach. The 11th-century Abergavenny Castle is built on a mound. In 1177, William de Braose invited the notable Welsh leaders to dine and then, suddenly, had them all murdered in order to secure control of the surrounding area.

This section of canal is unstable and had to be drained for several years after a serious breach in 1975. New cracks appeared after heavy rain in December 1994 and a temporary trough was placed in the canal until 1999. The whole section from Talybont to Llanover was closed in 2007 for repairs because of leakage problems.

Llanfoist Wharf is home to Beacon Park Boats' hire centre, which is arguably the most attractive in the country; it has no advertising hoardings, just lawns, rambling roses and stacks of logs for the winter in a beautiful setting. The wharf buildings were built for a tramway, the remains of which can be seen by the boathouse. This was the setting for Alexander Cordell's *Rape of the Fair Country*.

Below the Punchbowl nature reserve are a pigeon loft and a golf course but the line is generally completely natural. Flowers proliferate, buttercups, foxgloves, vetch, herb robert, primroses, stitchwort, violets, yellow arch-angel, ferns and succulent lichens. Notable near Llanellen is a hillside with a film of blue from bluebells in the spring.

Llanellen has a three-arch stone bridge over the River Usk and a 19th-century church. This is also the location of the *Lord William de Braose* cruising restaurant, a rather inappropriate choice of name.

Views over the border country are extensive. Swallows hunt spiders and insects over the canal and herons fish in quiet corners. It is hard to believe that this rural

The boat hire centre near Llanfoist.

Mynydd-y-garn-fawr stands beyond Blaen-Ochran. Blaenavon is only a kilometre beyond the ridge.

The Folly watchtower breaks the skyline beyond the trees.

The portal of Cwmbran Tunnel hides in the shadows.

The beam from Halfway Bridge at Cwmbran.

spot is only 4km from the Welsh pit village of Blaenavon.

The mountains begin to pull back at Goetre. There is a spur to the right which served Goytre Wharf with its limekilns, now home of Red Line Boats, a boat hire marina. The Waterside Rest offers meals and displays with information on the local industrial heritage.

The canal crosses a small brook flowing down from a holy well and would hardly be noticed were it not for the wooden roller for the drain plug chain beside the canal.

At Mamhilad the canal passes near the Star Inn and a church surrounded by massive yews. In the distance is the former ICI Nylon factory and a water tower. The more conspicuous tower is the Folly watchtower on top of the ridge to the west side of the canal, the ridge being far enough back to allow fields of hay and rape at its foot, although close enough to dominate the view.

The end of the national park, at Jockey Bridge, and the beginning of **Pontypool** arrive suddenly, together. Houses back onto the canal, one with a treehouse perched on top of the remains of a tree, and the canal passes through a gloomy cutting.

Yellow irises grow along the banks approaching the aqueduct over the Afon Lwyd, the valley of which the canal is to follow. The A472 also descends the valley and crosses over the canal to meet the A4042, which follows the canal here. Pontymoile Basin has a day hire boat company and a boat-shaped tea room. The round-ended junction toll cottage is now private, by a roving bridge and a stop lock. On the far side of the A472 is a basin with private moorings. The former Monmouthshire Canal, the lower section of the Monmouthshire & Brecon Canal, known as the Snatchwood Branch, continued up the river valley to Pontnewynydd.

Pontypool, Welsh for the bridge at the pool, stands on the rim of the coalfield and has been industrial since Roman times. In 1720 it produced the first tinplate in Britain and in the 18th and 19th Centuries it was an iron town. It was a japanning centre in the 19th Century but also a market town. The Italianate Grade II listed town hall of 1856 was donated by ironmaster Capel Hanbury Leigh, to celebrate the birth of his only son. Pontypool Park House was built for the Hanbury ironmaster family and is now a school. Its Georgian stable block houses the Valleys Inheritance Centre, with a leisure park featuring tennis, bowls, the third longest dry ski slope in Britain at 265m, a triple hydroslide and a grotto summerhouse built of shells and bones. The wrought-iron entrance gates were given to John Hanbury by Sarah Churchill, Duchess of Marlborough. A more recent industry has been the Pilkington glassworks.

The disused railway Skew Bridge passes over, the line heading up the river valley. A picnic table is located on the left bank almost opposite a hospital. An aerial stands on top of the ridge that confines Griffithstown next to the canal. One of the houses has a cypress hedge clipped into large turrets along the canal bank while the Open Hearth, on the towpath side, is more welcoming.

For many years the navigation ended at a 2.4m diameter culvert at Sebastopol. Crown Bridge has been rebuilt and powered boats are now able to continue along a broad reach – with duckweed, arrowhead and plantains, home to plenty of frogs – to Five Locks.

The 80m **Cwmbran** Tunnel, with no towpath, lies below a golf course. The water after the far portal is shallow at first but deepens towards **Pontnewydd** Locks. Five Locks Moorings were opened in 1998. Full restoration is planned to here with preservation to Newport as the current objective. The canal fell 103m in 14km through 31 locks. The Five Lock restoration scheme after the Cross Keys has been designed. Town planners of the Cwmbran new town were guilty of closing this section of canal as recently as 1954.

Below Five Locks, the canal is controlled by anglers who claim that any boats other than their own on the navigation create disturbance and so are not permitted.

As a landscaped water garden, the ten locks of the Cwmbran flight begin very attractively. As a canal they have less to recommend them. The long walk down to Forge Hammer passes the Old and New Bridgend Inns and a selection of modern industrial units. Launching is possible among the shopping trolleys in front of the gas holder but the water comes to an end again where 500m of canal has been covered by a roundabout and a busy road. A vertical cutting for the road means that the line of the canal cannot even be followed on foot. Instead, a diversion is required up the footpath to the right from the roundabout, along a residential road, left past some shops and right at the bridge crossing the new road.

The canal begins again where an iron beam from the Halfway Bridge of 1847 has been erected across the end of the reach as a feature. A swan's nest was visible at a surprising site in the middle of a housing estate with only 300m of clear water. A low bridge crosses halfway down this reach. Another short walk is required where a housing estate road has been laid along the line of the canal. Further walking is needed at the Waterloo, two Oakfield Locks and then, from Ty-côch, ten locks dropping away in open country.

Some of the Tycock lock flight can be canoed but the rest forms a pleasant portage after Cwmbran. Dragonflies hover around the reedmace and some picnic tables are well placed. A cuckoo might be heard in the distance. Three sets of powerlines cross to an electrical substation.

Towards the bottom of the flight, reeds and shallows begin to make paddling difficult, even when the locks

The fully restored Gwasted Lock with its bridge.

The final reach into Newport.

Residents on the Tycock lock flight.

are far enough apart to justify it. Eventually the water runs out altogether in most conditions at Malpas. Youths from a local housing estate are making a good effort to demolish a traditional, arched canal bridge. They have already removed one parapet but their attempts to prise out stones from the underside of the arch with scaffolding poles suggest there could be a sudden end for all parties involved.

As the water deepens again it becomes clearer. Wildlife includes a goldfish that is obviously thriving and has grown to a considerable size in the canal, which was dredged here in 2000 and 2004.

The restored Bettws Lane Lock awaits water. Gwasted Lock and its stone bridge, with lock gates and fittings smartly painted, comes as an encouraging sign for the future and is regularly used by the canal society trip boat.

The canal continues, to pass under the west end of Junction 26 of the M4, just outside the Brynglas motorway tunnels. A portcullis drop door prevents flooding.

Beyond the motorway, the main line of the canal is joined by the Crumlin Branch at the foot of Barrack Hill.

The canal comes to a sudden end at the ivy-covered Barrack Hill Tunnel, where a screen of mesh confronts the boater and the noise of water can be heard tumbling away into the darkness. The canal used to serve **Newport** Docks but sections were closed in 1879 and 1930, making any further restoration extremely difficult. Newport was new in the 14th Century. In 1796, the year the canal opened, the town of just 750 people shipped 3,500t of coal, increasing to 150,000t in 1809 and becoming the third coal port in Britain by 1830.

There are plans to make the Bettws Brook navigable with two locks to reach Crindau, where there will be a canal boat and nautical marina, possibly by 2015.

Distance
66km from Brecon to Newport

Navigation Authorities
Torfaen Council/ Newport City Council

Canal Society
Monmouthshire Brecon & Abergavenny Canal Trust www.mon-brec-canal-trust.org.uk

OS 1:50,000 Sheets
160 Brecon Beacons
161 Black Mountains
171 Cardiff & Newport

24 Monmouthshire & Brecon Canal, Crumlin Branch

One of the best lock flights in Britain

The Valleys can only be those steep valleys in south Wales which run south-eastwards, regardless of the geology, formed by previously overlying rocks, possibly Mesozoic. Most importantly, they run across the formerly rich Welsh coalfield, resulting in a grim industrial setting, with the narrow valley floors being given over to mining while terraces of houses cling to the sides of the valleys. Only as the mines have been closed in recent years has the black begun to be replaced by green as nature is allowed to return.

Thomas Dadford Jr was appointed engineer for the design of the Crumlin Branch of the Monmouthshire Canal, also known as the Risca Canal, which ran down Ebbw Vale from Crumlin. It crosses Gwent in a south-easterly direction to meet the main line of the canal on the northern edge of Newport, its purpose being to transport coal and iron to Newport docks. Thus, it was not connected to the main canal network in Britain. It was not completed until 1798 as priority was given to connecting up with tramways from the mines rather than on building the branch itself. While the financial importance of the tramways was a significant realisation, it was not appreciated that these tramways could easily be connected together and the canal bypassed. The canal carried no significant freight after 1870 and in 1880 it was bought by the Great Western Railway with the usual tale of neglect, being closed by Act in 1949. It was partly filled in during 1962 yet much of it remains in water and it is planned to preserve from Mill Street to the M4, with water but not locks, as an amenity. The change of name to Monmouthshire & Brecon Canal came relatively recently as a result of work on the main line, fittings on the canal still bearing only the Monmouthshire Canal name. Restoration has begun at two points.

Over the greater part of its length, the current position of the branch is very clearcut. It either has deep water or it is dry. For the most part, the obstructions are easily portaged.

From **Crumlin** to Newbridge the canal has been filled but at Newbridge, by the end of the A472, the canal makes a false start with 300m alongside a rugby field. On the west side, the towpath is of black slate chippings. The far side resembles a swamp where dead trees stand in water that is clearly over its original bank. The trees bear number boards at close intervals. The water ends as suddenly as it began.

It begins again equally abruptly 3km south at Pontywaun. Restoration has been started and a slipway built. The lack of space results in transport routes tripping over each other in the confined location of the valley side. Cwmcarn Forest Drive is an 11km route through the 16km^2 Ebbw Forest, with picnic and barbecue sites and views over eight Welsh and English counties.

With the backs of houses and small gardens squashed on the uphill east side, the canal is soon joined by a freight railway line on the right. The B4591 squeezes over the canal and railway and the uphill bank becomes free of housing at **Crosskeys**, a field of horses clinging to the mountainside between woods. The canal has lilies and reeds which soften the edges without obstructing passage. The edges are also protected by a flexible lattice material which breaks up any wash and would allow a swimmer to climb out but, being a soft rubbery material, does not mark boats.

Various angling club notices ban everything from boats and swimming to litter although the latter suggests that the angling officials do not carry a great deal of authority as bags of rubbish and beer cans detract from

The isolated section at Newbridge is more like a swamp with its dead trees.

The canal at Crosskeys, surrounded by vegetation but with clear passage on the water.

what is otherwise a beautiful stretch of canal, well kept and running through deciduous woods of willows, alders, birches and, predominantly, sycamores. Owners Caerphilly County Borough Council allow boats. Not satisfied with the trees, a squirrel runs along a telegraph line above the towpath. Interestingly, the litter problem improves later in the more built-up area.

On the far side of the valley, a grey scar on the side of Mynydd y Lan draws attention to a disused quarry above the confluence of the Sirhowy with the Ebbw. At intervals there are striking views across the valley and down onto traffic on the dual carriageway that dominates the valley bottom. A transmission aerial tops Mynydd Machen, below which runs the Rhymney Valley Ridgeway Footpath. At **Risca**, the valley cuts through the rim of the Mountain Limestone basin containing the coal measures and the valley and its various transport routes become further constricted.

The water comes to an abrupt end under a bridge, its reappearance being far from obvious 200m to the southeast on the far side of the steeply climbing road, opposite the foot of a 600m long quarry.

The canal winds along the hillside, the drop being indicated by the view down onto a church spire at one of the breaks in the canalside hedgerow.

A portage is necessary past a bridge at Pearhiw but it is worth getting out again at the Prince of Wales, if for no other reason than to look at the view down onto Pontymister and the roofs of Risca or up the valley to landscaping of the valley side. The following bridge, among newer housing, also needs to be portaged; the pipes under the road look almost large enough to get through with a small boat but there is a grill on the downstream end. This next reach has a dense growth of weed but it ends at another blockage by a playground and an Air Training Corps hut. Another short reach leads to a break in a section of embankment at Ty-Sign, where an aqueduct seems to have been removed. The portage down and up again passes an Indian takeaway. Massala Junction is in an area of no man's land at the centre of a housing estate, where a new slipway has been installed.

Normal service is quickly resumed and the canal continues its wooded way for over a kilometre until it reaches an unexpected heavy growth of reedmace. The required portage is only for about 100m and is well worthwhile for giving the last extensive view over the hedge, down into the valley, before the canal begins to move away from the Ebbw River.

Beech trees shelter the canal and provide a habitat for magpie, kingfisher and wagtail.

The pipes under the next road by a golf course look the most tempting yet but lumps of masonry in the water prevent a hole in one. Little by little, the water depth reduces between the neatly trimmed banks. Behind a row of houses a clay stank has been built across the canal to provide access between Cefn and Rogerstone.

From the next bridge is a section which is relatively dry and so needs to be portaged but water returns as the B4591 comes alongside for the last time, accompanied by the mouthwatering smell of fresh bread wafting from a bakery close by.

A short paddle reaches the top lock at Cefn. Below it is a large basin in water, by which is the 14 Locks Canal Centre, a facility with tea rooms and conference centre, opened in 2008 and run by the Monmouthshire, Brecon & Abergavenny Canal Trust, with a display on the canal and its part in the Industrial Revolution in south Wales. There is a parking area, picnic site, toilets and a panoramic view for several kilometres to the east as far as Brynglas.

To the north-east are two reservoirs, which were built to supply not the canal but **Newport**. James Simpson built the 10m high earth dam with its clay puddle core

The Sirhowy valley joins Ebbw Vale, seen over the towpath. Above is Mynydd y Lan with its disused quarry.

Beech trees shroud the canal at Rogerstone as it hugs the side of the valley.

Mynedd Machen, seen from Crosskeys, makes a substantial backdrop to the canal.

Looking up the line of the canal from Rogerstone towards Mynydd Machen and Mynydd y Lan.

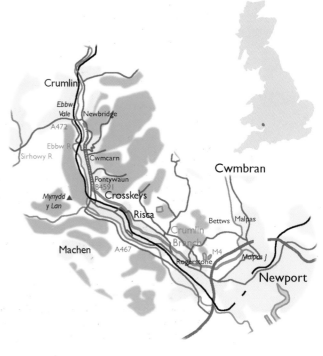

Two locks in the Cefn flight or Little Switzerland with a side pond on the right.

The 14 Locks Canal Centre. From the car park there is a panoramic view past the trees on the far side of the basin.

The traffic noise steadily increases down towards the M4 and is to remain for the rest of the route. Water is absent from all but one pound of the lock flight and the line of the canal under the motorway is capped with a continuous full width concrete cover. This bridge is just north of Junction 27 and from here the motorway curves round to follow the canal along the foot of a wooded hill. The hill is topped by a viewpoint at Ridgeway but is scenic even at its foot, with extensive views to the north past Bettws.

The canal soon passes under a footbridge, beneath which is a barrier with a drop of some 200mm, another lock following soon after. The five Allt-yr-yn locks are being restored. The motorway comes alongside again. Gradually, it begins to lift onto a low viaduct, clearing the main line of the Monmouthshire & Brecon Canal, which emerges from the left at Malpas Junction. The viaduct rises to Junction 26 and then to the twin Brynglas motorway tunnels, hidden from the canal.

Distance
14km from Newbridge to Malpas Junction

Canal Society
Monmouthshire Brecon & Abergavenny Canal Trust www.mon-brec-canal-trust.org.uk, Islwyn Canal Association www.islwyncanal association.com

OS 1:50,000 Sheet
171 Cardiff & Newport

and stone pitching on the upstream face to hold Ynys-y-fro Reservoir of 323,000m^3, to supply the 19,000 population of Newport. A second dam, which now carries a road, was built upstream in 1881 to increase capacity and trap silt that discoloured the water.

It is not just the view that justifies having an open lock centre on a closed canal. The flight of locks, at the end of a dredged reach used by the trip boat, drop the canal 51m in 800m; it is the most spectacular flight in south Wales and one of the best in Britain. The top lock has been restored, the next four will follow in 2008/9 and the rest should be completed by 2018. The wooded route is used as part of a series of signposted walks through the trees. The lock chambers, in good condition if a little overgrown, are very deep. Side ponds to save water have been cut into the hillside between them.

Halfway down the hill the canal emerges from the wood to give an extensive view along the M4 to Brynglas.

Looking northwards across the M4 to the route of the main line of the Monmouthshire & Brecon Canal, from the foot of Barrack Hill on the edge of Newport.

25 Neath Canal

The Neath Canal was opened in 1795. It runs southwards down the Vale of Neath, across Neath Port Talbot. It follows the River Neath, itself following a straight line of crushed rock between two large blocks of mountain. For the first half of its route it passes through a steep valley with woods up each side, including oaks and other deciduous trees.

Designed by the Dadfords, it was intended to use the course of the river, formerly navigable for ships, but never did so. It was expensive to build but eventually proved a financial success through the 19th Century, carrying timber, coal, limestone, silica, copper, iron and cannon balls, traffic eventually ceasing in 1934. It survived as it is used for water supply. It has bream, carp, roach, tench and eels so it attracts anglers. It is privately owned and the towpath is not a right of way although there is a public right of navigation. When built, there were 19 locks.

The original terminus was in **Glyn-neath**. This has been lost but the dry channel is in good shape opposite the leisure centre at the end of the village. Below a lock chamber it comes into water, shallow with reedmace and quickly crossed by a low bridge at the entrance to Aber-pergwm House, sitting below Aber-pergwm Wood. There are coal measures and the name on the gate is that of a coal mine. On the Blaengwrach side of the River Neath are the remains of mining spoil.

The water only continues to the next corner before becoming shallow rapids festooned with brambles as it drops down through a couple more lock chambers and disappears into a small stone culvert under the B4242, the downgraded former Heads of the Valleys Road, which follows the canal to Aberdulais.

A bend in the road has been cut off to leave a lay-by. The water reappears in the resulting traffic island with a bridge under the lay-by. However, now it is deep, wide and fully navigable with a clear towpath. It quickly reaches Ysgwrfa Lock, which is restored. Many of the structures on the canal have interesting detailing. In this case, a mini cast-iron aqueduct carries water over the end of the lock chamber. A house brings a lawn and daffodils down to the lock. The towpath side is wooded with ramsons and celandines in the spring.

Maes-gwyn Lock follows, with a limekiln alongside for easy supply. The A465 Heads of the Valleys Road in its current form becomes relatively intrusive down to Aberdulais but it could have been worse. The plan had been to route it down the bed of the canal.

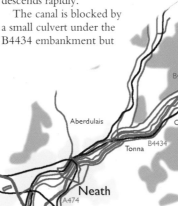

Canal bridge at Blaengwrach.

An aqueduct takes the canal over the Rheola Brook in a flurry of gorse to Rheola Lock, alongside which a large aluminium factory has been dismantled. It looked as if it had been burned out but it seems the black was just part of the normal working environment.

Life for workers along the Neath and Tennant Canals was described in Alexander Cordell's *Song of the Earth* and references to events from the book are quoted at intervals, including at a restored 18th-century lengthsman's cottage where the canal society moor their trip boat. The canal is forced hard against a substantial stone retaining wall, secured with 27m rock bolts. The left bank is open and almost parkland across to the river. Resolven lies at the foot of 383m Mynydd Resolven, from which the Clydach Brook descends rapidly.

The canal is blocked by a small culvert under the B4434 embankment but

Bringing down coal for export

Water first appears in the canal to a limited extent at Glyn-neath.

Blaengwrach Ysgwrfa Lock with an aqueduct below the bridge.

Limekilns by Maes-gwyn Lock.

Above the lock at Rheola.

a temporary terminus has been made with a slipway, a feature met at intervals on this canal, which is not linked to the general canal network.

The water continues from the other side of the culvert but only as far as Farmer's Arms Lock, just out of sight of the road. For the next 700m the canal is more silt than water, with frequent trees fallen across it. The towpath is little better. Then it disappears into a bramble patch and the footpath emerges onto a large area of crushed stone, obstructed on the far side by security fences and barking dogs. There is still much work to be done here. A better proposition at present is to walk along the B4242 to Abergarwed and go down the muddy footpath, facing across the valley to Melin Court Brook. The brook has the notable 24m Melincourt Waterfall over a Pennant sandstone ledge, accompanied by an early ironworks from 1708. It is necessary to turn right before reaching an arched metal bridge over the river, right again by an angling notice banning canoeing on the river and past a small sewage works to

Crugau Wood rises above Crugiau Lock.

The canal centre and trip boat at Resolven.

Ynysbwllog Aqueduct. The 34m steel trough gives the longest single-span aqueduct in Europe.

132

Ynysarwed Lock, where rabbits scamper about. The canal is now in water, not always full depth, for the rest of its journey.

Views across the valley to the south are extensive and up towards the cairns and former Carn Caca Roman camp, later towards a row of wind turbines looking over a ridge like so many Martian fighting machines. On the north side the hill rises steeply to the Sarn Helen Roman road.

Someone with a small boat at Ynys-dyfnant has some interesting carvings for those passing, including a large wooden frog squatting on a wall.

The local mine closed in 1903. When British Coal finally stopped pumping, there was a problem with inflowing orange iron pollution, as on a few other canals, and a treatment plant has been constructed on the bank.

The A465 finally crosses and an underpass has been built for the canal with a slipway next to it.

The original Ynysbwllog Aqueduct, a five-arch structure, collapsed in floods in 1979. In the past, floods sometimes flowed over the aqueduct. For three decades there was an argument about which takes precedence, the right of navigation or the Environment Agency's requirement to pass a 100-year flood on the river, during which time the canal water was passed over in three pipes. In 2008 a single-span steel trough was placed to clear the entire gap, the longest single span aqueduct in Europe. Angling club notices on the bank ban dogs and even their own members without rods.

At Clyne, Gitto and Whitworth Locks have been rebuilt and landscaped and Lock Machine also has been subject to restoration. The canal runs at the foot of a steep wooded bank climbing up to the Blaen-cwmbach Roman camp site. Meadows open up to the right, towards the river. The views must be impressive from the hospital at Tonna, on the ridge beyond the powerlines.

Aberdulais has its own share of views as the Dulais joins the River Neath. Aberdulais Falls have been generating power since 1584, when a copper smelting works was established, and the site was a corn and flour mill in the 17th Century, was a tinplate works from 1831 to 1890 and today has Europe's largest electricity generating waterwheel. Most visitors come for the view however, Turner among them in 1796. Penyscynor Wildlife Park, with its ski lift and alpine slide, is a further attraction these days.

Beyond the junction with the Tennant Canal, the Neath Canal winds round past a Calor Gas depot and the steep hillside becomes vertical cliff. The final lock is Lock House Lock, with restored workshops and stables.

The squat 13th-century church of St Illtud, whitewashed and standing next to the canal at the foot of the hill, is a striking start to **Neath**, site of the Roman fort of Nidum, and what should become a linear park at the end of the canal. As new housing builds up, so does the birdlife, not just the small ones with birdbox accommodation by the canal but also swans and herons.

A basin is located in front of the ruins of the Norman castle although Neath Museum is further back. The edge of Morrisons' supermarket car park, approached on foot over a lattice footbridge, allows nesting swans maximum scope for scowling at everybody.

The Fishguard to Paddington railway passes over and then there is a road bridge that has been lowered to water level and needs to be raised again. By now the River Neath has become tidal. The A474 passes over the canal. It had been planned to link the canal to the river at Melincryddan but the expected river improvements did not come so the canal was continued to Giant's Grave. While the views across the estuary might be expected, there are also extensive views to the south across low land at Penrhiwtyn, the edge of the canal so low that sandbags have been added in places to prevent

The junction to Aberdulais Basin at Tonna.

St Illtud's 13th-century church beside the canal.

Neath Castle and another trip boat.

The canal leaves Neath at Penrhiwtyn.

Horses on the hillside at the start of the Briton Ferry Canal.

Running above the tidal estuary of the River Neath.

Distance
19km from Glyn-neath to the Earl of Jersey's Grave & Briton Ferry Canal

Navigation Authority
Company of Proprietors of the Neath Canal Navigation

Canal Society
Neath & Tennant Canals Trust www.neath-tennant-canals.org.uk

OS 1:50,000 Sheets
*160 Brecon Beacons
170 Vale of Glamorgan*

flooding. Moorhens and mallards dabble among the lilies.

Another railway bridge crosses the estuary and powerlines fan out. A low drawbridge has been added to take a cycle path past a crane yard. Removal of a badly positioned piece of associated fence assists those trying to pass.

A stone roving bridge gives notice of arrival in Zoar or Giant's Grave, where a low bridge needs to be portaged past a house wall battlemented with white horses' heads. Horseboxes stand around and stables dot the slope ahead as curious horses come down to investigate. This final reach below Briton Ferry is the Earl of Jersey's Grave & Briton Ferry Canal or Giant's Grave & Briton Ferry Canal, a feeder for factories. In the distance are the A48 Neath Main Road and M4 viaducts.

A crane depot at Briton Ferry.

Rocky corner on the Briton Ferry Canal.

26 Tennant Canal

The alternative names of Neath & Swansea Junction Canal, Red Jacket Neath & Swansea Junction Canal, Neath & Swansea Red Jacket Junction Canal, Glan-y-Wern Canal and Mr Tennant's Canal point to the confusing history of this short waterway. It began in 1788 when Edward Elton dug a line across Crymlyn Bog to link Glan-y-Wern colliery with Redjacket and the River Neath, no mean feat. It was taken over in 1817 by George Tennant, who could see the potential of giving better loading facilities in Swansea for Neath Canal traffic. However, he failed to get a river crossing to meet the Neath Canal. Undaunted, he extended his canal right up to Aberdulais to make the connection, opened in 1824, and built new wharves at Port Tennant, by Swansea Docks. Commercial traffic finished in 1934.

The canal remains in the ownership of the Tennant family and there is no right of way on the towpath or on the water. However, most of the canal is in surprisingly good condition for a waterway that probably has only two boats on it and there are plans to link it through to the River Tawe above the tidal barrage to connect with the Swansea Canal, which should join up these various sections of waterway that are currently isolated from each other as well as from the rest of the canal network.

The canal survived as it was used for water supply. It has pike, rudd and tench, making it popular with anglers.

It runs south-west across the coal measures of **Neath** Port Talbot, following the River Neath and then the coast of Swansea Bay.

The Tennant Canal begins in striking fashion. It leaves the Neath Canal under a skewed towpath bridge with some intricate detailing and heads into the banana-shaped Aberdulais Basin, with a slipway on the site of a former drydock. From here the line is straight across the spectacular Aberdulais Aqueduct, a scheduled ancient monument. It is not just the structure itself that is impressive at 104m long with ten arches but its position across the River Neath just below a full width natural weir. In spate, the aqueduct can cover.

The line then passes through the canal's only remaining lock.

The route on foot is a little more complex. From the far end of the basin it is necessary to go left under

Brunel's railway, which was the downfall of the Neath Canal as it served the Vale of Neath, past the Railway public house, over the river, right back under the railway and left to join the water below the lock. In places there are panels giving information relating to Alexander Cordell's *Song of the Earth*, which was set around the two canals.

A recent boathouse is located where the water starts, water which is reasonably clear of weed, wide and deep enough. At the end of the first straight it passes below a

Aberdulais Aqueduct and Brunel's railway viaduct.

Overgrown bridge at Cadoxta-Juxta-Neath.

Aberdulais Basin as the Tennant Canal leaves the Neath.

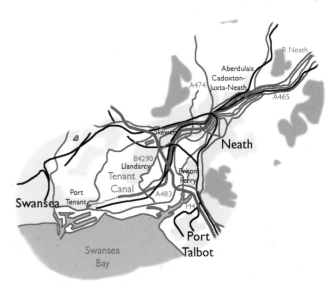

135

Passing the grand ruins of Neath Abbey.

More a wall than a bridge at Skewen.

the river having been a swing bridge until fixed in 1988. Railway tracks are now running on both sides of the canal. The M4 crosses and some apparatus on the bank draws attention to the former oil refinery at Llandarcy. Watercress grows in the canal as if it is a chalk stream. Railway and powerlines are each to cross and cross back.

The canal increasingly runs in marshy areas as it turns to cut between hilly outliers. The marsh and canal originally both continued to the River Neath along the Red Jacket Pill, which floods to the canal at high water. The river, sea and chemical works at Port Talbot all remain hidden from the canal, just as most of Neath did, and it is predominantly a marsh landscape, overlooked by high hills. Cairns have been built on Pant-y-Sais. Pant-y-Sais Fen wetlands Site of Special Scientic Interest occupies what was an old course of the River Neath. It consist primarily of reeds, a boardwalk having been laid out through them in a loop that meets the towpath. Greater spearwort, narrow buckler fern and royal fern are present and the reeds encourage such birds as reed buntings and reed and sedge warblers. It is one of the most important sites in south Wales for dragonflies, with 11 species being recorded. Limited wet woodland of willow and birch is present in places.

Pant-y-Sais Fen with Jersey Marine beyond.

The B4290 crosses Briton Ferry Road Bridge at Jersey Marine, the last reliable road access point. A boathouse is built in the canal and southwards is a massive stone tower.

The reeds hide the remains of a chapel as the hillside gives way to Crymlyn Bog. The Ford motor works are hidden to the south. Between them a pipe bridge crosses the canal with not just a single pipe but about thirty of them together, an oil industry legacy.

Having flowed wide and clear with clumps of weed below the surface, the width suddenly narrows just before the junction with the original Glan-y-Wern Canal. Disused since 1910, there is still a steady flow from it as it helps to drain the bog. The main line deteriorates quickly, first a short length of significant reeds and then a dense wall that extends all the way to Port Tenant, somewhere in the middle of which was the former Tir-Isaf Branch Canal to the north.

Beyond the carpet of straggly brambles which pass for a towpath on the south side is a muddy cycle route that soon becomes a surfaced single-lane road crossing from Neath Port Talbot into **Swansea**. It can be reached by vehicle if the gate is open at the Port Tennant end. If not, it is a further 1.6km walk, partly alongside a railway yard. At the end, the canal does reappear in a very derelict state with some walls and dumped rubbish before submerging into the embankment of the A483, across the road from Swansea Docks. Locks and wharves have been filled in and the red brick Vale of Neath Inn stands with its toughened glass windows all damaged.

The establishment of the final two kilometres through to the River Tawe will be difficult but canal enthusiasts have overcome greater obstacles and there has already been public funding for restoration work.

Both ends of the A465 bridge span are sealed.

Distance
13km from the Neath Canal to Port Tennant

Navigation Authority
Tennant family

Canal Society
Neath & Tennant Canals Trust
www.neath-tennant-canals.org.uk

OS 1:50,000 Sheets
159 Swansea & Gower
170 Vale of Glamorgan

complex interchange of the A465 Heads of the Valleys Road with a series of flyovers and a tunnel, claimed to be the shortest canal tunnel in Britain, all in concrete.

Quickly it is into an apparently rural area lined with reeds, an area for herons and mallards. Brambles trailing off bridge arches suggest little boat use. Only the Dulais valley railway separates Cadoxton-Juxta-Neath from the canal, to be joined by the Vale of Neath line which crosses to follow the right bank of the canal to Llandarcy. Immediately, the A465 sweeps over on a long low viaduct, the canal runs along the bank of the tidal River Neath, the Fishguard to Paddington railway crosses, the A465 crosses back and then the A474 crosses the other way.

A Tesco store stands back from the canal and gorse begins to appear around the aqueduct over the River Clydach. Adjacent to both the river and the canal is Neath Abbey, now lacking its roofing and quite a few of its walls yet still startling for its enormous size. Founded in 1130 by Lord Richard de Granville, it was of the Cistercian order.

Into cutting, the canal is crossed by a bridge at Skewen that is more like a wall with a mousehole in the bottom. The canal itself has a stone-lined section that is a scheduled ancient monument, the stone having been brought by tramroad and laid across quicksand.

The A465 passes over the canal for the last time and causes the only blockage except at the ends of the canal. A corrugated iron door has been fitted across each end of the bridge and it is necessary to go through the adjacent accommodation arch where water and mud lie calf-deep. Powerlines cross, as does a railway line from Briton Ferry to Llangennech, the adjacent bridge over

27 Stratford-upon-Avon Canal

The Stratford-upon-Avon Canal has often been considered as two different canals. The northern section was opened in 1802 to provide an alternative route to the Grand Union Canal into southern Birmingham by linking across to the Worcester & Birmingham Canal at King's Norton. The resulting canal also happens to be more attractive than the Grand Union Canal.

At King's Norton, a modern industrial estate has many vacant buildings and access to the Worcester & Birmingham Canal.

A turn off the Worcester & Birmingham brings a pair of fine guillotine stop gates, now rusted solid but originally designed to be dropped to save water in the event of a breach in either canal. The initial stretch is industrial with much graffiti but the numerous hawthorn bushes on the canal banks soon begin to block out the industry and the housing estates that follow. Factory effluent is pumped into the canal just before a swing bridge, the scene of protest cruises by Sir Peter Scott and L. T. C. Rolt when the bridge was fixed and supporters were trying to get the canal reopened.

The only tunnel is at Brandwood. Its 320m length is barred to unpowered craft. The walking route is over a trodden down chain-link fence and through a housing estate. The tunnel portals were fitted with Shakespeare plaques and niches for statues at a time when tourism was a thing of the future.

In **Birmingham**, the water is dark and dirty with numerous pieces of flotsam.

The Horse Shoes public house is adjacent to an arm that is approached under a brick arch bridge. This has been relieved of its parapet. Many of the older bridges on the canal are of red brick with surprisingly large arches but it is encouraging to see that even one or two of the newer concrete structures have lengths of rail at the corners to prevent towropes rubbing and have raised lines of brickwork on the towpath to allow horses to grip. Sadly, the towpath is in a poor state throughout and is only really used by boatowners at locks.

The drawbridge at Yardley.

In 1936 the Government tested a bridge at Yardley to destruction. The 150-year-old structure held 125t for 45 minutes before collapsing.

Beyond a stretch of high embankment, under which a road passes, there is a modern drawbridge adjacent to the equally modern Drawbridge public house.

Passage under the railway at Major's Green brings an end to the built-up area, not met again until Stratford.

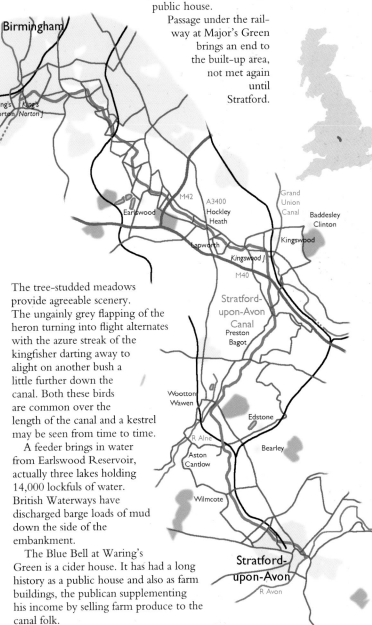

The tree-studded meadows provide agreeable scenery. The ungainly grey flapping of the heron turning into flight alternates with the azure streak of the kingfisher darting away to alight on another bush a little further down the canal. Both these birds are common over the length of the canal and a kestrel may be seen from time to time.

A feeder brings in water from Earlswood Reservoir, actually three lakes holding 14,000 lockfuls of water. British Waterways have discharged barge loads of mud down the side of the embankment.

The Blue Bell at Waring's Green is a cider house. It has had a long history as a public house and also as farm buildings, the publican supplementing his income by selling farm produce to the canal folk.

At Illshaw Heath comes a bridge for the M42.

The model for other canal restoration schemes

The guillotine gates at the head of the canal.

Just before the A3400 at Hockley Heath is a small canal arm that once served as a coal wharf, giving the adjacent Wharf public house its name.

At Lapworth is a most unusual bascule bridge that gives access to a farm. The end of the top pound is reached with the first of the 55 locks down to Stratford. Although some are in flights, especially at Lapworth, the spacing, typically, is at about 400m intervals.

A house between the first two locks has a lorry wing mirror mounted outside the window so that the occupant can see down the flight without going outside. Another house has railway relics, including a railway signal. Squeezed between the canal and the road is a timber-yard, so narrow that timber has to be pulled out of the stacks into the road.

A kilometre to the left of one of the bridges stands the timber-framed Packwood House. Built about 1560, it is opened by the National Trust and contains interesting tapestry, needlework and furniture. More notable, perhaps, is its fine Carolean garden and yew garden clipped to represent the Sermon on the Mount.

Two bridges over the Lapworth flight at Kingswood are the first of many split bridges on the canal. These have slots in the centre to allow the towrope through without unhitching the horse and are built as pairs of cantilevered arms, the handrailing being in the form of a distinctive double cross on each side. The locks themselves have large side pounds between them, so large that some are used as moorings. The locks have small paddles, making them slow to operate. Just below the Boat public house and a nearby grocer's shop, one has an overflow which discharges a cascade 2m high and nearly as wide rather than having the usual discharge channel running through the undergrowth around the back of the lock.

Group of canal buildings by the first southern section lock.

Parting of the ways at Lapworth Junction. The southern section of the canal passes under the split bridge, typical of those on the canal, while another lock to the left of the house connects with the Grand Union Canal.

A large side pound discharges into a basin, from which lead the southern section of the canal and also a branch to the Grand Union Canal which lies just the other side of the railway. Despite the fact that the two run parallel for some distance, the Stratford-upon-Avon falls rapidly through a whole flight of locks while the Grand Union is level from Rotton Row to Hatton, a distance of some 14km.

The southern section, linking up with the River Avon at Stratford, was completed in 1816, 14 years after the northern section. Unusually, it was owned by the National Trust and its recovery from disuse was a personal triumph for David Hutchings. Volunteers, services and prison labour restored it to use in three years and it was reopened by the Queen Mother in 1964, a model for other canal restoration schemes and the first restoration led by volunteers.

Immediately, an attractive set of canal buildings face

One of the barrel-roof houses. The cut in the foreground reduces locking to the Grand Union Canal.

the first lock of the southern section while just below, on the other bank, stands the first of half a dozen houses unique to this canal. They have barrel roofs, each house being built like a single bridge arch with ends. A couple of them now act as canal gift shops.

Above the house is a cut to the short connection to the Grand Union Canal, added to prevent boats coming up from Stratford and heading for the Grand Union needing to lock up to Kingswood Junction and immediately down again to the connecting arm with the waste of time and two locks of water.

A house with high brick chimneys stands away to the right of the canal at Copt Green. Half-timbered brick buildings become more prominent. In places the canal is the newest item in the landscape.

A notice bans anglers because of their bank erosion. At Finwood a shop offers home-baked bread. The Fleur de Lys has a lawn which comes down to the canal, with lengths of birch log acting as seats at the tables.

Yarningdale Common has a short cast-iron aqueduct over a stream. Like the two other aqueducts to follow but unlike almost any others on the canal network, the towpath is at the level of the invert of the trough so that the horse had to climb down onto the aqueduct towpath and up off it at the other end. Because it leads to a lock, the canoeist must portage onto the towpath, one of the few occasions he will ever have to climb down out of his boat.

A footpath from the last lock for a while, at Preston Bagot, runs to All Saints' church, which has a Norman nave and details. The Warwickshire countryside now becomes more rolling. It impressed E. Temple Thurston, who wrote about it in his book *The Flower of Gloster*, about a narrowboat journey in the early years of the 20th Century.

A canal basin and boatyard at Wootton Wawen, formerly a temporary terminus, is served by the Navigation public house. A boatyard gantry stands in front of the reason for the delay in extending the canal, an aqueduct over the A3400. Simple railings stop the tow horse falling over the edge on one side but there is no such refinement on the other. It is possible to sit in the centre with only 100–200mm of freeboard separating the boater from the tops of lorries going underneath.

Aqueduct at Wootton Wawen.

Just visible over the trees down the road is the roof of the magnificent five-storey 18th-century brick watermill on the River Alne. Beyond it stands Wootton Hall and then St Peter's church, which dates from Saxon times. It is the oldest in the county and one of the oldest in England. It also boasts an 11th-century sanctuary, a Norman nave, a superb 14th-century east window and a lady chapel like a barn.

Up the hill, Austy Manor is seen and, beyond it, Austy Wood, the largest remaining piece of the Forest of Arden that once covered the whole area.

The 230m Edstone or Bearley aqueduct is the longest and most impressive on the canal, crossing the railway, a minor road and a stream. Once again, the views are magnificent with a tinge of drama.

From here the canal is more overgrown with bushes for a while, adding to the canal's attraction as there is always plenty of room for passage.

Wilmcote is much visited by tourists seeking the house of Mary Arden, Shakespeare's mother, and its attendant farming museum and cider mill. Less often noticed but worth looking at are the school and vicarage by Butterfield, dating from the mid 1840s, and the old fashioned railway station.

The Wilmcote Locks start the final drop down into Stratford, the tallest blocks of which now come into sight. Alongside, tree planting has been to disguise a rubbish tip at Burton Farm.

Stratford is approached the back way, past industrial estates and a gasworks. Narrowboats are moored at Western Cruisers, who invite all boats to use their shop and showers. A café stands on the A3400.

The next lock is hard up against a minor road bridge and so the downstream balance beam has been replaced by a metal framework which is at right angles to the normal line of the balance beam and has to be pulled to open the lock. The portage involves a tight S-bend under the bridge which might stop a large boat being carried through. A nearby grocery store calls for custom. Down the road to the right lies Stratford Motor Museum, covering cars, fashions, scenes and music from the 1920s and 30s.

To most people, however, Stratford is Shakespeare. His birthplace lies just beyond. Stratford has more historic buildings than any other town of comparable size in the country and must be visited.

Bollards carry the names of benefactors of the canal in the area which is now residential, some of it recent, as the canal wanders to the back of the bus station and a café before bursting out into the splendour of the Bancroft Gardens, set around the canal basin. On one side, the Bard sits on a pedestal, surrounded by flowerbeds and statues of his characters. Opposite is the Royal Shakespeare Theatre, built in 1926 in dubious cinema architecture to replace the 1879 one destroyed by fire. The Royal Shakespeare Company are in residence from April to December. The tourists are always in residence, especially in the summer, and all the facilities here face the gardens, from McDonalds to a chance to sample life in Shakespeare's day.

Significant tourist dates are April 23rd, Shakespeare's birthday, and October 12th, Mop Fair, the time when farm workers were traditionally hired.

The final lock down to the River Avon is a broad beam lock. Originally, the whole canal was to have been broad gauge but when the Worcester & Birmingham Canal and Warwickshire canals were constructed as narrow canals this idea was abandoned.

The footbridge over the canal is being replaced by a much longer one with viewing platform over the canal.

Downstream can be seen the 15th-century Holy Trinity church in which Shakespeare is buried. Upstream is a brick bridge built in 1823 to carry a horse-drawn tramway and, behind that, Sir Hugh Clopton's bridge, started in 1480 and finished ten years later, two years before he became Lord Mayor of London. The tower at the north end was the toll house and the bridge still carries the full weight of traffic on the A3400.

The Bard and Prince Hal in front of Stratford basin.

Distance
41km from King's Norton Junction to the River Avon

Navigation Authority
British Waterways

Canal Society
Stratford upon Avon Canal Society
www.stratfordcanal society.org.uk

OS 1:50,000 Sheets
139 Birmingham & Wolverhampton
151 Stratford-upon-Avon

28 Worcester & Birmingham Canal

The country's longest lock flight

Gas Street Basin in the centre of **Birmingham** might be considered the central point of the British canal network. This was certainly the case in 1791 when the Worcester & Birmingham Canal Act was passed. The Staffordshire & Worcestershire Canal Company obtained an injunction to prevent the Worcester & Birmingham coming within 2.1m of the Birming-ham Canal Navigations, resulting in the Worcester Bar, over which goods had to be manhandled in an effort to prevent loss of business and water to the new canal. When the Worcester & Birmingham was opened to traffic in 1815, a stop lock was opened through the bar and this provides a route through with a footbridge over the top.

The Worcester & Birmingham was acquired by the Sharpness New Docks Company in 1874, nationalised in 1948 and kept in use by Cadburys until 1964. Use by pleasure craft keeps it free of weed although the water is dark at the Birmingham end and muddy elsewhere when holiday traffic is active.

Opposite the Wharf office development are various adult premises with Central TV adjacent. The canal crosses Holliday Street on a cast-iron viaduct. It turns the right angle Salvage Turn in front of the former post office centre, now the Mailbox development of shops, restaurants and BBC studios, which are approached across the canal by a footbridge surrounded by cobbles in which narrow gauge rails are set. Opposite the Kinnarec Thai restaurant a three-storey building on the left, used to stable towing horses which were taken up ramps to the higher storeys, is being replaced by the 17-storey Cube, so complicated that the designers had to erect it themselves because no contractor would tackle it. Breach protection stop gates are passed at intervals. Birmingham University halls of residence occupy the Davenports brewery site. At Five Ways, the railway pulls alongside the canal. This is the Cross City Line with a University station opened in 1978. It runs beside the canal until beyond Bournville.

The canal passes some new houses and then enters a long section through Edgbaston, where it is secluded by trees and the city disappears. Someone's private retreat on the left bank is protected with much barbed wire but the quietness is little disturbed.

Ornate ironwork on the Holliday Street aqueduct.

Ivy lines the cutting leading into Edgbaston tunnel.

Bridges are of black brick and many have red-painted trapdoors at road level to allow fire hose access.

Edgbaston Tunnel is just 96m long and has a towpath. The railway goes through an adjacent bore and a heavy arched retaining wall flanks it at each end. The other bank comprises red sandstone and supports Birmingham University campus, with its brick clocktower landmark in the centre and silver, domed roofs. Surprisingly, the dirty canal water supports newts.

At Bournbrook, a renovated viaduct carries the railway across the canal. The smell of paint wafts from a car body repair shop on the left bank, hemmed in by a former builders' merchants' stockyard and an electricity substation, protected from a children's playground by a very high and substantial fence. Dudley Canal No. 2 line used to enter on the right but was abandoned after the 3.6km Lappal tunnel collapsed in 1917. A Sainsbury's store is about to be built and the Number 2 line restored to Harborne Lane.

At Bournville, in 1879, the Cadbury brothers set up their chocolate factory as a garden city on what was then a rural canalside site. This pioneer community now has 3,500 houses. Old wharves front the canal and a plank footbridge, lifted by a crank handle, stands sentinel on the left bank. A housing estate occupies Cadbury's former wharf, opposite the factory.

After another substation and timberyard, a modern industrial estate is passed through and there are views between the trees across the rooftops as the canal is at a high level here. A heavily protected wall at Lifford surrounds nothing more than a refuse and sewage

The Worcester Bar at the west end of Birmingham's Gas Street Basin. Broad Street tunnel is just visible beyond the stop lock.

works while graffiti on its surface shows a better grasp of spray cans than of the English language.

Just after the first brick arched bridge is a fine brick double-fronted canal toll house of 1796, facing the junction with the Stratford-upon-Avon Canal.

The octagonal spire on the 15th-century tower of St Nicholas' church at King's Norton is a prominent landmark.

Wast Hills tunnel is the major obstacle on the canal. At 2.3km long it is one of the longest still in use in Britain. The walking route involves going up a flight of steps on the right and then avoiding the modern housing estate of Hawksley and Birmingham University playing fields with their private notices. In horse-drawn days it boasted a steam tug and, later, a diesel one.

The tunnel emerges in deep cutting in open countryside, the city having been left behind. The banks are muddy. The scenery is now rolling fields and just occasionally reeds encroach into the canal on the right, the side away from the towpath.

At Arrowfield Top is the Hopwood House Inn. Barnt Green sees boats moored alongside Lower Bittell Reservoir, built to compensate mill owners for their loss of water after the construction of Upper Bittell Reservoir, which is a canal feeder. The reservoirs are popular with anglers and birdwatchers. There are few anglers on the canal.

Houses here have large gardens beside the canal, one with a duck house and wooden open canoe as lawn features, while flower tubs on top of the aqueduct over a road are painted in ornamental canal style.

Wheeley Farm overlooks the canal on top of its hill, which is steeper than most in this part of the country. The Crown is passed and the small Weighbridge offers real ale and food behind Alvechurch marina.

Shortwood tunnel is 560m long. The footpath over the top once followed the line of the canal almost exactly but the farmer has ploughed up half of it, resulting in people making a detour round his field to the wood at the bottom where the canal emerges. It is possible to see daylight dimly through this tunnel and the following one, the 520m Tardebigge tunnel, which is crossed by three roads including the dual carriageway A448. Standing on the central reservation while trying to keep a small boat parallel to the carriageway in the wind blast from passing vehicles is not one of Britain's safest boating experiences. A detour past the prison and Tardebigge public house and under the A448 is less dangerous.

Natural ventilation by the wind causes exhaust to rise silently from the downwind ends of tunnels, an eerie sight in the evening light. Tardebigge tunnel has moorings at both ends and points straight towards the spire of the church in Tardebigge. In this area the fruit orchards are surrounded by high fences, cranked at the top, of rusting barbed wire.

The land starts to drop away as the top lock of the Tardebigge flight is reached, marking the end of the 27km level pound from Birmingham. This flight is the longest in the country, dropping through 30 locks in the next 3.4km, with a further 28 to go down to the River Severn.

The top lock is one of the deepest narrow locks in the country with a fall of some 4m. It was originally built as

The top end of the Tardebigge flight of locks.

a lift lock but was converted to a conventional lock in 1815 after it was found not to work properly in its original guise.

A three-storey pumphouse by the second lock has stood empty for several years. The reservoir on the left bank is, at first, below the level of the canal and then above, giving an idea of the rate of fall of the canal.

It takes a narrowboat about four hours to lock through the first flight but there can be few pleasanter places to pass the time if the weather is right. Apart from the occasional lock cottage and distant farm, there is nothing to break the solitude of the countryside. The canal drops past hawthorn bushes and fields of hay and barley. Swallows drink from the canal while on the wing. From lock 42, the BBC's massive Henbrook transmission aerials come into sight.

Lock 33 has a split footbridge.

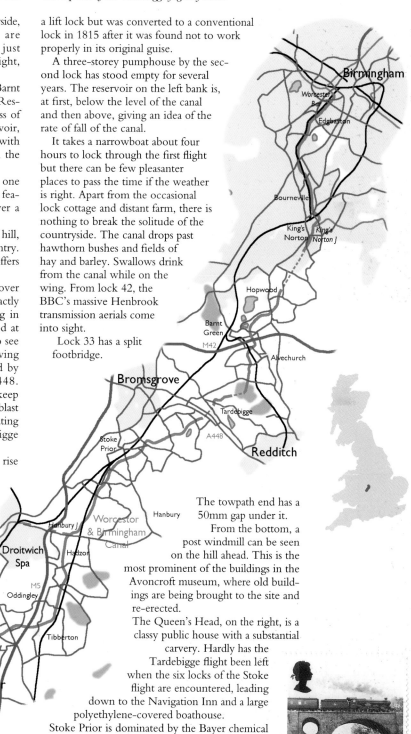

The towpath end has a 50mm gap under it.

From the bottom, a post windmill can be seen on the hill ahead. This is the most prominent of the buildings in the Avoncroft museum, where old buildings are being brought to the site and re-erected.

The Queen's Head, on the right, is a classy public house with a substantial carvery. Hardly has the Tardebigge flight been left when the six locks of the Stoke flight are encountered, leading down to the Navigation Inn and a large polyethylene-covered boathouse.

Stoke Prior is dominated by the Bayer chemical works. The canal runs through the middle between high walls and security fencing but the smells leave no doubt as to the activities that are being pursued here. Brine was pumped to the

The bridge at lock 50 on the Tardebigge flight.

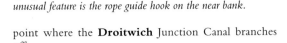

Split footbridge at lock 33 on the Tardebigge flight. Another unusual feature is the rope guide hook on the near bank.

Narrowboats working through the Tardebigge flight. The building is one of the lock keepers' cottages on the flight.

salt-producing factory here from 1828 and did much for canal trade.

The Boat & Railway, with a carvery instead of its former famous pickled eggs, draws attention to the high-speed trains running just west of the canal. A few doors along is a village hall with a dramatic diamond-patterned roof.

The Astwood flight brings another six locks and lawns down to the water's edge. The cottage at lock 18 is unusual in having a vegetable garden along the far bank of the lock in place of the conventional grassed area.

A footpath runs left from lock 17 to Hanbury Hall, a fine Wren style red brick mansion owned by the National Trust and open to the public.

Surprisingly few orchards are seen, considering the part of the country concerned, but one is passed on the left and it is followed on the right by low ground giving views right across to the Malvern Hills. The cream block of Hadzor Hall dominates the hill on the right.

The Eagle & Sun offers food on the line of the Roman Salt Way. On the other bank, a garish blue and yellow lighthouse marks the Hanbury marina at the point where the **Droitwich** Junction Canal branches off.

The 70m Dunhamstead tunnel has an irregular brick roof. Dunhamstead Wharf, beyond, boasts the Fir Tree Inn and garden and Forge Studios.

Hedgerows are generally fully grown to shelter the canal from wind and visual intrusion but a neat piece of hedging has been undertaken between the canal and the railway at Oddingley. On the other bank, a squat stone church overlooks a fine half timbered building, forming a very attractive group.

The canal has been of generous width for much of its length but, after Tibberton, it achieves river-like proportions and passes through a deep cutting as it approaches the M5 and the six locks of the Offerton flight, overlooked by the cream bulk of the house at Hindlip Park.

Increasing use is made of galvanised sheet piling, which gives the canal a well kept look. Tollandine Lock gives the first view of the northern outskirts of **Worcester**. After Blackpole Lock come another substation and a modern industrial area, followed by extensive playing fields and a sports centre on the right.

At lock 5, the towpath switches to the right bank. The surroundings become poorer in quality. A football ground is surrounded by old factory premises about to become a housing estate.

The Cavalier Tavern offers food and coffee with views to a squat, towerless brick church one way and an architecturally dramatic brick railway bridge over the canal the other. A children's playground with slide running down from the top of a cobbled mound adds a note of interest among the drab lines of old terraced houses. A final marina offers moorings near the city centre.

Blockhouse Lock gives a close view of Worcester cathedral. Kings Head Lock is alongside fine half-timbered tea rooms. The 15th-century Commandery was used by Cromwell during the Battle of Worcester.

The canal skirts almost unnoticed round most of Worcester and slinks into Diglis Basin the back way. Housing is steadily replacing industrial premises. Narrowboats and seagoing vessels are found together as the final two locks down to the River Severn are both broad locks.

Diglis Basin offers a telephone box and the Anchor, in an area through to the river which has been heavily redeveloped.

Distance
48km from the Birmingham Canal Navigations to the River Severn

Navigation Authority
British Waterways

Canal Society
Worcester & Birmingham Canal Society
www.wbcs.org.uk

OS 1:50,000 Sheets
139 Birmingham & Wolverhampton
150 Worcester & the Malverns

The rural setting at Tardebigge.

29 Staffordshire & Worcestershire Canal

Running southwards through the West Midlands, the Staffordshire & Worcestershire Canal was an early James Brindley canal, built 1768–72 to join the Trent, Mersey and Severn as part of the Grand Cross route which was also to have a link with the Thames. It was one of the first canals of the Industrial Revolution to be completed and was the main route north from Birmingham until the Shropshire Union Canal opened. The Worcester & Birmingham also took away some traffic later on a more direct line to the Severn but it remained a considerable commercial success and was one of our most prosperous canals. It was constructed as a narrow beam contour canal so views are constantly changing and it is frequently above the level of the surrounding fields. Despite its location, it is largely rural throughout and forms part of the Four Counties Ring as far as Autherley Junction so it is heavily used by narrowboats with the consequent disadvantages of murky water, chopped weed and occasional oil film.

The Staffordshire & Worcestershire Canal leaves the Trent & Mersey Canal at Great Haywood Junction, sandwiched between the River Trent and the Rugeley to Stoke-on-Trent railway. A farm shop has toilets and a stock of last minute food supplies. Anglo Welsh Narrowboats are in one of several narrowboat marinas down this canal.

The towpath bridge over the junction has an unusually wide bridge arch of 10.8m with two courses of bricks overlain by sandstone slabs. All bridges on the canal are clearly named and numbered and the aqueducts say which rivers are being crossed. Trent aqueduct in stone carries the canal over its river and another arch clears a mill tailrace as the canal starts up the valley of the River Sow. Great Haywood Wharf has mellow red-brick warehouses but the attention is drawn more to the tiny brick toll office on the other side of the canal with its iron-lattice windows, now used as a craft shop.

To the left of the canal lies Shugborough Park.

On the other side of the canal the large building with the long arched front is merely Tixall Farm but the four-storey ruins of a gatehouse and stables point to something grander. Tixall Hall and its replacement have both been demolished. The first house was built in 1580 by Sir Walter Aston with Mary, Queen of Scots being a pris-

The old toll office at Great Haywood, now a craft shop.

oner for a fortnight in 1586. Thomas Clifford added an 18th-century replacement alongside. The owner felt the view from his windows would be spoiled by the canal so an artificial lake had to be constructed on the line of the canal to placate him. Today Tixall Wide is surrounded by willows and reeds and has moorhens, mallards, herons, kingfishers and mink.

Tixall Lock is the first of the dozen that lift the canal 30m to its top level at Gailey. From the lock can be seen the largest engineering structure on the Trent Valley railway line, the 708m Shugborough Tunnel curving under Shugborough Park to follow the line of the canal.

A rural canal serving the early Industrial Revolution

Tixall Hall gatehouse, seen across Tixall Wide.

Shugborough Tunnel dives below Shugborough Park.

The unusually wide bridge arch at Great Haywood Junction.

In keeping with the locality, it has battlemented tower portals with additional towers at this end.

The art of grand living is not dead, however, and a modern house at Milford Bridge with its swimming pool, greenhouse and landscaped gardens sweeping down to the water cannot be matched along the rest of the canal.

Roving bridge and stoplogs at Milford.

The River Sow is crossed on a massive low stone-arched aqueduct while the next structure is a classic brick roving bridge at Milford. Close by is a bridge under the railway, leading to a farm shop and stained glass studio.

Plovers tumble over the fields sloping down from the railway. With a south-westerly wind the large sewage works across the valley from Weeping Cross will be passed largely unnoticed.

For smaller birds there are numerous nestboxes in a canalside garden at Baswich. A long line of chalets have their sun verandas facing out onto the canal, an idyll that is spoiled by the fact that they are on the south side of the canal so the verandas are almost permanently in shade.

There is a parting of the ways as the canal turns southwards to follow the valley of the River Penk. The railway crosses to head for the centre of **Stafford**, as did the former Stafford branch of the canal which continued up the River Sow until abandoned in the 1920s. Stafford's name comes from the Old English *staeth ford*, ford landing, when the waterway was shallower.

A nesting swan may be the largest gathering of white near what was the Baswich Salt Works. Beyond the Trumpet there is a salt well at Bickerscote. Lawns and a play area top the left bank of the canal on its way up to Deptmore Lock.

Two other major transport routes are to follow the canal's general line, the Crewe to Birmingham railway line and the M6, both now closing on the canal and adding a distant roar of traffic which is to come and go for a considerable distance. Junction 13 of the M6 is just before Acton Trussell. But for the traffic noise, the village retains its old charm with beehives beside the canal and the 15th-century church looking down across mown lawns to the Moat House with its bars and restaurants.

As the motorway closes on the canal after Shutt Hill Lock the noise is less intrusive than that from cars using the busy minor road running along the canal's right bank. If the wind was right for the last sewage works it will be wrong for this one although this one is rather smaller. Finally, the wood on the left bank is marked out with ropes, tapes and number boards as people roam around it with guns.

The mansion at Teddesley Park has been demolished but Park Gate Lock acts as a reminder. Directly above the lock is a boatyard.

The M6 passes over and the noise recedes sufficiently quickly for a heron to have taken up residence. Before Longford Lock someone has used the entire abutment of a bridge to write out a verse from a Motörhead number in crimson paint, not exactly decorative but slightly more thought provoking than the usual lines in bridge graffiti.

Traffic on the M6 races past Rodbaston Lock.

The approach to the Boat (a 1711 millhouse which became a public house in 1779 and sold the second strongest beer in England) and Penkridge and Filance Locks in the centre of **Penkridge** is past a line of chalets. Perhaps the most notable building on the west side of the village is the 12th-century church, restored in 1880 and with a Dutch wrought-iron screen.

A kingfisher streaks away up the pound beyond the Cross Keys. Beside the canal is a sign proudly proclaiming that the adjacent field is the site of the new Otherton Boat Haven but ducks dabbling in a large puddle at the edge after rain are the only indications of such activity as paint peels off the sign. In the distance are the buildings of a mine at Huntington and a transmission mast at Pye Green on Cannock Chase. The M6 returns and from Otherton Lock it runs along an embankment next to the canal for over a kilometre, high enough for a freight railway line to have passed over the canal and under the motorway. The agricultural college at Rodbaston tries to present a more rural atmosphere.

Rodbaston, Boggs, Brick Kiln and Gailey Locks take the canal away from the motorway and up to its 17km summit pound. The canal thus avoids Junction 12 of the motorway which is surrounded by the Gailey Pools reservoirs which not only have a heronry but large flocks of ruddy ducks introduced from North America and being seen as pests by ornithologists in this country.

The road from the motorway to Gailey wharf is the A5, built by the Romans as Watling Street, here just short of Water Eaton, the Roman burga of Pennocrucium, together with a legionary fortress, two forts and two temporary camps.

The bridge carrying the A5 over the canal is accompanied by a narrow arch built for tow horses. The top lock is overlooked by a magnificent castellated round house. The setting is completed by a yard of narrowboats and a hand-operated crane.

A large chemical complex seems totally out of place in the woods at Gravelly Way but its stainless steel flues and flare stacks are accompanied by some iron-banded brick chimneys that show that it has been here for many years and the rhythmic popping and hissing sounds might have had their ancestry in the Industrial Revolution.

The canal turns past a gravel pit at Four Ashes and heads towards the M6 again with a sharp change in direction at Hatherton Junction where the Hatherton Branch leaves. Formerly connecting to the Churchbridge Branch of the Birmingham Canal Navigations and then to the Cannock Extension Canal, it was largely abandoned in 1954 after subsidence near Cannock. A picnic area, Misty's and the Calf Heath Marina front the junction.

The Staffordshire & Worcestershire Canal moves away from the M6 for the last time, crossing a tributary of the Penk and passing a moat, to be followed by another moat among the network of powerlines on the way to Slade Heath.

At Slade Heath the railway line finally crosses and makes its departure. The A449, which has followed a similar line and is a Roman road here, passes close to the Anchor at Cross Green and then also crosses and departs in the direction of the railway. Not far away is the prison at Featherstone but Coven Heath has a name which suggests the activities of witches. The problem today is a powerful jet of sewage works effluent issuing across the canal to a point where the sheet piling has collapsed. Whether this is the result of scour or of unsuspecting narrowboats being pushed onto it is unclear but it would be worth avoiding any craft which look likely to be deflected here.

The canal now passes under Junction 2 of the M54 and into a progressively more built-up section of **Wolverhampton** at Fordhouses. Although the worst excesses of the built environment are avoided there is little to show of the *hean tune*, the Old English for high village, of which the lady of the manor in 985 was Wulfrun. Stacks of large tyres for earthmoving equipment are passed at a depot before the canal reaches the first of the sandstone cuttings that characterise much of its route to the south. This one is particularly narrow and lined with a rich growth of lichens as the canal leaves Staffordshire for the West Midlands.

School playing fields lined with poplar trees lead towards Autherley Junction with the Shropshire Union Canal. Massive tolls were imposed on the section from here to control traffic using the faster Shropshire Union Canal route to the Mersey, only being reduced when the Shropshire Union threatened to build a bypass route.

A large sewage works follows the right bank as far as the Shrewsbury to Wolverhampton railway bridge with a golf course on the other bank at Oxley.

Aldersley Junction quickly follows, the bottom end of the New Main Line of the Birmingham Canal which climbs steeply up the side of a horse race course from where the pounding of hooves on turf may be heard. A rifle range and a stadium complete the sporting facilities.

This is now part of the Stourport Ring. A Norman church tower with battlements is hidden by the trees at **Tettenhall** where the A41 crosses. The oblique railway bridge beyond it no longer carries trains but the line between Oxley and Castlecroft has become the Valley Park nature trail and cycle route.

Compton Lock brings the end of the summit pound. This is where Brindley began building his canal and, until recently rebuilt, the lock was reputed to have been the oldest lock in the West Midlands.

Compton Wharf Lock is the first of 31 dropping the canal 81m down to the River Severn. Between Wightwick Mill Lock and Wightwick Lock there are school playing fields on the left while trees and bushes on the right make a valiant effort to screen the housing. At the latter lock the cottage is boarded up but a house by the next road bridge provides contrast with a beautifully kept garden and a cheerful display of daffodils in the spring.

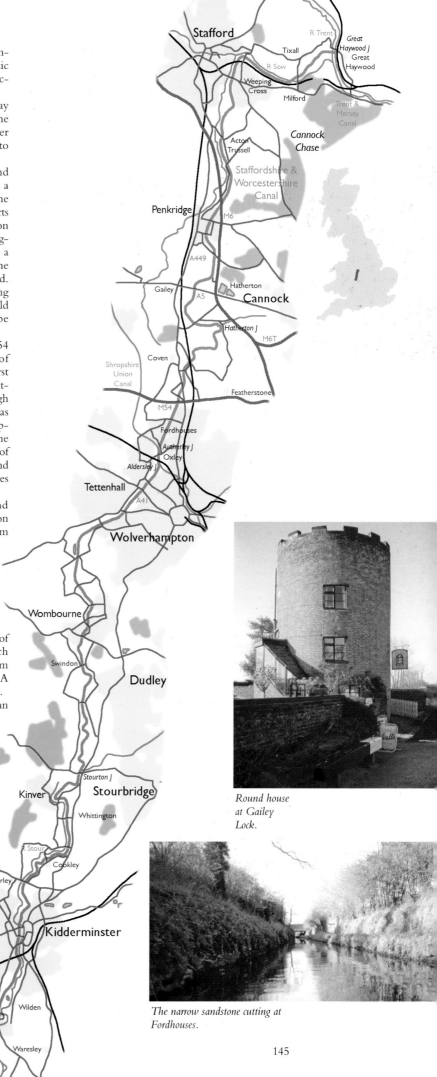

Round house at Gailey Lock.

The narrow sandstone cutting at Fordhouses.

145

An interesting bridge design near Aldersley Junction.

Bilston waterworks beyond Bratch toll office.

It looks across to the 29ha of Victorian/Edwardian gardens with topiary, yew hedges, terraces, ponds and woods that surround Wightwick Manor. Ornately timbered in Jacobean style, the house dates from 1887–93, was furnished in Arts & Crafts style and contains Kempe stained glass, original William Morris wallpaper, tapestries and fabrics, paintings, de Morgan ware and other pre-Raphaelite works of art and there are also stables, pottery, studio workshop and antiquarian bookshop.

The canal returns to Staffordshire and Wolverhampton is suddenly left behind, a predominantly rural situation which is to prevail for most of the rest of the canal's route, ridges of higher ground to the east providing views of open countryside while hiding the second largest conurbation in Britain. Gorse grows on the banks and there is reedmace, frequently chopped up by narrowboats. Herons quietly fish and Canada geese move onto the towpath from reservoirs hidden behind the hedge after the derelict Pool Hall.

Powerlines converge on an electrical substation which follows Dimmingsdale and Elstree locks. Many of the lock weirs on this canal are of an unusual design: a circular weir protected by a circular brick wall and a birdcage arrangement in the centre to keep out people and debris. The example at Awbridge Lock is typical of a number of these weirs.

Approach to the Bratch is past a cricket pitch with a notice warning about low flying cricket balls. Perhaps it is just a ploy to prevent boats from mooring here as there is little that passing boats can do to avoid the problem.

The three Bratch Locks were originally built as a staircase but extra gates were later added to give 1.5m gaps between the locks. A side pond curving away to the right tempts the unwary although, thinking about it, there is no way narrowboats can reach it. The correct route is down the flight and through the delightfully complex brick access stairways at the bottom lock. Watching over everything is a strategically sited 18th-century octagonal toll house. The area is one of the gems of the canal system, no less so for the restoration work carried out by British Waterways in 1994. The conical slate towers and turrets of Bilston waterworks top what looks like a castle beyond the bottom lock.

A red sandstone cliff on the right at Bumble Hole Lock is a typical feature which is to be seen on this canal with increasing regularity.

The Round Oak and the Waggon & Horses mark the two ends of the jig through Giggetty, the westward extension of **Wombourne**. There are also some large industrial units by the canal, giving way to a wildlife centre at Smestow before the two Botterham Locks.

From here the canal closely follows the Smestow Brook down to the River Stour.

At Marsh Lock the towpath switches to the left. The bridge carrying the towpath has had additional metalwork added to strengthen it. Originally it had two cantilevered halves with a gap down each side and across the walkway so a towrope could be passed through without unhitching the horse. Although this is no longer possible, the slots remain and the uprights have been

Circular weir at Awbridge Lock.

The full height of the unique Bratch toll house.

sawn almost to comb profiles by the wear from towropes over the years. Richard Thomas & Baldwins' 19th-century steelworks were built here.

Hinksford waterworks, below Hinksford Lock, have a red-brick Victorian pumping station of castle proportions.

Between Swindon and Greensforge locks the Romans sited two forts and six temporary camps and the canal crosses the line of a Roman road at Greensforge in the vicinity of the existing bridge and Navigation public house.

Moorings in Ashwood Basin occupy what was formerly a wharf owned by the National Coal Board.

Exquisitely laid out grounds around a house and nursery greenhouses include a conspicuous rockery and waterfall. The next lock is Hockley or Rocky which has a cave nearby in the rock.

What appears to be an isolated round weir opposite a section of wooded hillside well away from Gothersley Lock is the remains of Gothersley Roundhouse, rebuilt as a picnic area. Originally three-storeys high with a

Rocky Lock is carved out of the red sandstone.

Gothersley Roundhouse, now a picnic site. The balance beam is used as a seat.

A low, narrow, lock footbridge with no handrails.

The footbridge at Marsh Lock, formerly split. Below, a rare working barge enters the lock.

The church at Kinver occupies a highly conspicuous position.

rectangular extension, this tower was built in 1805 to house a wharfinger. The adjacent ironworks were closed about 1890 and the roundhouse blew down in a storm in 1991.

The wood is populated by pheasants and perhaps polecats.

A cave cut into the red cliff, the Devil's Den, can really only be approached from the canal but at a point where the water is too shallow for large craft to get close.

An aqueduct carries the canal over the River Stour, which the canal now follows for the rest of its course. From Stourton Junction the Stourbridge Canal climbs away to the left to reach higher up the Stour valley.

Stewponey Lock is overlooked by another octagonal toll office and the Stewponey & Foley Arms. The name is not a reference to a way of disposing of unwanted barge horses but came from Estepona in the Peninsula War.

Below Dunsley Hall is Dunsley Tunnel, Britain's shortest official canal tunnel at 23m. The towpath goes through but, these days, horses look down from the field over the entrance rather than pulling boats.

Red sandstone cliffs continue frequently and at one point they are so undercut that a series of brick piers have been built up out of the water to support an overhanging stratum.

A rope swing hangs out over the canal at Hyde Lock, site of an ironworks.

Before **Kinver** Lock there is another large pumping station. The dominant building for the village, however, is the church high up on a ridge above the canal. The ridge extends westwards and runs south-west as the gorse- and heather-covered Kinver Edge. On the west side of the village is Holy Austin Rock where a series of caves had brick fronts built and were occupied from the 16th Century by up to 12 families, some 80 people, until they were moved out by health officials in the 1960s. One is again occupied. Some of the older houses in the village have rock cellars. The area was noted for its witchcraft.

The pound south from here is exceptionally pretty

Weathered sandstone near Whittington, yet to be hidden by the year's horse chestnut leaves.

The canal runs along a steep hillside at Kinver.

with cottages set in the side of a high wooded ridge, arguably the most attractive section of the whole canal. The canal is shallow on the east side, as witnessed by local boys riding their bicycles along in the water.

Whittington Horse Bridge, the lowest and smallest on the canal, lost its northern parapet to a herd of cows crossing in 1953. The railings that replaced both parapets were removed and brick parapets restored in 1994.

The lock cottage at Whittington Lock is also attractive but the most interesting building is the timbered Whittington Inn, the former manor house of 1300, up the hill on the left. At one stage it was owned by the grandfather of Dick Whittington, the boy who went to London and became Lord Mayor after he heard the streets were paved with gold. The resident ghost is that of Lady Jane Grey, who lived here, and Queen Anne stayed in 1711. There is a 300m hidden tunnel to Whittington Hall and priestholes are built into it. A final odd building is the wooden shed on stilts on the embankment after the lock, originally fitted with wheels which have not been removed but remain attached, along with the supporting piles. Another rope swing hangs out over the canal for use by the local youth.

As the canal moves into Worcestershire the A449 briefly pulls alongside and there is significant traffic noise from this busy road.

Austcliffe Rock is marked with a series of reflectors and old oaks, being noted for the way it overhangs the canal on a corner, making life difficult for narrowboats.

Caravans and chalets follow before the canal reaches the 59m Cookley Tunnel, again with the towpath going through. A minor road passes over the tunnel, together with housing so that there is a sheer drop from the rear

wall of the end house down the high sandstone cliff to the water at the mouth of the tunnel. An ironworks beyond the tunnel formerly had its own spur.

Cookley or Debdale Lock has a large cave cut into the sandstone next to the lock chamber. Probably used as a store or keeper's shelter in the past, it has only a small doorway but could easily hold a couple of cars inside. The lock is followed by rock cutting with walls up to 6m high.

Alongside Wolverley Lock is the Lock, an inn with a fine collection of horse gear and harness and with a large model narrowboat on the wall facing the lock and a message extolling the virtues of life in the slow lane. The red Italianate church dates from the year the canal was opened while the grammar school was built in 1629. An open sandstone cavern was fitted with an iron gate as a pound for stray animals. More recent is a triple arch of steel pipelines from 1904, carrying water from the Elan

Houses directly above Cookley Tunnel.

Entrance to the large cave in the sandstone at Cookley Lock.

Impressive sandstone cutting near Wolverley.

Welcoming message on the wall of the Lock alongside Wolverley Lock.

Church and wharf crane at Kidderminster.

valley, and a modern octagonal conservatory on top of a house beside the canal. The lock itself is in a cramped position next to the road so that the balance beams on the lower gates have to be L-shaped.

From Wolverley Court Lock with its formerly split cantilevered footbridge, the canal works its way towards **Kidderminster**, the environment becoming less attractive as the tower blocks rise ahead. Industrial sites passed include the Stourvale Ironworks and a relatively recent warehouse surrounded by charred paper bales. The newest construction is a large Sainsburys development which may or may not be a good thing, the next kilometre of towpath and canal being littered with dozens of silt-coated shopping trolleys. Kidderminster, Old English for Cydela's monastery, is not a town that makes best use of its canal.

One of the better spots is just before Kidderminster Lock where a Perpendicular church tower stands on top of a bank of cultivated shrubs above a wharf with a hand-operated crane. When the wharf was removed there was a collapse of coffins and skeletons down the canal bank and it had to be reinstated. Close by is the spire of a late medieval red sandstone church. The footway is on the left and into a concrete tunnel leading down below the ring road. The River Stour is crossed over by the lock.

A large brick mill stands on the left bank. Close by is a museum that features the town's carpet industry and an art gallery with Brangwyn etchings. On the right bank are a timber yard and the Old Parkers Arms.

A common and unobtrusive way of repairing canal sides is with sacks of concrete used like sandbags. Here, the job has been done with blue plastic fertiliser sacks which are a complete eyesore. If they were used as an emergency repair then the appearance could be much improved by someone taking a box of matches and burning off the plastic rather than leaving it flapping in the wind.

The canal passes under the far side of the ring road. Despite the surrounding of industrial compounds, the kingfisher may still be seen. Caldwell Lock has the remains of a cottage that was built into the adjacent cliff until the 1960s.

Kidderminster's old brick mill buildings.

The Falling Sands Viaduct over the canal carries the Severn Valley Railway with its steam trains.

Oldington Lock is also known as Falling Sands from the instability of the red sand-

stone. The remains of a disused arm lead off on the left side, feeding through Pratt's Bridge Lock to Wilden Ironworks until 1949. There was a junction with the River Stour via Pratt's Wharf Side Lock. On the hill above Wilden is a conspicuous race scrambling circuit.

The Upper Basin at Stourport and the decorative clocktower.

Derelict canal buildings and the obligatory shopping trolley at Stourport.

Distance
37km from Great Haywood Junction to the River Severn

Navigation Authority
British Waterways

Canal Society
Staffordshire & Worcestershire Canal Society www.swcs.org.uk

OS 1:50,000 Sheets
127 Stafford & Telford
138 Kidderminster & Wyre Forest
139 Birmingham

At Upper Mitton the canal passes between the Bird in Hand and a graveyard. Although the name Lower Mitton, formerly a Saxon settlement, may be seen on a bridge, the village's name was changed to **Stourport-on-Severn** when the canal was built. One wharf has a series of arches highlighted with courses of blue bricks. The increase in canalside public houses such as the Rising Sun, Black Swan and Bell give indication that the terminus is being reached. York Street Lock is the final lock into the basin. Alongside the lock is a Gothic cottage of 1854 while upstream are interesting wharf buildings that deserve a better fate than their present situation of standing derelict with all their windows smashed.

Being on foot involves crossing the A451 which can be busy at times. If the gate on the right of the bridge is open, its private notice will not be visible. A roller in the bridge parapet reduced towing line wear. A hand operated wharf crane stands below.

The town was built by the canal company to serve what would become a 7ha complex of basins at the terminus, which was in use by 1771. When the Upper Basin was dug it was only 230mm deep in the centre but deeper at the outsides. In 1956 it was deepened right across. There are Georgian warehouses, one with an overhanging roof on cast-iron columns and one, originally a grain store, with a prominent clocktower of 1812 which gives its name to the Clock Basin, also known as the Middle Basin despite being on the upper level 9m above the Severn, clear of floodwater. Two of the three original drydocks in the Lower Basin survive and there are flights of both wide and narrow locks down to the river with a cast-iron footbridge crossing. An aerial ropeway led from the Furthermost Basin to a power station which was supplied with Cannock coal along the canal until the 1950s but was demolished in the 1980s, although the basin has been restored with new housing to be added around it. Buildings on the left at the bottom in the site of the gasworks are on what was formerly another basin. There are boatyards, chandlery and souvenirs in the former toll office and a hire cruiser base in the former offices of the canal company. The Grade II listed Tontine hotel, the most prominent building at the terminus, was built by the canal company but is being converted to housing; a tontine is a funding arrangement whereby the longest surviving shareholder receives all the assets. The area also makes carpets and has the largest chain works in Europe. Times are changing, though, and part of the site is now a funfair. Below the inland port is the River Severn, a much older and more major navigation route.

The River Severn above its usual level at the foot of the canal.

30 Shropshire Union Canal

The canals that were united to form the Shropshire Union Canal were very different in many respects. The Chester Canal was built between 1772 and 1779 and connected the Dee at Chester with Nantwich. The Wirral Line of the Ellesmere Canal followed in 1793–7 and was intended to run on past Wrexham to the Severn. They were built to broad gauge to take Mersey flats. A branch was built to Nantwich but the Trent & Mersey Canal refused a connection to the Middlewich Branch for fifty years, afraid that they would lose their northern trade to the Chester route. As a result, the Chester Canal and the Wirral Line became semi-derelict and it was to be the first canal to fail.

The Birmingham & Liverpool Junction Canal was built by Telford, one of his greatest canals, completed in 1835. It was built to narrow gauge as the Birmingham Canal Navigations were narrow, so there was no point in wasting money and water by building wide. In order to challenge the railways it was built with deep cuttings, high embankments, grouped locks and straight lines, so that it was shorter than the Trent & Mersey Canal and had thirty fewer locks. It was the last canal for Telford, who died six months before it opened. Problems caused by two landowners resulted in collapsing embankments, deteriorating health for one of our greatest engineers and technical problems that remain to this day. The link with Birmingham increased through traffic.

Steam tugs were introduced in 1842 to tow trains of barges but they proved uneconomic and horses were brought back. The Shropshire Union Canal was formed by amalgamation in 1845, the following year becoming the Shropshire Union Railways & Canal Company (although the company also used 'Railway'), with an Act to allow laying of railways along its canal beds. The canal was leased to the London & North Western Railway and later subsidised by them in order to reach traffic within Great Western Railway and Cambrian Railway territories. The Manchester Ship Canal added more business in 1894 and the route was profitable later than most. The SURCC had its own fleet of narrowboats until 1921. Flyboats ran a twice-weekly 30-hour service between Birmingham and Ellesmere Port. Commercial traffic, including oil products and metals, operated until the 1960s.

The canal runs north-west from Wolverhampton to the River Mersey across New Red Sandstone, through the 30km wide passage of the Cheshire Gate between the Shropshire and Staffordshire hills and then across the Cheshire Plain, with its dairy farming. The canal was needed in an area with few large rivers. It has long straights, sweeping curves, stone-arch bridges and extensive views. It is one of the prettiest canals, rural for 80km to Chester, often tree-lined. It is not crossed by a single B-class road. What was once a transport artery is now largely forgotten except by boaters. It has been proposed for World Heritage Site designation. As far as Barbridge Junction it forms part of the Four Counties Ring, the most popular canal cruising ring. The water is muddy and there is quite a lot of canal traffic.

It leaves the Staffordshire & Worcestershire Canal at Autherley Junction in **Wolverhampton**. The junction is reached by a track running past the large Barnhurst sewage works. The canal passes under a fine brick towpath bridge and runs through Autherley Stop Lock, which has only a 150mm fall but is intended to protect the Staffordshire & Worcestershire Canal's water at its 94m level. Unusually for a British Waterways canal, gate

balance beams are painted grey and white, the canal company colours of 1918–64, rather than black and white as elsewhere. The Shropshire Union Canal had its own good supply of water from the sewage works. There used to be a toll office here and there is now a canal shop for provisions. Autherley Boat Club are based in one of a pair of red-and-blue brick horse stables.

There are herons, mallards, coots, swans and greylag geese in Pendleford, a modern housing estate with mown lawns to the water. The houses on the right side are built on the former Wolverhampton Aerodrome, where the Bolton & Paul Defiant was manufactured.

There is nothing to indicate crossing the line of the Roman road from Swindon to Water Eaton. The following turnover bridge, accompanied by Wolverhampton Boat Club with its crane, is important not so much because it takes the canal from the West Midlands to Staffordshire but because it moves immediately from a built-up area to open country, a situation that lasts almost unbroken to Chester.

Near **Codsall**, the canal passes over the young River Penk, fed by another sewage works and passing a wood with a nature reserve and trail. The canal has extensive views to the north-west before narrowing into a rocky cutting, a taste of things to come. Upper Hattons bridge has a keystone with the date 1802. This end of the canal was built between 1826 and 1835 but a mason made an error in the 1960s.

The M54 crosses and the canal moves towards overhead powerlines with no more disturbance than the squawk of a pheasant among the larches. The high balustraded Avenue Bridge carries Lower Avenue to Chillington Hall, the Georgian house by Francis Smith and Sir John Soane in 1760s gardens by Capability

The roving bridge at Autherley Junction.

Wolverhampton Boat Club's premises.

The ornate Avenue Bridge leading to Chillington Hall.

Narrow section where the canal embankments have slipped.

financed. The village did have a real fort, built on Beacon Hill by the Romans to defend Watling Street.

There are caravans at Shutt Green and celandines, primroses and dandelions on the towpath, oaks and ashes helping to screen Belvide Reservoir, which feeds the canal and is noted for winter wildfowl.

Watling Street was the Roman road from Water Eaton to Wroxeter, now the A5. Here, one of his transport arteries crosses another on the Stretton Aqueduct with its cast-iron trough and railings, ornamental circular stone pillars in the corners and curved brick wingwalls, the 30° skew adding to the complexity.

The canal quickly leads into Lapley Wood Cutting, the wood including sycamores, hawthorns and ashes with violets in the spring and a wide variety of summer flowers. There has been slipping. A rock in the water has been painted with crocodile features.

The two Wheaton Aston Locks have a ruined lock cottage. A bomb dropped by a German Second World War aircraft onto a narrowboat loaded with aluminium on a moonlit night exploded less than 100m from one lock. A bench provided here by the Brownies may be an old church pew.

The 27km pound heading north begins by passing Wheaton Aston, with the Hartley Arms and a church rebuilt in 1857. The canal crosses the line of the Roman road from Water Eaton to Whitchurch at the start of an airfield that had a squadron of Oxfords and was one of the RAF's biggest training units during the Second World War. One runway was close to the canal and almost parallel with it, a 2.5km straight here. The canal perhaps was more visible in some conditions than the runway and could have been why an American Thunderbolt ended up in it on one occasion. The airfield was abandoned in 1947 and is now a pig farm.

A hall with moat stands between the airfield and Little Onn. Owls, kingfishers and bluebells are found in the woods beside the canal. High Onn Wharf, with its warehouse, follows St Edith's Well and then someone has improvised a drydock for building a cruising boat. The following corner has first-rate hedging beside the towpath to show that some traditional crafts have been retained in the countryside.

The unlined Cowley Tunnel is 74m long and has a towpath through, the entrances sometimes covered by a curtain of ivy. It was designed to be 640m long but a series of collapses during construction mean that most of it is now rock cutting, home to mosses, ferns and young trees.

After the Boat, which is located in former stables and has a curved wall, the A518 crosses at Gnosall Heath, next to the Navigation Inn. The following bridge used to carry the Stafford to Wellington railway, which was built by the SURCC. To the east it passed **Gnosall**, where the church of St Lawrence has some Norman details, mostly 13th-century. To the west it passed

Brown, home of the Gifford family since the reign of Henry II. Sir John Gifford assembled one of Britain's first zoos. In 1513 a leopard escaped and he shot it with a crossbow at Giffard's Cross as it was about to pounce on a woman and child. Moorhens are less fearsome wildlife these days. Of continuing concern are slips: kinks in the line of the edges of the canal show where the sides have slumped. Bridges

may have towline abrasions on the brickwork.

Brewood has a landscaped basin with the stump of a wharf crane. The canal looks down on the village. Conspicuous is the 13th-century red sandstone church of Sts Mary & Chad, its tower with spire and turrets. It has a 16th-century font and effigies and 17th-century monuments to the Giffords. It also has the grave of Colonel Carless, who helped Charles II escape. This is one of the venues where he was claimed to have hidden in an oak tree to escape Cromwell's troops. The 18th-century Gothic Speedwell Castle, among the Georgian houses in the square, is named after the racehorse that brought the winnings by which it was

Brewood seen from the canal.

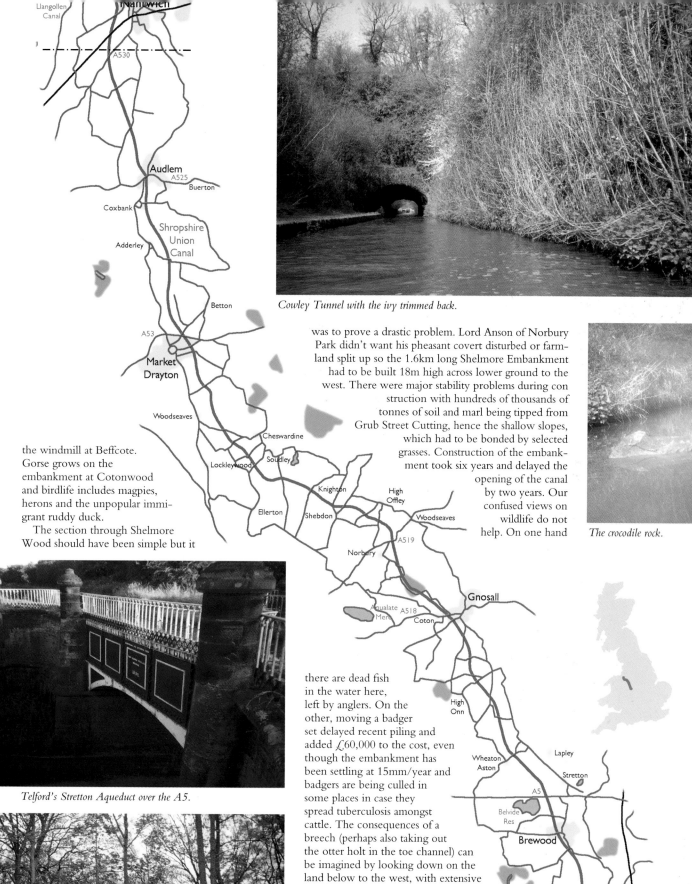

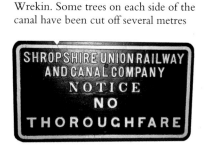

Cowley Tunnel with the ivy trimmed back.

The crocodile rock.

the windmill at Beffcote. Gorse grows on the embankment at Cotonwood and birdlife includes magpies, herons and the unpopular immigrant ruddy duck.

The section through Shelmore Wood should have been simple but it was to prove a drastic problem. Lord Anson of Norbury Park didn't want his pheasant covert disturbed or farmland split up so the 1.6km long Shelmore Embankment had to be built 18m high across lower ground to the west. There were major stability problems during construction with hundreds of thousands of tonnes of soil and marl being tipped from Grub Street Cutting, hence the shallow slopes, which had to be bonded by selected grasses. Construction of the embankment took six years and delayed the opening of the canal by two years. Our confused views on wildlife do not help. On one hand there are dead fish in the water here, left by anglers. On the other, moving a badger set delayed recent piling and added £60,000 to the cost, even though the embankment has been settling at 15mm/year and badgers are being culled in some places in case they spread tuberculosis amongst cattle. The consequences of a breech (perhaps also taking out the otter holt in the toe channel) can be imagined by looking down on the land below to the west, with extensive views across Aqualate Mere to the Wrekin. Some trees on each side of the canal have been cut off several metres

Telford's Stretton Aqueduct over the A5.

Strangely cut trees in Shelmore Wood.

SHROPSHIRE UNION RAILWAY
AND CANAL COMPANY
NOTICE
NO
THOROUGHFARE

What remains of the Newport Branch at Norbury Junction.

The Junction and basin at Norbury.

High Bridge with its telegraph pole.

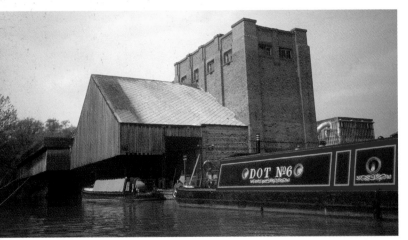

The conservation area around Cadbury's wharf at Knighton.

above the ground, the stumps being cut into strange crown shapes.

Norbury Junction was at the end of the Newport Branch of the Shrewsbury & Newport Canal, of which the first 200m remains in water. Restoration plans are being discussed. British Waterways have a maintenance yard here but the crane is not original. The Junction is a popular public house.

Manor farm and its moat face across the canal to Norbury.

Grub Street Cutting is 1.6km long and 27m deep, not to be confused with Woodseaves Cutting despite the proximity to the first village of Woodseaves. It is crossed at High Bridge by the A519. Telford built a number of very tall arches on this canal and this one is braced across, halfway up. Canals, like railways, became popular routes for lines of telegraph wires. The bracing has the top of a telegraph pole, now without wires, left standing on top, a surreal location for it. Another peculiarity of the cutting is a black creature like a monkey that has haunted it since a boatman was drowned in the 1800s.

North of the end of the cutting is High Offley with its large 15th-century church. The Anchor is an unspoilt boatman's public house. After the arched Shebdon Farm, the Wharf Inn has a crane at the start of Shebdon Great Bank, this one 1.6km long and 18m high, again with a history of having constantly shifted, partly collapsed and consistently leaked. Skylarks are heard.

Knighton is a conservation area around the wharf which served the 1911 factory where Cadburys received milk from dairy farms by canal, being one of the first users of motorised canal craft. The community has a distinction of having been absolved from paying rates by Charles II, a benefit that was not rescinded until 1990.

The canal passes into Shropshire, a county in which it spends little time despite its name. Knighton Reservoir supplied water to the canal.

Once again there are extensive views to the west over farmland. Ellerton Farm has an impressive dovecote.

At Little Soudley, a minor road crosses Fox Bridge. From here, with fine views westwards, the canal runs between oaks, ashes and sycamores with bluebells, primroses, celandines and dandelions in the spring to give colour. There are bream, chub, carp and roach in the canal. There is a moat at Goldstone Wharf although the Wharf Tavern is not until the Lockleywood to Cheswardine Goldstone road bridge.

Woodseaves Cutting is through friable rock, 2.4km long and 27m deep, damp at the best of times and still affected by frosts. Boats are warned to keep their speeds to 3km/h for their own safety. Ferns, creepers and nettles grow in the damp conditions, trees can close right across the canal and there are mallards, magpies, rooks, bats and a selection of bugs to feed them. On a canal noted for its tall bridges, one of those across the cutting is exceptionally high and narrow.

For 3km the canal is at the county boundary but is on the Staffordshire side as if trying to dissociate itself from its name. Beyond Old Springs Hall is a wharf with stone and brick buildings of 1837, having stone mullioned windows.

The five locks of the Tyrley flight include two set in a red sandstone cutting, not unique but unusual and attractive. Pine trees and gorse flank Peatswood. Opposite is Tyrley Castle. An aqueduct crosses Coal Brook and another passes over the River Tern, both as streams here.

Beginning with the Talbot and including a garden with a clever Bill & Ben feature, the wharf area at **Market Drayton** is particularly fine, forming the north-eastern extremity of the town. Many of the town's houses are black-and-white half-timbered buildings from the 17th and 18th Centuries after a great fire in 1641. It has two food claims to fame. It is said to be the gingerbread capital of the world, using a recipe containing rum. Using local trees, it produces everything involving damsons, damson jam, damson cheese relish, lamb and damson pie and damson gin.

Robert Clive of India was born here in 1725. Among his exploits were carving his initials in a school desk that

An exceptionally tall bridge in Woodseaves Cutting.

Tyrley top lock and a lack of enthusiasm for the EU.

The Tyrley lock flight enters its wooded cutting.

The mooring area in Market Drayton.

remains here, building a dam to flood a shop which refused to pay protection money and climbing the outside of St Mary's church tower and perching on a gargoyle. There were probably some residents who thought India was not too far away for him.

The church, built in the 14th Century, was founded in the 12th Century and has a fine Norman doorway. There has been a street market since 1245 and an annual fair, which used to be a horse fair with dealers coming from Wales. In the 19th Century the town prepared horsehair for furniture.

The town is left with the A53 crossing Lord's Bridge before Tunstall Hall and Victoria Wharf. The canal crosses the route of the North Staffordshire Railway's former line from Market Drayton to Stoke-on-Trent and then is followed by the route of the GWR's line from Wellington to Crewe, the canal having outlived both railways.

Hawthorn, blackthorn and cowslips are found beside the canal. At Betton Copse there may be a shrieking ghost near the roving bridge.

The five locks of the Adderley flight get increasingly close as they descend, dropping through a cutting lined with larches and pines. In the 1980s the lock keeper had the flight looking like an ornamental park. Adderley has its own ornamental park and a red sandstone church rebuilt in 1801 in neoclassical style. A canal breach in 2000 flooded the appropriately named Pool House, collapsed a road bridge and marooned a fire engine. Perhaps the adjacent motte stayed above it all.

After all too short a visit, the Shropshire Union Canal leaves Shropshire for Cheshire, the English county with a greater length of canals than any other. The fifteen locks of the **Audlem** flight, the Thick, drop the canal 28m. Two flyboats were excavated from Coxbank in 2003. The third lock of the flight is accompanied by a converted stable block.

The A525 crosses at Audlem, where there is a Bridge Inn and old buildings at Audlem Wharf. Audlem Mill is a craft centre and restaurant, originally a 1916 grain and animal foodstuffs mill, with a cantilevered gallery over the canal on the second storey for loading narrowboats.

The Talbot say their supply of bottles of pickled small girls is unreliable.

The wharf crane came from the local railway goods shed. The Shroppie Fly has the replica front half of a narrowboat as its bar. The town has timber-framed houses, a Butter Market, a market cross, which has become a stone-pillared oak-beamed bus shelter, and the Gothic 12th-century church of St James the Great, with its well

Scoured bridge protection in Market Drayton.

Audlem with the Shroppie Fly and Audlem Mill beyond. The railway crane is a recent import.

Nantwich Basin was the terminus of the Chester Canal. Telford's Birmingham & Liverpool Junction Canal comes in from the left, not the orginal plan.

preserved oak ceiling. There is a windmill to the east of the village at Buerton.

The bottom lock of the flight is by Moss Hall, where a stable is now a private house. Swans have nested here and there are a beehive and Canada geese to add to the wildlife. There are no more deep cuttings from here to Ellesmere Port, which may or may not be why the plant life in the canal improves. An embankment takes the canal onto the Moss Hill Aqueduct over the fledgling River Weaver. During the Second World War there were plans to make the river navigable from here for 100t barges and to replace the locks with barge lifts so that they could continue up the canal to the West Midlands.

The Wellington to Crewe railway line finally crossed at Coole Pilate. Mickley Hall had a moat but there are slightly more conspicuous defences at Hack House, a

Narrowboats at Hurleston junction wait for entry to the Llangollen Canal.

The former transhipment dock at Barbridge, now without its overhanging roof.

35m radio mast and a radar scanner. Mostly below the ground is the 1950s £32 million secret nuclear bunker that was prepared for Cold War use, 3,300m² of rooms with blast doors, Government headquarters, a decontamination room, a missile early warning system and a BBC studio. On show are nuclear weapons and military vehicles and there is a NAAFI style cafeteria.

Hack Green Locks have an old stable block for flyboat horses. This point is just 4km from the Llangollen Canal but the two lines do not meet for another 8km. A picnic table is provided beside the canal here.

The A530 crosses over the canal by Baddington Bank Farm. In 1995 British Waterways staff repairing a canal breach were troubled here by a nest of angry bees. Baddington railway bridge takes the Cardiff to Crewe line over Edlestone Bridge, next to another moat.

Nantwich, a salt town until 1856 because of the underlying rocks, means trading settlement of famous buildings. In 1583 there was a twenty-day blaze, following which there was a national collection and Elizabeth I had the town rebuilt with timber from Delamere Forest. The 14th-century St Mary's church in red sandstone was greatly restored in 1885 and has an octagonal tower, fine stone carvings and carved choirstalls, with grotesque oak misericords of knights, monks and dragons and other beasts. Nantwich Town Museum covers the Civil War and fire service and has a dairy. By the River Weaver is a Victorian Gothic savings bank. Old Piggott, a 19th-century coach driver, was grandfather of racing jockey Lester.

Telford had planned to join his line end on to the Nantwich spur of the Chester Canal on level ground but he reckoned without Dorfold Hall. This is one of the two best Jacobean houses in the county, built in 1616 for Ralph Wilbraham, with a drive and lodge added in 1862, landscaping by William Nesfield, including an avenue of ancient limes and a Spanish chestnut that is said to be a thousand-year-old relic of Delamere Forest. During the six-week Siege of Nantwich in 1644 it was occupied by each side. The Royalists camped in the grounds. The siege was lifted on January 25th so Nantwich people wear holly on Holly Holy Day, the nearest Saturday to this date.

The hall's owner demanded that the canal should not go too close and so the 800m Dorfold Bank had to be built with many stability problems. The last slip was in 2000. Despite the risk of collapse, the town has now been extended onto the embankment's flood plain. Concern was enhanced by the discovery in 1999 of 500 rabbit burrows in the embankment. A school has been built next to Telford's fine aqueduct of 1828, of similar design to his Stretton Aqueduct, carrying the embankment across the A534 Chester turnpike. Daffodils enhance the embankment's spring appearance.

Nantwich Basin, with its marina and restored quality cheese warehouses, was formerly the terminus of the branch from Hurleston. The narrow but direct Birmingham & Liverpool Junction Canal passes through an automatic stop gate and makes an oblique junction with the broad Chester Canal, now a vetch-edged contour canal. The Chester Canal could not afford to pay for the land so the landowners blocked the canal twice, once for a year, until rent was handed over.

St Mary's church in Acton was established by the monks of Combermere Abbey. It has the oldest tower in Cheshire, with Saxon stonework, rebuilt in 1757 after a partial collapse, rebuilding of the church body following in 1879. Another collapse is better remembered. Buried in the graveyard is Albert Hornby, who captained the English cricket team to an eight-run defeat by Australia, described in the press as the ashes of English cricket. So began the Ashes series, Albert had an honour he would probably have preferred to have foregone and many is the time since then that England would have been delighted to have been beaten by Australia by a mere eight runs. Much missionary work was carried out by the church at the canal basin, including marriages and funerals.

Other village properties include the Star Inn, with steps on the outside for mounting horses, and the site of Whitening House where bones were ground in an early fertiliser works. Beyond the village are a moat and a windmill.

The A51 crosses over by Henhull Wharf, used for roadstone. Other services crossing are powerlines and a pipe which carries Dee drinking water for mid Cheshire. The water comes from a reservoir at Hurleston Junction.

Barbridge was briefly a foundry site. It has the Old Barbridge Inn and then the Jolly Tar beyond Barbridge Aqueduct at Barbridge Junction. There was a transhipment dock with a roof spanning the canal where it narrows by a former stop lock. A long low building included a boatmen's mission church with a wooden cross on the end. The blind exit from the Middlewich Branch has resulted in many accidents over the years and these continue today. Much of what traffic has not gone to Llangollen now takes this branch as the Four Counties Ring leaves.

At the very point where the canal traffic leaves, road traffic arrives in the form of the A51 alongside until Calveley, the first time there has been significant traffic alongside on the whole canal, and the Crewe to Birkenhead railway is closing fast.

Although the short Wardle Canal is at the far end of the Middlewich Branch, Wardle is on the main line here. It boasts a radio telescope but this is well hidden, certainly not in the Jodrell Bank league. Forgetmenots on the bank are more conspicuous, with a small industrial estate above.

Calveley has warehouses that were built for Cheshire cheese and had a canal – railway transhipment wharf.

From Calveley, the canal follows the River Gowy, initially to Huxley, sheltered by oaks, ashes and hawthorns and disturbed only by the occasional train on the Crewe to Birkenhead railway. Coots are seen.

Beyond Alpraham, Bunbury Locks are met, a staircase of two in a wonderful heritage setting at Bunbury Wharf, resulting in the intricate Bunbury Shuffle if two descending boats pass two ascending boats in the middle of the flight. The stables have now become steel narrowboat workshops and there is a shop in the lock keeper's cottage, a pump in front, a tiny Edward VII letterbox round the back and a drydock on the other side of the canal. The place has tremendous atmosphere.

In the village of Bunbury is Bunbury Mill, a notable 19th-century working watermill, damaged in a storm in 1960 but restored and working. There is also the 14–15th-century church of St Boniface.

The railway crosses over immediately and is to follow the canal to Moston. Bluebells and marsh marigolds grow in the woods alongside the canal.

Tilstone Lock has a lock hut that consists of a single circular room with a bowed wooden front door to maintain the curve of the brickwork. At Tilstone Bank, the bank is used by snowdrops, primroses and bluebells and the mill is used by the Scouts. A prominent heavily timbered hotel on the south side of the valley is decorated with a range of flags that suggest American tourists might be a significant part of their customer base.

Beeston Stone Lock has another of the round lock huts. It has an overflow channel with a low arch and a high step. The hillside above is a blaze of gorse. Mink live here, perhaps with an eye on the mallards and black ducks.

Beeston Iron Lock was built on a new line in 1828, in the aftermath of a collapse in 1787, using diagonal iron plates as it is based on unstable sand, but the sides have bowed in. The lock cottage is set back from the current line of the canal.

The A49 crosses Beeston Brook Bridge at Tiverton, a village with a cattlemarket. On the other side of the railway are hidden wartime underground oil dumps, the number of entrances suggesting that it was a very large reservoir of fuel.

Reedmace and larches grow beside the canal and herons, swifts, curlews, owls and rabbits live around it. There is a deer farm above the canal.

Wharton's Lock, where the Sandstone Trail footpath crosses, shows Beeston Castle at its best. The castle was built from 1226 by Earl Ranulph of Chester after his return from the 5th Crusade, where he had been inspired by the impregnable hilltop strongholds in the Holy Land. Built on Beeston Rock, it has a precipice 150m to the Cheshire Plain with the best castle views in England, from where can be seen eight counties, the Welsh mountains, the Pennines and the Wrekin. Richard II was supposed to have hidden a vast treasure in the 110m well but it has not been found. The castle was taken by Simon de Montford in revolt against Henry II, then by the Roundheads, followed by the Cavaliers in the Civil War, including by eight men who

Crop circles: the farmers fight back? Not the Wardle radio telescope but a huge straw sculpture.

Former stables at Bunbury Locks

Shop and pump at Bunbury Bottom Lock.

One of the circular sheds at Tilston Bank.

Canalside dwellings in the warehouse at Egg Bridge, Waverton.

Riparian residents at Waverton.

Beeston Castle from Wharton's Lock.

appeared to be part of a large army at night. Earlier, the site had been used by Neolithic farmers, as a Bronze Age hill fort and by Iron Age warriors. The castle has a large outer bailey, D-plan towers, an inner bailey using cliffs on two sides and an early 13th-century curtain and gatehouse.

Peckforton Castle, behind on the Peckforton Hills, is a Victorian folly, built in 1850 for the 1st Lord Tollemache, strange that he should have chosen to build it so close to an obviously superior model.

By Bates Mill Bridge is the Shady Oak public house. Less shady is the towpath hedge where a long section of it has been systematically smashed with a machine to show that hedging skills are no longer available on this part of the canal, unlike further south. Blackberries grow on the undamaged side of the canal beyond the River Gowy Aqueduct.

Over the railway, Cheshire Farm makes cheese and one of the widest ranges of dairy ice cream flavours in the country, over 40, including both striped Tiger and Monkey Poo for Chester Zoo.

Crows' Nest Bridge has a pipeworks and a mast. It has also been a swans' nest bridge.

Waverton's church was much restored by the Victorians and has a Jacobean farm beside it. Sports activities range from a golf course next to the canal, to the Crocky Trail, a 2km fun and adventure course. The village had Waverton Mill and there was a warehouse beside Egg Bridge.

The canal sidesteps Rowton Moor where, in 1645, the last major battle of the Civil War was fought. Charles I watched from the city walls at Chester, 4km away, as his soldiers were beaten. The city withstood a two-year siege until starved out.

After the Cheshire Cat, a reach of weeping willows leads into Christleton, probably an early Christian settlement. The 14th-century almshouses are by John Scott and the 19th-century church has a 15th-century tower by Butterfield. There is a village green, a pump house and a pond that was a fertiliser marl pit. In the Civil War it was an outpost for the Royalists, who started a serious fire here. Christleton College of Law is in an 18th-century hall. Riparian owner David Wain played an important role in our inland waterways and set up the canal museum in Llangollen. Back in 1935, a local boatyard may have been the first hirers of canal pleasure craft in Britain. By Quarry Bridge is a 19th-century warehouse with a canopy.

The A55 **Chester** ring road crosses before Christleton Lock, locks now having take out and launching platforms for small craft. A blockage of the bywash channel here in 2000 resulted in flooding and a slip of the railway embankment. The railway passes under before Greenfield Lock, which has a peacock and an assortment of exotic hen breeds wandering about. The A41 crosses and the built-up part of Chester begins.

In front of a cricket pitch, Tarvin Road Lock has a further cylindrical tool shed. Tarvin Road, the A51, is a Roman road from York, beside which is the Bridge Inn.

Much redevelopment has been going on in this area. A major occupant has been the water treatment works. The Victorian water tower is a substantial structure beside the canal. Water has been pumped from the River Dee 200m away, treated since 1853 and stored in the 3.7m deep 1,200m^3 cast-iron tank on its 20m x 21m diameter tower. In 1889 the tank was jacked up 6m and more brickwork added to meet increased demand.

There are two Cow Lane Locks, Chemistry Lock and Hoole Lane Lock. Between them is Steam Mill, home to Club Globe, and the Old Harker's Arms, trendy city bases in old industrial buildings at Boughton. The Mill Hotel is tucked in beside the low, inner ring road, St Oswalds Way bridge with the Fortress on the opposite corner.

The canal uses the moat at the foot of the city walls. Prominent is the King Charles Tower, from which Charles I watched the battle of Rowton Moor, now a Civil War museum. One of the bridges crossing the red sandstone canyon by Northgate Street is a strange footbridge that reaches between two blank stone walls.

Can anywhere pack so much heritage into so small a space as the centre of Chester? It is based on the Roman street pattern, Deva being named after the Dee, and was a fortress completed in 79 for the 20th Legion by Agricola, the modern name coming from the Old English ceaster, a Roman fort. The 3km circuit of the walls is almost complete, the most complete in Britain, being broken only by the council offices, and is one of the best walled

Steam Mill in Chester.

The bridge to nowhere.

The King Charles Tower from where Charles I watched the battle of Rowton Moor.

the ghost of a Roman centurion who passes and returns after 20 minutes, walking through brick walls. Chester has more ghosts than any other English city. After 1,600 years, the changing of the Roman guard now takes place again at noon on summer Sundays.

There were Viking raids in the Dark Ages but the Vikings were driven out by Aethelflaeda by the 10th Century and the walls extended and strengthened. Amongst the wall's features are the Wishing Steps, which require a person to walk the walls and then run up, down and up again without drawing breath to have a wish come true.

The Saxon minster was rebuilt in the early 10th Century as a Benedictine monastery, the most complete medieval monastery complex in the country, parts remaining from 1092. It became the Norman St Werburgh cathedral from 1541 and has 11–18th-century architecture of Keuper sandstone. The 1380-carved choirstalls, with the Tree of Jesse and monastic buildings, are the most complete in England, notably the

Roman cities in Europe and the best surviving Roman wall fortress in northern Europe. The 1769 Eastgate is on the site of the Roman East Gate. Spud-u-like in Bridge Street is built over a Roman bath house. The half-excavated amphitheatre, seating about 7,000, the shrine of the Greek goddess Nemesis, is the largest stone amphitheatre in Britain and may have been used for gladiatorial combat. The Roman Gardens have stone and pillars from various sites and a reconstructed hypocaust. The Roman fortress houses the Dewa Roman Experience including a Roman galley. The George & Dragon Inn is on the site of a Roman cemetery and has

Part of Chester's Roman walls.

159

13th-century chapter house, magnificent refectory and cloisters. The massive gateway to the square is the venue for Mystery Plays, the original texts being the most complete in existence, and there is a copy of Handel's marked score for the *Messiah*, first rehearsed here in 1742. There is a 200-year-old cobweb picture, wonderful stained glass and a Renaissance-style font in black marble. There is also a shrine to St Werburgh, a Mercian princess with miraculous healing powers. In 2002 the cathedral launched Chester Pilgrim Ale. The cathedral is the most popular free entry destination in the UK.

The market dates from 1139. Katie's Tea Rooms are Britain's largest in one of the oldest buildings, even older than the market. At the time of the Conquest, brewers of poor ale were fined four shillings or ducked in the town pond. At noon on summer Tuesdays to Saturdays there is a town crier at High Cross, where there were bear baiting, stocks, a whipping post and sedan chairs on hire at various times.

Following the Conquest, Hugh Lupus d'Avranches, the cruel nephew of William the Conqueror, was made 1st Earl of Chester and built the first Chester weir to power the Dee mills. Chester Castle of 1069 was built on a Saxon fort site but only the gatehouse remains, the Agricola Tower, dating from the 13th Century, now with canal documents. The current building is in Greek Revival style, by Thomas Harrison, and contains the Cheshire Military Museum, with 300 years of Cheshire regiments plus Ypres trenches and the Victoria Cross of Todger Jones, who captured 120 Germans single handed.

The 12th-century church of St John the Baptist has fine Norman pillars and arcades plus the ruins of the choir and collapsed tower of 1573.

The Three Old Arches of 1200 is England's oldest shopfront. In a city of many half-timbered black-and-white buildings, the Rows are exceptional. These unique 13th-century shop galleries on two levels through half-timbered Tudor buildings are like being on a galleon.

The expression 'There's more than one yew-bow in Chester' made to jilted girls is a reference to the large number of local archers lost at Agincourt, Crécy and Poitiers, Chester archers being the best in the country. Not all girls were losers, however. A medieval mayor attempted to force his daughter to marry Luke de Taney. In a game with him she hit the ball over a wall, sent him to fetch it, escaped through the Pepper-gate while he was gone and eloped with a waiting Welsh knight and the gate has been locked ever since. Because there were so many border brawls, Henry IV banned Welshmen from carrying any weapons except eating knives and instituted a sunset curfew for them.

Bishop Lloyd's house of 400 years ago has a richly carved front with biblical scenes and animals. Chester Heritage Centre is in the old St Michael's church. St Mary's centre has exhibitions in a 15th-century church with a Tudor nave roof, two 17th-century effigies, medieval stained glass and a wall painting at the top of one of England's steepest streets.

Chester Visitor Centre, the biggest in Britain, is in a Grade II listed building with a history of Chester, including the Rows and a recreated 1850s Victorian street. The Grosvenor Museum has exhibits from Roman times to the present day, including a Roman graveyard, a period house, a Victorian schoolroom, Chester silver, Anglo-Saxon coins and natural history. The Chester Toy & Doll Museum has 5,000 items from 1830 on, including the biggest collection of Matchbox toys in Europe. There is a Broadcasting Museum and the Gothic-style Victorian town hall with its 49m tower containing the Chester Tapestry. The Victorian Eastgate jubilee clock of 1897 is claimed to be the second most photographed in the world after the Houses of Parliament.

There are festivals for summer music, jazz, blues, folk, literature, fringe, street processions and cheese rolling, an international horse show since 2002 and an international church music festival begun in 2003. The city is used for filming Channel 4' *Hollyoaks* series and was the birthplace of the BBC's Barnaby Bear.

The Northgate Locks were hewn out of solid sandstone. The Chester to Holyhead railway passes low over the flight so that boaters look down onto the tracks from above, with a background of Welsh hills, the Welsh border being less than 2km away at this point. It was the collapse of Stevenson's nearby wrought-iron-braced, cast-iron railway span over the Dee in 1847 that hastened the use of straightforward wrought-iron then steel-plate girders. The Dee is 200m ahead as the canal turns sharply right under a road bridge with some complex girders supporting a tight corner. The connection, straight ahead originally with two further locks, goes through a hairpin arrangement that adds about 500m to the journey down through three locks.

In the Middle Ages, Chester was the most important port in northern England, exporting cheese, candles and salt, but it gradually silted up. Although part of the Roman harbour wall remains, the Roodee is now part of the site of Chester races, Britain's oldest horse races, running since 1539. The Water Tower was a port defence structure, built in 1322.

The Northgate lock flight, the railway and the Welsh hills.

The Old Dee Bridge, built in the 14th Century by Henry de Snelleston, has had various rebuilds but was the only bridge across the Dee in Chester until the 19th Century. Thomas Harrison designed the Grosvenor Bridge in 1802 but delays meant it was not opened by Princess Victoria until 1832. At 61m it was the longest stone-arch bridge in the world. It remains the longest in Britain and the fourth longest in the world. Chester weir is Britain's oldest surviving mill dam and is where an Environment Agency report first recorded that canoes do not disturb fish. Chester rowing regatta, is the world's oldest, running from 1733.

Telford built two magnificent warehouses on the canal; sadly both were burnt down in recent years by arsonists. The one in Chester has been rebuilt as the Telford's Warehouse public house. By the junction is Tower Wharf, filled in during the 1950s but re-excavated in 2000. A development of housing, offices and restaurants has taken place around it. Harvest House was the canal company headquarters and is still the maintenance centre, with Georgian office, small warehouse, manual crane and neat cast-iron roving bridge. By the towpath is a plaque to Tom Rolt, the local man who used his Shropshire Union boat, *Cressy*, to inspire leisure use of the canals and who is the main figure responsible for the canal revival. Taylor's Yard is a dry dock in the fork with stable block, forge, former steam sawmill and workshops. There is the dilapidated Shropshire Union dockyard on the west side of the junction, the largest surviving. The restoration potential here is enormous. A century ago there were 448 houseboats on the canal in Chester. It is a very different canal today.

From here the former Chester Canal became the Ellesmere Canal or Wirral Line. There are few pleasure boats on the canal from this point, surprising as there is still significant rural scenery to come.

The Deva Aqueduct is crossed just before the A5480 crosses over, followed by the former Mickle Trafford to Connah's Quay railway bridge at Abbot's Meads. After Blacon there is open country, partially wooded, with swallows but hardly a building in sight for 6km, Chester completely hidden. The most garish intrusions are cyclist signposts that look like gaudy sculptures. There are powerlines. The A540 crosses. The Crewe to Birkenhead railway with Merseyrail, on a splendid 1839 red sandstone viaduct with 11 skewed arches, makes its final appearance between Moston and Mollington. The view north-west between the railway and the A41 is unbelievably rural. Somebody picked a prime spot for the county offices at Backford.

Even Chester Zoo, up Butler Hill at Upton Heath, is virtually invisible. The UK's largest and most popular zoo, it is one of the finest in the world, carrying out much conservation work with endangered species. Covering over 50ha, it has more than 520 species, over 7,000 animals including jaguars, black rhinos, red pandas, marmots, Britain's only komodo dragons and birds of paradise, a bat cave, the world's largest elephant house, buffy-headed capuchin monkeys and macaws in forest close to the canal. They have the biggest monkey house in Europe, the largest social group of chimpanzees in the UK, orchids, 80,000 plants including Roman and South American gardens, a monorail and a waterbus running on mini canals.

A cast-iron arch bridge crosses at Caughall and there is a straight reach with reedmace and burr reeds, herons and many moorhens. It even remains relatively quiet while passing through the middle of the freeflow interchange linking Junction 15 of the M56 with Junction 11 of the M53 Mid Wirral Motorway although Stoak must suffer with motorway alongside the hamlet on two sides. The canal is now following the River Gowy again. It passes under powerlines, slips under the A5117 New Stanney Bridge and is at one end of the vast Stanlow refinery complex started by Shell in 1924. Surprisingly, not too much is visible from the canal, even at night, although British Oxygen do have a large plant next to

Elegant footbridge at Tower Wharf.

Britain's largest surviving canal dockyard, sadly dilapidated but with scope for restoration.

Taylor's Yard with the main line, left, and the link to the River Dee on the right.

Glimpse of the tanks, chimneys and stacks of Stanlow.

British Oxygen's canalside plant.

from Netherpool by the Ellesmere Canal Company in 1796 to be the port for the Shropshire town with which they hoped to connect directly. A transhipment port, it is the finest canal port in England and is now the home of the three hectare National Waterways Museum, Ellesmere Port, Britain's premier canal museum with the world's largest collection of inland waterways craft, over 60 of them, including a 1912 tunnel tug and an Iron Age dugout. Other exhibits include dock workers' cottages of the 1840s and 1950s, a power hall and working pumphouse with steam, gas and diesel engines, painted boat ware, tools, plans, documents, stables, a working forge, a café and Ellesmere Port Pottery. It was opened in 1976. A popular attraction, the museum is visited by many groups of children during term time. An entry fee is likely to be requested if following the canal through the site.

The Whitby Locks are paired with broad and narrow locks. Halfway down the flight the towpath peters out and, if on foot, it is necessary to find a route to the right around various buildings. Telford's winged warehouse was to have been the site's central exhibit but was destroyed by an arsonist in 1970 ahead of a preservation order being served. Opposite is the Holiday Inn, voted their best worldwide. By it is the Waterways Conference & Leisure Centre.

There are basins to the left of the lock flight and then the canal turns right past a lighthouse of 1795. This was to guide shipping in from the River Mersey although, even at high water, that is at least 500m away across marsh now and the Manchester Ship Canal directly fronts the Shropshire Union Canal. It is closed to small boats and even narrowboats have to raft up in pairs. Nearby are a ro-ro ferry terminal, a container terminal and large ocean tankers and other craft passing down the Manchester Ship Canal. For many, over the decades, it has not been the end of the journey so much as the start of something much bigger.

the towpath. From the towpath or overbridges it is easier to see 141m Helsby Hill, Runcorn, Widnes, Fiddler's Ferry and the flare stacks of Stanlow although £75 million is being invested in clean fuel technology that should reduce some of the impact.

Little Stanney has the Blue Planet, the UK's largest aquarium with over 2,500 fish, more than a score of sand tiger sharks in Europe's largest collection, the world's longest moving underwater walkway and Europe's largest indoor rockpools. Close by, at Wolverham, is the Cheshire Oaks McArthur Glen designer outlet village, Europe's largest with 140 stores. 2003's local tourist board guide said it has 'Children's play areas, easy access to motorways.' There is also the Iwerks Extreme Screen Attraction 3D cinema, the first in Britain. The Coliseum is a 15-screen multiplex cinema with Megabowl, two nightclubs, restaurants and the UK's largest single storey book/music/film store.

Gradually everything gets squeezed together, powerlines crossing from side to side of the canal as the M53 moves in and then crosses. The Runcorn to Birkenhead railway goes under the motorway and over the canal to a fine Victorian station although a signal box has to be protected with mesh over its window.

The M53 crosses back at Junction 9 and ahead lie **Ellesmere Port** docks of 1833. The name was changed

Telford's lighthouse by the Manchester Ship Canal.

Distance
107km from Autherley Junction to the Manchester Ship Canal

Navigation Authority
British Waterways

Canal Society
Shropshire Union Canal Society www.shropshireunion.org.uk

OS 1:50,000 Sheets
117 Chester & Wrexham
118 Stoke-on-Trent & Macclesfield
127 Stafford & Telford

The paired Whitby Locks at Ellesmere Port, the final locks.

31 Llangollen Canal

The names do no more than reflect the confused history of a canal that bears little resemblance to what was originally planned. Built by Telford and Jessop, it was opened below Pont Cysyllte in 1805 and from Llangollen three years later, not that Llangollen featured in the original plan. It was to be a canal running north–south from Chester via Wrexham, Chirk, Ellesmere, Frankton and Weston to Shrewsbury and the River Severn, taking coal from Ruabon and bringing lime to numerous kilns. A change was made to a cheaper east–west link via Whitchurch to the Chester Canal, to become the main line of the Shropshire Union Canal. The plan to link the Mersey, Dee and Severn was not to happen as a direct line. Had the change been made sooner, the section above Frankton would not have been built and the feeder from the Dee would have been along the south side of the river. However, the magnificent Pontcysyllte Aqueduct was complete, leading up to what was intended to be a 4.2km tunnel to the north of Trevor. Cutting of this never began.

Signs show it as the Llangollen Canal and it has also been known as the Llangollen Branch of the Shropshire Union Canal and the Welsh Canal or Welsh Cut. The line from Carreghofa via Frankton to Hurleston was the Ellesmere Canal while the line from Llantysilio to Frankton was the Pontcysyllte Branch. Above Trevor it was intended as a navigable feeder, shallow, narrow and fast. The line has been nominated as a World Heritage Site from the falls to Chirk Bank.

To supply the feeder, Telford built Horseshoe Falls, the falls actually being J-shaped, despite the name. 140m long and 1.2m high, of masonry with a cast-iron capping, it provides a reliable source of 5,500m³/day from the River Dee, supplying water for domestic use at the far end of the canal at Hurleston Junction and canal water as far apart as the northern end of the Montgomeryshire Canal and the Manchester Ship Canal. A dam increased the depth of Llyn Tegid at Bala to enhance supplies.

A footpath to the weir leads down over the canal and along the path, which is covered by the Chain Bridge Hotel entrance.

Above the weir is Llantysilio Hall. Parts of the interior of the Victorian church at Llantysilio come from Valle Crucis Abbey. The glass in the south window is 16th-century. Buried here is Exuperius Pickering, who built a private canal from Trevor towards Cefn-mawr.

The canal is quickly crossed by the long stone bridge carrying the B5103, itself crossed by the Llangollen Steam Railway as it emerges from a tunnel at Berwyn station, built for the old Barmouth–Ruabon line of the Great Western Railway. Beside this is another of Telford's notable projects, the A5 Holyhead mail coach road. From below this a large stream discharges in a waterfall to the river. The area is a network of stone bridges. The other important bridge is the chain footbridge which has been in disrepair since the 1970s and is closed, waiting for £70,000 of restoration work to be undertaken.

The canal begins cut into the rock wall of the valley, overhung with trees (as it is for much of the Welsh section) and with the wall smothered in ivy. Canal and river follow a syncline and here the canal runs directly above the Serpent's Tail rapid.

Below the Horseshoe Pass is Valle Crucis Abbey. The Cistercian abbey was founded in 1201 and fell into disrepair after the Dissolution of 1535, when it was the second richest Cistercian monastery after Tintern. It has been partially restored. It has sculpted memorial slabs and includes parts of the church, 14th-century chapter house and cloister. It has a notable west front with carved doorway and rose window. The chapter house has a rib-vaulted roof near the original fishpond. The name comes from the valley of the cross. The 2.4m inscribed stone cross, just up the valley, was probably 9th-century and known as Eliseg's Pillar after the Prince of Powys who built Castell Dinas Bran and was killed in battle in 603. The top third of the cross has been lost, having been pulled down in the Civil War. Owen Glyndwr met the abbot one morning and remarked that he was up early. The abbot replied that it was Glyndwr who was early, by a century, perhaps a reference to the rise of the Tudors in 1485, to whom he was related. Glyndwr was not seen in public again after this meeting.

A tramway ran from the hills past the pillar to the canal at Pentrefelin, where the motor museum has now been joined by the canal museum from Llangollen.

Alders, sycamores and oaks hang over the canal while red clover, buttercups and foxgloves emerge from the grass and mallards search for smaller prey than the roach, dace, gudgeon, bream and pike that frequent these waters. The A542 passes over on what is almost a concrete tunnel and the canal comes alongside a field that has been the home of the International Musical Eisteddfod

Horseshoe Falls on the Dee, feeder for the canal.

The canal begins cut into ivy-covered rock wall above the Serpent's Tail.

163

Repaired section clinging to the hillside in Llangollen.

Another narrow section cut out of the rock.

since 1947, now with a permanent pavilion. The Eisteddfod is an international gathering of music, song, dance and poetry each July. The inaugural National Eisteddfod was also held in **Llangollen** and returns here from time to time but it changes its venue around Wales every year.

An aerial tops the high wooded Geraint's or Barber's Hill on the far side of the valley.

Beyond the cottage hospital comes the head of navigation for powered craft, where an offline basin has now been constructed. The water becomes continuously muddy as this is the most popular section of canal in Britain, visited by 4,000 boats each year, busier than it was in the days when it was carrying agricultural products and slate, business which ceased before the Second World War.

Fifty thousand people per season take horse-drawn barge trips from Llangollen Wharf, the longest running trips boats on the canal system, having been established in 1884 and currently operated in three boats pulled by three Welsh cobs. The wharf building, with its crane, is now a restaurant but was previously the canal museum, set up in 1972. A Tardis of a place, it always seemed bigger inside than out and won founder David Wain a Prince of Wales award.

There is a folk museum in the town and, on the far side of the valley, a museum in the magpie Tudor- and Gothic-styled Plas Newydd which was the home of the Ladies of Llangollen, Lady Eleanor Butler and the Honourable Sarah Ponsonby, Irish wits and entertainers whose guests included Sir Walter Scott, the Duke of Wellington, William Wordsworth, Edmund Burke and Josiah Wedgewood, in the period 1780–1831. In 1809 the house was bought by their maid, Mary Caryll, who subsequently left it to the Ladies.

The town is called after Collen ap Gwynnawg ap Clydawg ap Cowdra ap Caradog Freichfras ap Llyr Merim ap Yrth ap Cunedda Wledig, more a family tree than a name to roll off the tongue. St Collen's church has a splendid carved oak roof which may have benefited in the 13th Century from the skills of craftsmen working on Valle Crucis Abbey, the church being enlarged in 1865. Also enlarged was one of the Three Jewels of Wales, the bridge crossing the Dee just below Llangollen Town Fall. Begun as a stone packhorse bridge in 1282, it was rebuilt in sandstone about 1500 in its present style by John Trevor, Bishop of St Asaph, being lengthened to accommodate the railway in 1865 and widened to 6.1m in 1873 and 11m in 1969. The weir downstream was breached in the 1960s to reduce the depth at the bridge and prevent scour. Llangollen was an important staging post for the Holyhead mail coaches and now

Former packhorse bridge in Llangollen during a canoe slalom.

takes the form of a slated Victorian and Edwardian resort.

A large school is left on the uphill side as the canal moves away on a section rebuilt with a heavy concrete lining in 1985. This section is narrow and has a long line of moored narrowboats in the summer. It is at its best when the trees are in their autumn colours as, once again, it cuts into the grey rockface, high above the road. It passes the back of what was the Carter's seed factory, facing a combined Dr Who exhibition, the world's largest model railway exhibition and the Dapol model railway factory in the home of the original Wrenn and Hornby Dublo factories. Across the river is Gale's Wine Bar. When it was opened in the mid 1970s, the local paper protested it was not what Llangollen wanted as it would attract Hell's Angels and the like. With old church pews for seating and a harpist sometimes playing in the evening, since then it has done so much to improve the quality of the town that it has spawned a row of look-alikes.

The canal breaks clear of its rock cutting to reveal steep sheep pasture leading up to the 340m peak, topped by Castell Dinas Bran, the castle of the crow, the legendary castle of the Holy Grail. The home of Gryffudd ap Madoc, son of the founder of Valle Crucis Abbey, in the 12–13th Centuries, it was 88m x 43m and nearly impregnable but was burned in 1277 by the Welsh defenders in the war between Llewellyn ap Gryffudd and Edward I. On a clear day it has views of the peaks of Snowdonia. Behind is the grey carboniferous limestone escarpment of Creigiau Eglwyseg, packed with fossils and some shells dropped by the Germans heading for Liverpool until the RAF jammed their navigation beam.

The first of the lifting bridges with high Dutch-style beams, so typical of this canal, stands open before the escarpment.

After the A539 crosses over, the canal enters another narrow concrete-lined section. In 1945 the canal breached and washed a 15m deep channel beneath the adjacent railway, into which an 18-carriage mail and newspaper

Looking down on Llangollen from Castell Dinas Bran.

The first lifting bridge and the Creigiau Eglwyseg.

The other way, though, is one of the seven wonders of the canal world, the Grade I listed Pontcysyllte Aqueduct, a scheduled ancient monument, described by Sir Walter Scott as the greatest work of art he had ever seen, the longest and tallest aqueduct in Britain. Built of 3.8m wide cast-iron trough in dovetailed sections, it has 19 arches of 13.6m to give an overall length of 307m, striding 37m above the River Dee, from which it is about to depart. Built by Telford and Jessop between 1795 and 1805, it featured in a painting which used to hang in Telford's office but is now displayed in the library of the Institution of Civil Engineers in London. The masonry piers are partly hollow, built with a mortar of ox blood, water and lime. The trough was made waterproof with Welsh flannel dipped in boiling sugar. The towpath is built over the trough so that boats do not reach to both sides and water can flow past them, rather than a boat acting as a piston and pushing all the water across the aqueduct. The project brought together a construction team later to work on other major projects such as the Menai Bridge and the Caledonian Canal. The flimsy handrail is of slight consolation for people or horses with vertigo but on the right side the only protection is an upstand of some 300mm. The canal is dewatered periodically with a dramatic discharge from the centre of the aqueduct to the river far below. Upstream is Pont Cysyllte itself, a three-span sandstone-arched bridge of 1696 over the river. For those who are concerned about heights there is the added interest of a figure in crinoline sometimes seen on the towpath by moonlight.

Such is the grandeur of the structure that it is easy to overlook the earth embankment at the southern end, now obscured by mature trees. It is both long and high and was one of the greatest earthworks anywhere at the time it was built.

Just beyond the Aqueduct Inn at Froncysyllte is a drawbridge, which is normally lowered. Pen-y-Graig quarries

train fell, burning for seven hours. Views across the Vale of Llangollen towards the Berwyn hills are extensive, hence the picnic table. Here, the Dee formerly flowed south of Pengwern Hall to meander back to its present position at the golf course. Dog roses add splashes of colour.

Above Trevor Uchaf, a monument and fort site remain unnoticed but the large white block of Trevor Hall is seen up the hill from the canal as the views unfold northwards. The former line of the railway crosses over, getting clear of an area where there have been three breaches of the canal, the last in 1985. The sides remain vertical and sections of railway sleeper are chained along the edges of the canal at intervals, providing floating steps for wildlife. Below is Trevor Rocks, the site of the first British canoe slalom in 1938.

An old lattice footbridge carries the Offa's Dyke Path over the canal. While the path was opened in 1971, the dyke was probably built in the 780s to keep the Welsh at bay.

Narrowboat moorings at Trevor are sited in the remains of the **Ruabon** Branch on what was to have been the main line from Chester. Trevor Wharf is served by the Telford Inn and rather spoiled by the chimneys of Flexsys Rubber Chemicals' works at Acrefair.

reached up behind the hilltop aerial. Limekilns on the canal bank were closed in the Second World War because the glow they gave was too much of a marker for German bombers seeking the Brymbo steelworks.

Ahead is the Cefn or Newbridge Viaduct, carrying the Wrexham to Shrewsbury railway line over the river. With 21 openings, it is 470m long and was the longest viaduct in Britain when built in 1846–8. The

The aqueduct soars above the treetops with minimal protection.

Castell Dinas Bran looks down over the autumn colours along the canal.

B5605 crosses both railway and canal at Pentre on what is called the Irish Bridge, actually the name for a form of ford.

The 175m long Whitehouse Tunnel takes the canal under the A5. Opposite the site of Black Park Colliery, a branch has been excavated to form **Chirk** Marina, the dredgings being used to landscape a golf course. If the wind is from the east the odours will be a mixture from the sewage works, the large chipboard factory and Cadbury's chocolate factory. An industrial estate has been built on the site of the Glyn Valley Tramway transhipment sidings. This served the 711mm gauge

The River Dee and the Pont Cysyllte seen from the aqueduct.

Part of the Pontcysyllte Aqueduct striding across the valley.

1873 Glyn Valley Tramway which ran 10km from Glyn Ceiriog slate quarries until closed in 1935.

The cutting hides Chirk Castle. Built in 1284–1310 by Roger Mortimer for Edward I, it was bought in 1595 by Sir Thomas Myddleton, whose family have owned it ever since. It is the only border castle to have been occupied continuously since it was built. Restored, it has a 30m Long Gallery of 1678, a Great Saloon, a deep dungeon, walls up to 4.6m thick

round a central courtyard, large round corner towers and D-plan intermediary towers, all set in 1719 parkland and formal gardens by the Davies brothers, behind magnificent wrought-iron gates. There are elegant state rooms with fine ceilings and walls, Chippendale furniture, James I tapestries, an 18th-century staircase and a ghost. The family arms include a red hand. This is said to represent former misdeeds and that this may be removed when a prisoner can

Two magnificent structures, Chirk Aqueduct and the adjacent railway viaduct, crossing the Ceiriog valley.

survive in the dungeon for ten years. One prisoner almost managed it.

The 420m long Chirk Tunnel was built by Telford in 1801, with a slightly crooked bore and flared ends, which make the bore appear larger than it really is. All three tunnels on this canal have towpaths, unusual for the time, avoiding legging. This is the longest British tunnel regularly used by walkers.

The B4500 crosses just before the tunnel emerges into the basin that served Brynkinalt Colliery. From here the Chirk Aqueduct carries the canal 21m above the River Ceiriog on ten 12m arches to give a total length of 220m, the longest canal aqueduct in Britain when built in 1796–1801. The trough is lined with iron plates but the aqueduct is of stone, the thick walls reducing any vertigo effect, more so because, in 1848, Thomas Brassey built a higher 16-span railway viaduct alongside. Artists who have made this their subject have included John Sell Cotman. Swallows fly around the structures and there are fine views up the valley to Pont Faen, the original site planned for the aqueduct, and downstream over Telford's bridge carrying the B5070 towards the recent A5 viaduct.

The river marks the border between the Welsh county of Wrexham and the English county of Shropshire. This is Marches country. The Bridge was important as the first public house in England in the days when Sunday drinking was not allowed on the Welsh side of the border.

Mottes stand on both sides of the road by the post office village stores. A garden at Chirk Bank is filled with canal memorabilia. Monks Bridge, by the Poacher's Pocket at Glerid, carried the original Holyhead road. A Roman vexillation fortress at Rhyn was sited to protect this strategic route.

Rhoswiel is located between a timber yard and a concrete railway bridge of 1913, which served Ifton Colliery. A wharf was used by Moreton Hall Colliery. These days, Moreton Hall school hides behind a golf

course and a Little Chef as the A5 crosses back and heads away to the south. The canal turns east past Henlle Hall.

At St Martin's Moor a canalside house has a bad attack of giant insects such as butterflies and ladybirds. Despite the terrain, there are only two locks in the 51km between Llantysilio and Whitchurch. Ironically, they come in the much more gentle country at

Castell Y Waun / Chirk Castle, Clwyd, Cymru / Wales

New Marton, a village marked by a wind pump. Yellow irises, curlews and kingfishers make themselves known. Born at Marton Crest in 1656, Thomas Bray set up the Society for Promoting Christian Knowledge.

The Jack Mytton Inn is a prominent canalside building at Hindford. Just after the Oswestry to Whitchurch railway crossed, a neat brick roving bridge takes the towpath across before the Narrow Boat Inn.

Ellesmere Depot and Yard with its distinctive architecture.

167

There are extensive views southwards over the Perry valley, for the next 3km approaching Tetchill, to hills south of the River Severn. The Montgomery Canal line of the Shropshire Union Canal joins at Lower Frankton. Following the contours, the canal makes a near 180° bend before Tetchill.

Ellesmere, a town with Saxon origins, begins with Ellesmere Depot and Yard, joinery shops and forge, now including interesting private dwellings. The boat repair shop is still in use and British Waterways provide services for powered craft. On one occasion a floor collapsed under British Waterways top brass having a health and safety meeting. There is a picnic area as the Ellesmere Branch goes off to the town centre, the canal's eastern terminus initially. The motte and bailey castle site is now a bowling green in this 18th-century market town with its Georgian houses. St Mary's church has medieval origins, a medieval chest made from a solid oak block, a 15th-century octagonal font and many fine effigies. The town had an annual cheese market until the First World War. As the name suggests, it also has the Mere, a large lake, and there are plenty of herons.

Ellesmere Tunnel is 87m long, just long enough for a road junction on top. The towpath is carried on brick arches over the water.

The meres are kettle holes surrounded by kame moraines. Blocks of ice buried in the ground at the end of the Ice Age gradually melted without any drainage paths leading away. Local legend says that there was just one well in a field. A new tenant refused to let the villagers use it or they were charged by the bucket. They prayed for assistance and the well overflowed so that there was free water for all but the tenant still had to pay the rent for his now flooded land. Another legend has 16th-century highwayman Sir Humphrey Kynaston landing on

his horse after a 14km jump from Nesscliffe, perhaps assisted by the belief that the horse may have been the Devil. He also made other prodigious but more plausible jumps on horseback. These days the meres are dark pools surrounded by trees and even rhododendron bushes. The canal skirts Blake Mere and Cole Mere. The latter has the Colemere Country Park, wildfowl and information on the geology, flora and fauna of the meres and mosses and the remains of limekilns in this Site of Special Scientic Interest. White Mere is a cormorant roost. Three kilometres south, in 1860, Crose Mere yielded the Ellesmere Iron Age punt, 3.35m x 730mm x 400mm deep, estimated to weigh 410kg. It would have needed ballast so it must have been used for carrying cargo.

When the village school in Lyneal closed down it was bought by the Guides. Lyneal Wharf has open canoes and narrowboats and is used for trips by the Lyneal Trust, including those for the disabled.

Hampton Bank has another wharf, this time with limekilns and a sand quarry. A picnic table is sited to give a view along the embankment as the canal crosses the valley occupied by one of the headwaters of the River Roden. There was a breach in 2004, following excavation by badgers. At the far end of the embankment, where there was another breach in 1960, the canal passes back into Wales and turns sharply left to run along the side of the valley, from where there are extensive views.

An account by Sir John Hanmer of Bettisfield, written in 1876, suggests the medieval Hanmer dugout boat found wrapped in reeds actually came from near Bettisfield. There is plenty of swampy ground to encourage that theory. To the south of the canal is a windmill but it may well be hidden by the dense trees next to the canal.

For 1.6km the canal runs straight and wide, crossing back into England to a canal junction dominated by a red-brick house with circular insets. The Prees Branch never reached Prees but it did get to Edstaston, hence its alternative name of Edstaston Branch. These days, it only reaches Waterloo. Whixall Moss roving bridge is a wooden structure with lattice sides, adding to the character of the area. An adjacent car scrapyard undoes some of the good work.

Whixall Moss is the best of the remaining north Shropshire peat bogs with rich flora, snipe and many insects, including mosquitoes and 16 species of dragonfly. The more extensive Fenn's Moss to the north has nearly dried out to heathland with silver birch, yellow balsam and bracken, together with half a hectare of cloudberry, normally only found in the high marshes of the Berwyns. This site was chosen for commercial peat digging. The mosses form the largest raised peat bog in Britain. The canal crosses the moss on embankment, the banks having needed to be raised repeatedly.

Platt Lane has a typical slated, red-brick wharf and warehouse. There are lime kilns in a garden at Lower Tilstock Park and the site of a brickworks on the right before the former Cambrian Railways line from Oswestry to Whitchurch crossed over. These local industries are easily missed, unlike the views to the left over Wales, the border of which is close to the left bank of the canal

Lifting drawbridges are common along this canal.

Canalside buildings at Platt Lane.

for the last time. Gorse bushes add splashes of yellow among the oaks, ashes and hawthorns.

The Shropshire Way footpath crosses over and another footpath leads up past the Pan Castle motte and bailey site.

While this section has been lock-free, the same cannot be said of access drawbridges that lower almost to water level. Three come in a kilometre on the approach to **Whitchurch**. Major roads also arrive.

The partially restored Whitchurch Branch leaves at Chemistry. Houses have been built on the old line of the canal but it is hoped to extend the branch, using an inclined plane. Much of the town remains hidden from the main line. The town was the site of a Roman fort, Roman artefacts being shown in the local museum. The original Norman white church of St Alkmund collapsed in 1711 and was rebuilt two years later. The town is home to the Joyces, the oldest tower clock makers in the world, eight generations having run the firm for over 300 years; their work includes the church clock of 1849. The town's interesting architecture includes medieval, Tudor and Georgian houses and it was the birthplace of Sir Edward German, who composed *Merrie England*. A market town, it is a producer of Cheshire cheese.

While the canal doubles back round the head of a steep valley, the A41 strides straight across on an embankment and crosses over the canal again. It makes a final pass over at Grindley Brook, this time on the line of the Roman road from London to Chester. Since Horseshoe Falls there have been only two locks but that now changes. Grindley Brook Upper Lock is a staircase of three accompanied by a bow-fronted cottage and the Lockside Stores. Grindley Brook Lower Lock is three separate locks after the road. The Horse & Jockey stands back from the canal.

After the locks, the canal is crossed by a massive former railway embankment using a blue-brick arch with magnificent skewed brickwork. Here the canal meets the Cheshire border and this is a point from which various long distance footpaths strike out: the Shropshire Way, the Cheshire Way and the Sandstone Trail to Frodsham.

The canal now runs along the bottom of a deep valley with Hinton Hall prominent on the east ridge and the slope rising on the other side from Land of Canaan to Bell o' the Hill Farm. Hinton Bank Farm produces blue Cheshire cheeses.

There is a waterfowl sanctuary at Tushingham Hall, with its small lake.

The Willey Moor Lock public house stands next to its namesake. In 1983 the pub designed and had built a metal bridge over the canal to give access to the pub and to be used by walkers and emergency services rather than having to walk across the lock gates, not the easiest of moves even before starting drinking. British Waterways charged £50/year rent for the bridge site. When the figure came up for revision it had increased to £10,000/year. After some acrimonious exchanges, a compromise figure was reached.

Bulrushes make their appearance along the edge of the canal.

The A49 crosses and busy roads are left behind again.

The canal moves away through another band of meres, this time not so close to the navigation. Between the canal and Big Mere is Marbury with its 13th-century church and old timbered buildings.

Black-and-white timbered buildings are seen more, as are black-and-white Fresian cows in this dairy countryside. There are picnic tables around Marbury Lock, along from which is a farm with a large circular cowshed.

A lifting bridge at Wrenbury dates from 1982 and tilts to 80°, steeper than its listed predecessor, which caused accidents. There are two mill buildings, the older of which is 16th-century. The miller had his own fleet of narrowboats. The Dusty Miller by the canal is accompanied

The drawbridge and mill buildings at Wrenbury, now a canal shop on the left. The one on the right has been converted to a public house and restaurant.

One of the Baddiley locks with the bywash working well.

by the Cotton Arms down the road towards the battlemented church of St Margaret with its 16th-century west tower, 18th-century chancel and pulpit and fine 19th-century monuments. Wrens are still plentiful along the canal around the village.

The canal turns sharply north, at which point it is less than 4km from the Shropshire Union Canal although it is to be over 8km before they finally converge.

There are three Baddiley Locks.

Swanley has its swan to answer Wrenbury's wrens.

The A534 crosses at Burland and leads to an armless windmill.

The overflow for the final lock flight goes left into a 390,000m³ reservoir, fed by the canal and intended for domestic use. The reservoir is unusual for being contained in an embankment rather than a dammed valley. It was because the canal was used to supply water from the River Dee to the reservoir that the canal was saved from closure in 1944.

The four Hurlston Locks are fitted with experimental low-geared enclosed paddles introduced in the 1970s. The penultimate lock has been suffering with a wall moving in, making it too narrow for some boats to pass, otherwise the canal would be even busier.

At the foot of the flight the Llangollen Canal meets the main line of the Shropshire Union Canal at Hurlston Junction. Here it is just 2km from the end of the Middlewich Branch.

Reservoir and lock flight down to Hurlston Junction.

Distance
75km from Llantysilio to Hurlston Junction

Navigation Authority
British Waterways

OS 1:50,000 Sheets
117 Chester & Wrexham
118 Stoke-on-Trent & Macclesfield
125 Bala & Lake Vyrnwy
126 Shrewsbury & Oswestry

32 Shropshire Union Canal, Montgomery Canal

The first canal with a reopening Act of Parliament

It would be understandable if the Montgomery Canal had an identity crisis. The northern end as far as Carreghofa was opened in 1796 as part of the Ellesmere Canal, which started at Hurlston. The Montgomeryshire Canal (Eastern Branch) took the canal on to Garthmyl a year later but it was not until 1819 that the Western Branch continued the line to Newtown. The Montgomery Canal has also been known as the Newtown Arm, the Old Montgomeryshire Canal and the current Welsh Y Gamlas Maldwyn, in 1845 becoming part of the Shropshire Union Canal.

The height of the freight activity on the canal was in the 1830s, with limestone, coal, slate and timber being carried. It was not until 1852 that a passenger service was started from Newtown to Rednal, where it was possible to change for trains to Liverpool.

The Shropshire Union Railway & Canal Company, LNWR (which encouraged the flyboat and other services), LMS and GWR (which failed to maintain the canal) were successive owners. In 1936 there was a breach near the Perry aqueduct, which the owners did not repair, despite their legal obligations. The main instigators of collapse on this canal appear to have been water voles. One trader trapped at Welshpool Wharf was forced out of business. The owners managed to get the canal closed in the infamous Act of 1944, claiming that there had been no traffic on the canal for some years past, hardly surprising.

Restoration began in 1968 with a Big Dig the following year. In 1973 a committee, under the Prince of Wales, undertook to fully restore the Arddleen to Welshpool section. In 1987 it became the first canal to have an Act of Parliament for the reopening of a canal. The canal now exhibits all stages of repair and is unusual in that off-line nature reserves have been established on the branches. Navigation is permitted on the main line, subject to low speeds being imposed. In a number of places the canal has been culverted gradually to improve road alignment. These points will need rebridging at some stage. Consideration has been given to placing drop locks at such interruptions to allow the canal to pass under at a lower level. Williams Bridge, at Carreghofa, was demolished by Powys County Council and replaced with a 900mm-diameter culvert after restoration had started. This culvert installation was described as official vandalism.

Running north-east from Newtown across Powys and Shropshire to the Llangollen Canal at Frankton Junction, it is unusual in that it falls to a low point at the Wern, the sump, and then rises again.

Newtown basin was filled in and built on in the 1930s. The first 3km, owned by SevernTrent Water, is dry. Water first appears at the lock above Aberbechan. Next to this is a derelict lock keeper's cottage. Above the lock, the canal profile looks to be in good shape.

The canal is extremely attractive with an amazing variety of wildlife.

At first it is narrow with reeds reaching out from each side but always leaving a passage in the middle of the relatively clear water. However, there are fallen trees of various sizes across the canal from time to time. Over the Welsh section, the views to the right are limited but hillsides rise steeply to the left, wooded or grazed. The only sound, other than the black-and-white Fresian cattle, is the cracking of gorse pods in the sunshine. There are several kinds of dragonflies. Hazel, pine, hawthorn, alder, ash, willow, thistles and a selection of umbellifers make up some of the vegetation.

For the first half, the canal follows the River Severn closely although not obviously. The Cambrian Railway from Machynlleth to Shrewsbury and the A483 are never far away.

At Aberbechan there is a small aqueduct carrying the canal over the Bechan Brook. The canal now moves past sheep on a wooded bank, oaks, beech trees, lilies, ferns and bracken. Herons fish, as do the many families of swans nesting along the canal. In 1921, a head-on crash between an express and a local train killed 17 after staff broke operating rules.

Newhouse Lock is restored like the other locks where the canal is in water.

After Byles Lock and the first boat, the canal passes under the A483 in a tunnel. Dolforwyn Castle is hidden by a hill, from which the drilling of the woodpecker may be heard.

The next bridge, carrying a road over the canal at Abermule, is eclipsed by its neighbour, carrying the road over the River Severn. Cast in 1852 at Brymbo, it was the second iron bridge in the county and carries the road over in an elegant 34m span with just 3.7m rise between its stone abutments. The details of its construction are picked out in letters along the sides of the arch.

On the far side of the river are a motte and bailey between the bridges. Brynderwyn Lock follows. After the well kept lock there is a low swing bridge.

Below Pennant Dingle, an ornate bridge over the canal serves Glan Hafren with another motte and bailey on the far side of the valley.

At Fron, the canal is truncated by the A483, only to have to cross back across it around the next corner.

The first reach of the canal above Aberbechan.

The ornate bridge over the canal at Glan Hafren.

The canal loops round towards Plas Meredydd.

Cattle use the canal to cool down.

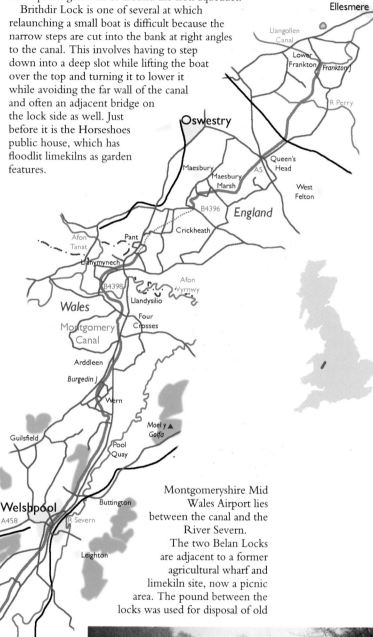

structure on the canal, carrying the canal over the River Rhiw. Completed in 1797, with two 9m river-arches and two smaller land-arches, it is surprisingly wide and has suffered major leakage problems, being rebuilt in 1889 and extensively repaired again in 1984.

Beyond Berriew, with its stubby-spired church, is Berriew Lock with flowers and apple trees.

Another motte and bailey and a settlement precede the Luggy Brook aqueduct. Blackberry, elder, rosebay willowherb, cranesbill, foxglove, meadowsweet and other plants grow on or around the iron aqueduct.

Brithdir Lock is one of several at which relaunching a small boat is difficult because the narrow steps are cut into the bank at right angles to the canal. This involves having to step down into a deep slot while lifting the boat over the top and turning it to lower it while avoiding the far wall of the canal and often an adjacent bridge on the lock side as well. Just before it is the Horseshoes public house, which has floodlit limekilns as garden features.

Berriew Aqueduct is the canal's second-largest structure.

A large angling lake has been excavated on the left of the canal, opposite housing for animals that include a llama.

The route at Garthmyl is not obvious, to the B4385/A483 junction and then northwards along the main road. After the Nag's Head Hotel the original road bears left and rises up over a canal bridge past the old malthouse, which now produces animal feed-stuffs. The water begins again underneath the bridge and this is most easily reached directly from the A483, opposite the bridge.

The canal winds away from Garthmyl, passes a builder's yard and moves out into fields again. On the left is a prominent cedar tree. In front, a low pipe crosses the canal. The canal is culverted under the B4385 at Refail, just around the corner. Below the lock is a growth of watercress.

Berriew Aqueduct is the second-largest

Montgomeryshire Mid Wales Airport lies between the canal and the River Severn.

The two Belan Locks are adjacent to a former agricultural wharf and limekiln site, now a picnic area. The pound between the locks was used for disposal of old

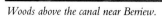

Woods above the canal near Berriew.

171

The much smaller aqueduct over the Luggy Brook.

Lifting bridge at the Moors.

Typical penstock gear at Belan Locks.

Powis Castle just visible through the trees.

working boats although these have now been cleared. A culvert passes under the A483. There are fine views to the Long Mountain and the churches at Forden and Leighton, nestling at its foot. Powys Estate sawmills were formerly powered by water from the canal.

Leading up the hillside on the left is the road to Powis Castle, built in stone about 1200 by the Princes of Upper Powys, who were allies of the English after Edward I's invasion of Wales. The architecture represents many periods. Red drums guarding the Inner Bailey are from the 14th Century. The Long Gallery is 16th-century. There are a Restoration State Bedroom and an early 18th-century Blue Drawing Room. Initially a military stronghold with military occupation, the protecting slope was made into four garden terraces with lead shepherdesses in the late 17th Century. There are an orangery and an aviary. Enormous clipped yews shelter rare plants in colourful herbaceous borders of high horticultural value, with a Braille guide to scented flowers. The ancestral home of the Herberts from 1587, it became the family seat of Clive of India as a result of a marriage of 1784. There is an illogicality about why the owner of a grand country estate should wish to leave it and live on another continent. There is a Clive of India Museum, covering the involvement of the two families with India, part of the finest country house collection in Wales, including notable furniture and paintings, the latter with two Romneys of Lord Clive's daughters in the Dining Room.

The next reach of the canal is wide, being used by tripping boats.

Welshpool is the largest town on the canal. It has had a market charter since 1263 and has the largest sheep market in Europe. Welshpool Lock is approached past an old stable block facing the canal. Just below the lock is the Shropshire Union warehouse, now the Powysland

Museum & Canal Centre. Horse-drawn narrowboat trips are run from the wharf.

The canal crosses the small 1836 iron Lledan Brook Aqueduct with stone-arched towpath. It passes under a footbridge that formerly carried a 762mm gauge railway, lost in the town but preserved to the west as the Welshpool & Llanfair Light Railway. The canal passes the backs of gardens and is quickly free of the town, a tunnel having been built in the 1990s under the A458 in front of the swimming pool when the bypass was constructed.

This reach is the one most used for cruising and is known as the Prince of Wales' length. Narrowboats used here include *Heulwen*, sunshine, Britain's first canal trip boat for the handicapped. A picnic area is a further amenity here.

At Buttington, the railway leaves for Shrewsbury. Formerly, another line branched off and followed the canal. Buttington Wharf has old limekilns that were burning 2,000t of limestone a year in 1830. Offa's Dyke, defining the boundary of Mercia, was another linear feature although the River Severn acted as the boundary for a while. The long distance Offa's Dyke Path comes down off the Long Mountain at this point.

The first of two lifting bridges comes at the Moors. Below is the riverside site of the former Cistercian Strata Marcella Abbey of 1170.

The canal pulls away from the wooded hillside, with its fort in one corner, and reaches the next lock at Pool Quay. Pool Quay is the legal head of navigation on the River Severn and it is surprising that there is no direct navigation connection from the canal to the river although there was a transhipment point. The river now turns eastwards. On the far side of the valley are the prominent Moel y Golfa and the dolerite Breidden Hill, with its roadstone quarries. The Breidden Hills were used by

Distance post by the road crossing. Pool was renamed to avoid confusion with Poole in Dorset.

Looking across the Severn valley to Breidden Hill and Moel y Golfa from Burgedin Locks.

The former salt warehouse by the wharf at Pentreheylin Hall.

Caractacus or Caradoc in his opposition to the Romans from 43, Middletown Hill with its earthworks being a possible place for his last stand. In the 18th Century they were planted to supply timber to Chatham shipyard. On top of Breidden Hill is a pillar to Admiral Rodney and at the end is a forest of radio masts.

A rural church is passed before dropping through three more locks, Crowther, Cabin and Bank, to the bottom pound. There is a picnic area at Cabin Lock. The channel that powered a 19th-century corn mill now carries excess water away through the Wern Nature Reserve to the River Severn. It is not noticed from the canal.

The climb back up begins with the two Burgedin Locks. Above the top lock, the main line is joined by the Guilsfield Branch, formerly 3km long, the remains used as a nature reserve, with a carved bird statue at the junction.

Algae and Canadian pondweed give way to duckweed. The height of the low A483 bridge at Arddleen is academic as wire mesh has been placed beneath it. Soon the A483 crosses back at another culvert point.

Near Four Crosses, the canal curves left past Clopton's Wharf, the old stone wharf building still standing beside the water, and away from the church spire at Llandysilio.

After passing the fort-topped hill at Bryn Mawr, the canal turns sharply onto the Vyrnwy Aqueduct, affording a close view of the old Pentreheylin Bridge. The aqueduct suffers bulging and leakage but is still standing, despite a collapsed arch during construction. Its floating water plantain has resulted in a Site of Special Scienic Interest designation being placed on it. Environmentalists have called for all boats to be towed across it by horse. The canal immediately crosses another aqueduct over part of a field before being cut by a culvert, a minor road having being lowered despite canal restoration already being in progress with royal support.

The following two Carreghofa Locks are accompanied by an 1825 tollhouse at the upper lock, because of the original ownership change at this point. The ownership aspect also resulted in complex water supply issues with a feeder from the Afon Tanat and some masterly brickwork spillways.

At Wern an aqueduct passed over the West Shropshire Mineral Railway. Wall's Bridge is the site of another culvert. St Agatha's church, in Norman style with a large clock, dates from the 19th Century.

The Wales/England border arrives as the A483 crosses with the Offa's Dyke Path for the last time, in Llanymynech, where the 57ha Iron Age or late Bronze Age hilltop fort is one of the largest in Britain. This might have been somewhere else that Caractacus could have made his last stand against the Romans. On the south side of the canal is a multiple ditched enclosure from this period.

In the 1820s this was the centre of activity on the canal, shipping up to 35,000t of limestone per year from the quarries on Llanymynech Hill, first worked by the Romans for copper and lead. Three barges per day were filled at the wharves. The whole area was a network of industrial activity, part of the 92 limekilns along 42km of canal, providing a peak of 57,000t per year of lime to transport. Railways were added progressively to help bring the limestone down to the wharves. The prominent 42m iron-banded square chimney is part of the renovated large Hoffmann limekiln, part of Llanymynech Heritage Site. It operated from 1899 to 1914 and was much more efficient than the old canalside kilns, still present, killing the canal trade. Everything was now transferred to the railway alongside. It is one of only two such kilns surviving in the world. The obvious activity today after the new Llanymynech bridge is a canalside pigeon loft, a restored stable block and the *George Watson Buck* trip boat. The railways have gone and, by Pant, so has the canal, the bed now being dry to all intents and purposes. A second arch in the bridge at Pant

Pool Quay church.

Carving at the start of the Guilsfield Branch, now a nature reserve.

The Vyrnwy Aqueduct, the canal's major structure.

Llanymynech Hill is riddled with mine workings. The Hoffmann limekiln chimney is prominent.

Old limekilns beside the line of the dry canal at Pant.

Timbered brick warehouse and roving bridge near Sutton. A semaphore signal under the eaves was to tell flyboats to stop for parcels.

Looking down the Frankton flight of locks towards Perry Moor.

Distance
*53km from
Aberbechan to
Frankton Junction*

**Navigation
Authority**
British Waterways

Canal Society
*Friends of the
Montgomery Canal
www.montgomery
canal.me.uk*

OS 1:50,000 Sheets
*126 Shrewsbury
& Oswestry
136 Newtown
& Llanidloes*

accommodated a horse-drawn tramway. Llamas are stabled beside the canal but a field in an old quarry contains the odd mix of a sheep, a goat and a noisy turkey.

Restoration is taking place at Crickheath Wharf and to Redwith Bridge. Trees have been removed from the canal past an animal feed mill. Water returns to a restored section at Redwith beyond the new B4396 canal bridge.

A tramway served Morda Colliery until the 1860s. A new post office with store, internet café and accommodation has been opened beside the canal near Maesbury Hall, a situation that bucks the trend in villages these days.

Maesbury Marsh was the transhipment point for Oswestry. The crane, wharf and other community buildings form the best remaining grouping of traditional buildings on the canal.

St Winifred's Well at Woolston was named after a 7th-century Welsh princess. It had healing properties but was closed to the public in 1755 because of bad behaviour.

There are three Aston Locks near West Felton. At the time of restoration, the required nature reserve by the top lock was the biggest and most expensive project ever tackled by the Waterways Recovery Group. At Queen's Head, the A5 passes high overhead on a new bridge, the minor road on the next bridge being the old line of the A5. Cart loads of sand pulled by donkeys used to feed boats through the building on the right.

By the roving bridge that follows, there is an interesting half-timbered brick wharf building. The severe canal edging results from dropping the water level half a metre and adding Graham Palmer Lock after the peat bog Perry Moor shrunk during the years of dereliction. Palmer was the founder of the Waterway Recovery Group.

Turning under the Chester to Shrewsbury railway, the canal passes Rednal Basin.

A recent aqueduct carries the canal over the River Perry. After another set of powerlines, the canal passes under the very steeply humped Lockgate Bridge and meets the line of the former 10km long Weston Branch, intended as the main line to the Severn.

Four immaculate locks lead up to the Llangollen Canal at Frankton Junction, source of water for this end of the Montgomery Canal. Features include a simple but charming footbridge and an interesting canalside arch. It was in the former drydock between the bottom two locks that L. T. C. Rolt had his narrowboat *Cressy* converted for steam power and a cabin added.

The former stables at Frankton Junction.

174

33 Shropshire Union Canal, Middlewich Branch

A short but important link

The Middlewich Branch of the Shropshire Union Canal was opened in 1833 by the Chester Canal Company and engineered by Thomas Telford as a short but important link between the Shropshire Union Canal's main line and the Trent & Mersey Canal. It provided a faster and less locked route than via the Staffordshire & Worcestershire and Trent & Mersey Canals. In 1888, there was an unsuccessful attempt to use narrow gauge steam engines to tow boats.

It leaves the main line at Wardle. Barbridge Junction, the branch point, is only 2km from Hurlestone Junction where the Llangollen Canal joins the main line on the opposite side and so traffic from Llangollen and Chester both feed through to the Trent & Mersey on the Middlewich Branch. It is part of the Four Counties Ring.

Despite the heavy canal traffic, which leaves the canal laden with sediment and not infrequently oily on the surface, the line is almost entirely rural as it reaches across the gently rolling Cheshire countryside, more often than not meadows with black-and-white Fresian dairy cattle. Studded with oak trees, it is a pleasantly pastoral scene which contrasts with the activity on the canal.

Winding holes are conspicuous even on this wide canal.

Swans are used to boats. For smaller birds, a bird table by the canal near the first lock, Cholmondeston, resembles nothing less than a complete dolls' house on a post.

Cholmondeston Lock, like other locks on the canal, is built of heavy red sandstone blocks. The locks are deep and narrow, daunting for narrowboat users, many of whom are novices emerging from one of the largest canal marinas in England and a caravan park, next to where the Chester to Crewe railway line crosses the canal.

A large slip has occurred on the left bank of the length of cutting on the way to Minshull Lock. The lock cottage here sells everything from confectionery to lettuces. Steps lead down to a wooden launching platform below the lock where lengths of timber help to funnel inexperienced narrowboat operators into the lock.

Oak trees on each side of the canal drop away as it runs out onto an embankment, carrying it high across the Weaver valley, crossing the river itself on a tall aqueduct which is inconspicuous from the canal. As it leaves the aqueduct, the canal turns to run along the side of the valley and is joined by powerlines that are to follow it as far as Clive Green.

Many of the bridges have been painted matt grey at some stage in their past, now peeling to reveal the original red brickwork. A bridge above Church Minshull shows how the canal is clinging to the side of the valley. The retaining wall on the downhill side forms a pair of ramps that connect to the left side of the bridge span. The village itself lies way below, a community of timber-framed houses powered by electricity from the miller's millwheel until 1960. Unlike these houses is the Badger public house beside the canal, a building with a suggestion of Dutch styling. A 147-berth marina is being created here.

The Weaver Aqueduct is hardly visible from the canal.

The canal curves on past a noticeable aerial by a wood. A stable block, formerly used to change flyboat horses, is now converted to residential use beside the canal. There are various large farms on the right bank until it is above the Top Flash. The view from here across the River Weaver far below and down its valley to Winsford and the start of the Weaver Navigation is phenomenal. To get to the river with a large boat, though, involves a detour around Northwich as far as the Anderton Boat Lift.

As the canal turns away from the Weaver it is crossed by the West Coast Main Line.

After Stanthorne Lock the canal crosses two aqueducts in quick succession, over the River Wheelock and the A530.

For the first time, the canal comes into a built-up area as it passes recent housing in **Middlewich**, complete with private landing stage serving all the gardens. The canal moves under several weeping willows and bridges, the last being close to the line of the Roman road which approached this important salt mining town from the south.

The Middlewich Branch leads away from the main line at Barbridge Junction.

A summer storm over Church Minshull.

Coming into the built-up area of Middlewich.

An old fly-horse stable block converted to housing near Wimboldsley Hall.

Extensive views over the Top Flash and River Weaver looking towards Winsford.

Much of the farmland is used for grazing dairy cattle.

Distance
16km from Barbridge Junction to the Wardle Lock Branch

Navigation Authority
British Waterways

OS 1:50,000 Sheet
118 Stoke-on-Trent & Macclesfield

The final lock belonged to the Trent & Mersey. Wardle Lock was named from the fact that it is at the west end of the Wardle Lock Branch. The branch, a mere 100m long, was opened in 1829 mainly as a convenient place to intercept and extract tolls from barges travelling from Chester on Shropshire Union Canal waters before they could get onto the Trent & Mersey Canal to reach wharves in Middlewich.

Wardle Lock and the entire Wardle Lock Branch.

34 Trent & Mersey Canal

Promoted by Liverpool corporation and the Staffordshire pottery owners under Josiah Wedgwood, the Trent & Mersey Canal was formerly the Grand Trunk Canal to link the ports of Liverpool and Hull. It was part of the Grand Cross to join the Mersey, Trent, Severn and Thames. Built between 1766 and 1777, it was James Brindley's most ambitious project and was one of the most successful canals. It was later bought by the London, Midland & Scottish Railway and regular freight traffic stopped in the late 1960s. A contour canal, it was wide beam as far as Middlewich, for salt barges, and again from Horninglow.

Despite the name, the canal never comes within sight of the Mersey but is to follow the Trent for the greater part of its length. It originally joined the Preston Brook Branch of the Bridgewater Canal at the northern end of Preston Brook Tunnel. This was subsequently extended to 1.133km, giving the unique situation of canals joining inside a tunnel. This was the first major tunnel to be built and is the ninth-longest canal tunnel still in use. Preston Brook, Saltersford and Barnton, all on this canal, were the first three British canal tunnels (except mine tunnels) to be built.

A fairly straight path leads over Preston Brook Tunnel, past a series of vent shafts and the Tunnel Top public house. The centre section of the tunnel had to be rebuilt in 1982 after a post office collapsed into it. To the west of the tunnel is an industrial estate with a large brewery. Running beside the tunnel is the West Coast Main Line,

to be joined at the southern end by the Liverpool to Euston line.

The tunnel is time-zoned with boats able to start south between 30 and 40 minutes past the hour with northbound half an hour later. A carefully timed departure can result in avoiding other craft initially.

This is part of the Cheshire Ring to Kidsgrove.

Dutton Stop Lock has a 150mm rise to isolate from the Bridgewater the water of this long pound, which runs from Middlewich. Beyond the lock is a dry dock with an ornate roof. It is an area with bluebells, celandines and red campion in the spring, gorse, hawthorn, ash and beech trees, set off by the occasional half-timbered black-and-white house.

By Dutton Hall, the canal is 15m above the River Weaver and Weaver Navigation, looking down on Dutton Locks and to the Dutton Arches of 1836, twenty 18m spans carrying the West Coast Main Line 20m above the river. This is followed on the Weaver by the Acton Bridge of 1933, a 560t opening structure on floating bearings.

Swans nest between stands of bulrushes.

Powerlines cross between Little Leigh and **Weaverham**, before a bank which is a carpet of bluebells in the spring.

At 129m and 174m long, Saltersford and Barnton tunnels follow in quick succession with a wide reach between them. Tunnelling methods were in their infancy

A canal built to end the pottery breakages

Bank of bluebells at Barnton.

Preston Brook, Britain's first major canal tunnel.

Looking south from Barnton tunnel towards the A533 crossing.

Dutton dry dock with the Mikron Theatre boat in for repair.

Anderton Boat Lift links the Trent & Mersey and the Weaver.

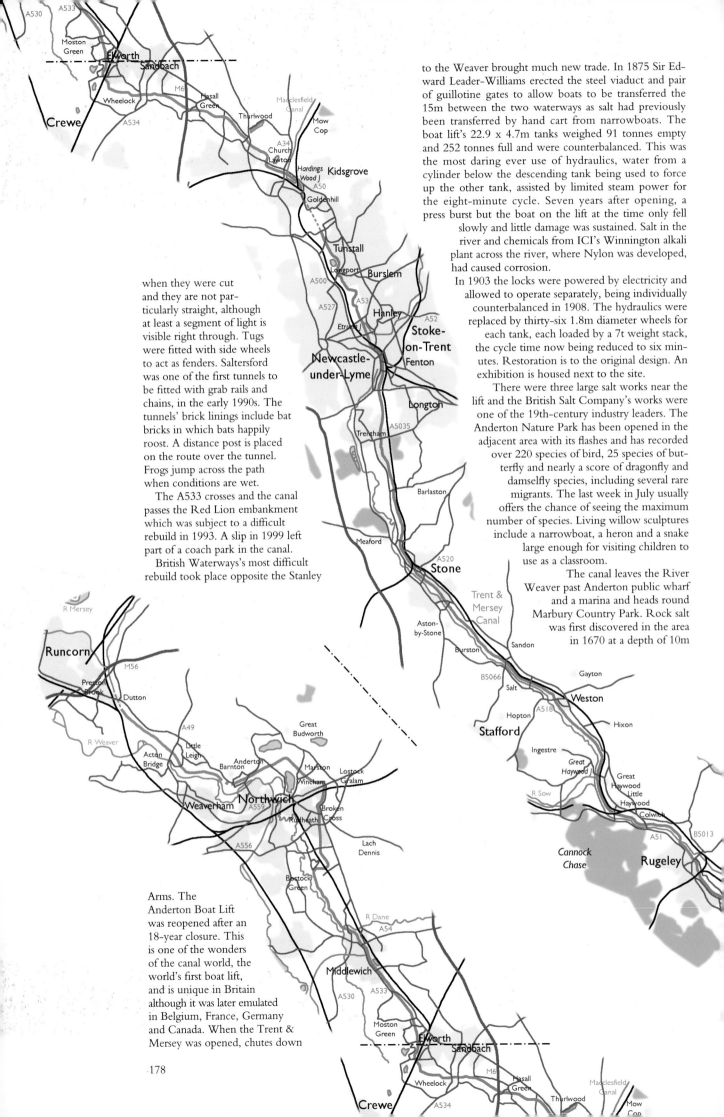

to the Weaver brought much new trade. In 1875 Sir Edward Leader-Williams erected the steel viaduct and pair of guillotine gates to allow boats to be transferred the 15m between the two waterways as salt had previously been transferred by hand cart from narrowboats. The boat lift's 22.9 x 4.7m tanks weighed 91 tonnes empty and 252 tonnes full and were counterbalanced. This was the most daring ever use of hydraulics, water from a cylinder below the descending tank being used to force up the other tank, assisted by limited steam power for the eight-minute cycle. Seven years after opening, a press burst but the boat on the lift at the time only fell slowly and little damage was sustained. Salt in the river and chemicals from ICI's Winnington alkali plant across the river, where Nylon was developed, had caused corrosion.

In 1903 the locks were powered by electricity and allowed to operate separately, being individually counterbalanced in 1908. The hydraulics were replaced by thirty-six 1.8m diameter wheels for each tank, each loaded by a 7t weight stack, the cycle time now being reduced to six minutes. Restoration is to the original design. An exhibition is housed next to the site.

There were three large salt works near the lift and the British Salt Company's works were one of the 19th-century industry leaders. The Anderton Nature Park has been opened in the adjacent area with its flashes and has recorded over 220 species of bird, 25 species of butterfly and nearly a score of dragonfly and damselfly species, including several rare migrants. The last week in July usually offers the chance of seeing the maximum number of species. Living willow sculptures include a narrowboat, a heron and a snake large enough for visiting children to use as a classroom.

The canal leaves the River Weaver past Anderton public wharf and a marina and heads round Marbury Country Park. Rock salt was first discovered in the area in 1670 at a depth of 10m

when they were cut and they are not particularly straight, although at least a segment of light is visible right through. Tugs were fitted with side wheels to act as fenders. Saltersford was one of the first tunnels to be fitted with grab rails and chains, in the early 1990s. The tunnels' brick linings include bat bricks in which bats happily roost. A distance post is placed on the route over the tunnel. Frogs jump across the path when conditions are wet.

The A533 crosses and the canal passes the Red Lion embankment which was subject to a difficult rebuild in 1993. A slip in 1999 left part of a coach park in the canal.

British Waterways's most difficult rebuild took place opposite the Stanley

Arms. The Anderton Boat Lift was reopened after an 18-year closure. This is one of the wonders of the canal world, the world's first boat lift, and is unique in Britain although it was later emulated in Belgium, France, Germany and Canada. When the Trent & Mersey was opened, chutes down

The Lion Salt Works with door tops now near towpath level.

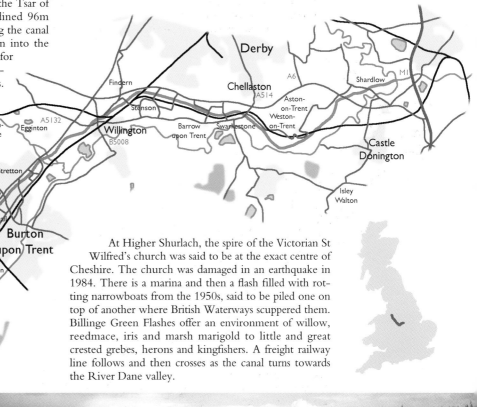

The ICI works bridges the canal.

Narrowboats stacked up at Lostock Gralam.

while exploring for coal. The first mine followed in 1675 with thicker seams below being worked later. Twin green pipes over the canal are the 1882 brine main to Weston Point. The hall in the park was derelict and pulled down in 1968. The park includes Budworth Mere. Salt profits paid for the 26m red sandstone tower with its peel of eight bells on the 14–15th-century church of St Mary & All Saints in Great Budworth.

The Marston Mine covered 14ha. In 1844 the Tsar of Russia and the Royal Society visited and dined 96m down in the mine. It collapsed in 1907, taking the canal with it. In 1958 the canal was diverted again into the 800m Marston New Cut, the first new canal for 50 years. Six months later, the old cut collapsed, leaving a conical hole that still exists.

Ingram's Lion Salt Works is now a museum, Britain's last works to produce salt by evaporating brine, operating from 1842 to 1986. It includes a horizontal steam engine, nodding donkey brine pump and 1900s pitched-roof railway salt wagon but the most interesting feature is right next to the canal where the arches at the tops of the boiler room door-ways are now at tow-path level, such has

been the settlement. A salt mine of 1833 – 90m below ground and only 23m from the canal – was abandoned in 1898 after the mineshaft, chimney and engine house subsided into it.

Beyond Wincham Hall, now a hotel, the canal crosses Wincham Brook aqueduct, which burst in the late 1800s. The A559 crosses on the line of the Roman road from Manchester to Chester and the Manchester to Chester railway also crosses. Between these are the Wharf and a wine bar, surrounded by narrowboats in the water and stacked up on the bank at Lostock Gralam.

Beyond the Wade Brook crossing is a chemical works, which has been in operation since the 1890s, including making First World War explosives. A bund contains lime waste from ICI's ammonia soda process for making sodium carbonate. This attracts lime-loving plants that are otherwise rare in the area, there being no natural limestone.

Powerlines cross and the adjacent A530 swings onto the line of the King Street Roman road. The canal passes the Old Broken Cross public house at Rudheath before reaching Roberts' Bakery, where Broadhurst's biscuits are made.

At Higher Shurlach, the spire of the Victorian St Wilfred's church was said to be at the exact centre of Cheshire. The church was damaged in an earthquake in 1984. There is a marina and then a flash filled with rotting narrowboats from the 1950s, said to be piled one on top of another where British Waterways scuppered them. Billinge Green Flashes offer an environment of willow, reedmace, iris and marsh marigold to little and great crested grebes, herons and kingfishers. A freight railway line follows and then crosses as the canal turns towards the River Dane valley.

The remains of boats scuttled in Croxton Flash.

Croxton Aqueduct spans the River Dane.

Middlewich Bottom Lock with a formerly split footbridge.

Traditional-style weather protection canopy in Middlewich, which actually only dates from as recently as 1995.

Looking towards the M6 and the Old Man of Mow beyond.

It passes round Whatcroft Hall, a Georgian mansion with a large copper cupola. On the far side of the valley is Bostock Hall, built in 1755, probably with profits from the slave trade. It was altered in the late 1700s and again in the 19th Century to make it more Italianate.

A former puddle clay pit for lining the canal has been converted to the Bramble Cuttings mooring area, laid out with picnic tables. A flash below Croxton Hall Farm contains more scuttled boats and acts as a heronry. Croxton Aqueduct carries the canal over the River Dane on a trough rebuilt after 1930 flood damage to the old wide beam trough alongside.

The canal arrives in **Middlewich**, the middle salt-works town between Northwich and Nantwich, *wic* being Old English for working place. Middlewich Big Lock gives access to the final broad-beam pound. The lock gate was operated by winch and chain because of the proximity of a bridge. The adjacent Big Lock public house has its name carved in a brick scroll over the door and on a fine day it is necessary to walk between tables of drinkers. Saltworks here included those of Murgatroyd's and Seddon's. There is a moat up the hill.

The A54 crosses next to the Perpendicular parish church and, after Middlewich Bottom Lock, the line of the Roman road to Whitchurch is crossed. The town dates from Roman times and is Cheshire's second-oldest salt centre with names like British Salt, Cerebos, Saxa and Stag. It was also a town where many of the former boatmen lived.

After Middlewich Top Lock, the short Wardle Canal leads off to connect with the Middlewich Branch of the Shropshire Union Canal. The Trent & Mersey Canal now becomes part of the Four Counties Ring to Great Haywood.

From King's Lock and its public house of the same name, the A533, which has not been too far away since Preston Brook, now runs alongside the canal for 4km to Stud Green, a contrast with the rural pound to Middlewich. There are waste lime beds from bleach and soda ash manufacture by Brunner Mond. The Ideal Standard bathroom fittings factory presents a smart front to the canal. Less salubrious on the other bank are an off licence and the Kinderton Arms.

After Rumps Lock is the British Salt works, extracting 2,000t of salt per day from brine and making four million packs of Bisto each week. Powerlines cross the site and the A533 changes to the other bank for the run up to the Hays Chemical works, producing caustic soda, hydrochloric acid, sodium hypochlorite, chlorine and hydrogen from salt.

The two Booth Lane Locks feature split footbridges, now fixed.

Stud Green brings Crow's Nest Lock and Moston Mill, which operated from 1826 to 1985 but is now a private house. The water in the canal is deep because of the necessity to raise the banks following subsidence.

In the Middle Ages a dragon used to live in Bache Pool and terrified the people of Moston. It was killed by Thomas Venables as it was about to eat a baby and the Venables chapel in Middlewich church has a crest showing a dragon with a baby in its mouth.

The Foden commercial vehicle works were sited by the canal at **Elworth**. By Elton Moss Bridge are **Sandbach** Flashes, which are important for waterbirds, especially in the winter. The apparently rather derelict Ettiley Heath has been active with herbicide works, steel fabricators, bone crushing mill and Sifta saltworks. The Manchester to Crewe railway follows and then crosses near Rookery Bridge, the first line to be electrified.

Wheelock may take its name from the Welsh *chwel*, to turn, because of the winding river. The A534, formerly the Sandbach to Nantwich turnpike, crosses by the Cheshire Cheese and Grade II listed Wheelock Warehouse, before the Commercial Hotel. Opposite the wharf was a

The rural setting of Church Lawton.

The large canal warehouse and wharf crane at Red Bull.

fustian mill, which was to become a phosphoric acid works during the First World War and then a printing works. A former forge site is now occupied by the Ensor concrete block factory. The canal leaves over the River Wheelock.

So far, the canal has been remarkably lock-free but that now changes. The two Wheelock Locks start Heartbreak Hill to Kidsgrove, 26 locks in 11km, often duplicated to remove bottlenecks.

The line of the former Elworth to Kidsgrove railway crosses and the canal arrives at Malkin's Bank, site of the Whitehall Salt Works. There are six duplicated Malkin's Bank Locks.

The environment of the six Malkin's Bank Locks is pleasanter for having a golf course alongside on what was the Brunner Mond chemical works site. The peace is briefly shattered as the M6 crosses but the roar fades away after the two Hassall Green Locks. The roar has also faded at the Red Lion, where the locals had their own opinion on their new pub sign, the establishment now formally renamed as the Romping Donkey. Other facilities include the Canal Stores and the Potters Barn. The hamlet's canalside is immortalised in the children's charity fundraising books *Tales of Hassall Green* and *Tales of the Hassall Green Lockopotamus*.

The two Pierpoint Locks lift the canal to cross the B5078 on the Chell's Aqueduct. The South Cheshire Way footpath crosses at the two Thurlwood Locks. The right chamber of the upper lock was built in 1958 in steel with a guillotine gate like a 32m long steel viaduct that could be jacked up as subsidence occurred but it was expensive, complicated to work and less reliable than the conventional chamber alongside and was demolished in 1987.

Rode Heath has the Broughton Arms and nurseries. It also had a saltworks until 1927, when the buildings subsided into the workings. A tall chimney served a steam-driven sawmill.

Above Rode Pool is Rode Hall, an 18th-century house with Georgian stable block, set in a Repton landscape with walled, formal and woodland gardens.

The A50 passes over and the canal immediately crosses a lane on Snape's Aqueduct at Lawton Gate.

The six Lawton Locks feature a wide variety of interesting details including split footbridges, a rope guide roller and a cutaway gate top. At one time there was a staircase on the north side.

Lawton was Saxon for hill farm. The church at Church Lawton dates from 1180, on a Saxon mound, and has a Norman doorway, 1540 spire and red-brick nave of 1803, following a fire four years earlier. The remains of St Werburgh rested here in 707 on the way to Chester for burial. Lawton Hall looks Gothic but is only 17th-century.

The Old Man of Mow, on the moorland ridge that has provided a scenic backdrop along this section of canal, gradually subsides behind the foreground activity.

The Crewe to Kidsgrove railway follows the canal, passing the six Red Bull Locks, one of which is accompanied by a large warehouse and wharf crane. The Red Bull public house is located at Red Bull Wharf. The short Hall Green Branch has confused many a boater by leaving on the south side of the Trent & Mersey Canal and then crossing over on the Pool Lock Aqueduct to join the Macclesfield Canal, taking the Cheshire Ring and its walk northwards with it from Hardings Wood Junction. In the vicinity of the Tavern, the Blue Bell and the former Albion Iron Works, the canal reaches its summit level of 111m for 9km. Originally, the summit level fluctuated considerably depending on the amount of pumping from the Harecastle mines. Thus, the stop lock at the end of the Hall Green Branch was fitted with gates facing both ways as the Hall Green Branch level could be either higher or lower than the Trent & Mersey summit level although officially slightly higher. The Hall Green Branch is now permanently higher and one set of stop gates has been removed. The water is bright orange from ironstone leaching in the tunnels but the colour flows only westwards from the tunnels.

Cheshire gives way to Staffordshire.

The Manchester to Euston railway crosses over the canal three times in quick succession, the third time in the tunnels, which are preceded by the Harecastle public house.

Harecastle Tunnel had water ingress problems during its construction to 1777 and Brindley abandoned his plans to connect it by other tunnels to the Golden Hill collieries as had been done with the Worsley mines. The 2.7km tunnel was a significant bottleneck and Telford added a second parallel bore in 1824–7, so that it could be used as a dual carriageway. Subsidence meant that the original bore was unusable by 1918. The first passage of this for 60 years found stalactites like rapiers reaching right down to the water, three roof collapses and one short complete blockage; in places the tunnel arch was only a metre or so above water level. Telford's new bore had a towpath but subsidence resulted in this being more than 300mm underwater in places so it was removed to assist navigation. Boats cannot pass so they operate to a timetable. Three extraction fans at the southern portal require doors at this end to be closed except when boats

The northern end of Telford's new bore for Harecastle Tunnel with Brindley's original bore to the right.

Harecastle Tunnel's southern portal. The owner of the first narrowboat in the queue when this photograph was taken was someone who, as a young teenager, had taken a kayak through. In those days it was necessary to buy a ticket for passage and, after he had bought one, the officials could not prevent him from using it.

are passing them, adding to the oppressive mood that makes many powered-boat crews nervous. There was an electrically driven tug in use through the tunnel until 1954, using a chain along the bottom, one of the last canal tugs in operation. Mine adits vent gases into the tunnel but an explosion in the tunnel in 2000, resulting in the tunnel's temporary closure, was thought to have been someone letting off a firework.

It is not just the length and size that add to the mood of the tunnel. Several tunnels around the country are said to have the ghost of Kit Crewbucket. In fact, Kitcrew meant **Kidsgrove** and a bugget was a ghost. Two men were said to have disposed of the body of a murdered woman in the tunnel. The ghost of a headless woman or a white horse has signalled impending disaster, sudden death or murder and a ghost boat with a spotlight has sometimes been seen. On one occasion the barrier at the northern end closed on its own and the strength of six men was not enough to force it open. The waiting boater lost his nerve and announced that he was not going through, whereupon the barrier opened itself.

The towpath goes over the hill and is not clearly marked although it is named Horseboat Lane at each end. It passes through an area of superior housing, a farm and the middle of a large and apparently prosperous permanent gypsy camp. Bathpool Park is where the Black Panther hid one of his murder victims in the 1970s.

The canal moves from Staffordshire to the City of Stoke-on-Trent. The city was formed in 1910 from the Five Towns of Stoke, Tunstall, Burslem, Hanley, Fenton and Longton where, presumably, mathematics was not the greatest attribute. The Potteries take their other collective name from the fact that this is the UK's fine china capital and the canal follows where the Etruria marls meet the coal measures, both needed for the pottery industry. Pottery has been made here since 1700 BC. Most of the brick bottle kilns have now gone. Arnold Bennett based *Anna of the Five Towns* in the Potteries. Minton, Moorcroft, Royal Doulton, John Beswick, Coalport and Wedgwood all have museums and factory tours from among the 40 pottery factory shops in the city. Another attraction of the location was being able to move porcelain by canal without the breakages that resulted from road transport.

The tunnel southern portal is at Chatterley.

At **Tunstall** the A500 runs parallel to the canal and the railway and is to follow for 10km. Attempts have been made to improve the canal environment, including the erection of distinctive metal information signs. Westport Lake has been given beaches of imported sand and proves popular with birdlife.

A wharf crane at Middleport with a pottery beyond.

At Longport is the Duke of Bridgewater public house, someone not directly involved with this canal. A bottle kiln remains among the buildings of the Middleport Pottery, where Burleigh's factory is open for tours.

At Burslem Junction the Burslem Branch was closed in 1962 following a mining collapse. Festival Waters is a 16ha redevelopment site from Middleport to Etruria, including the Shelton Steelworks and the restored Burslem Branch, with offices, housing and leisure facilities. **Burslem** takes its name from the Old English man Burgheard and the British Celtic *lyme* or *lyne*, elm tree district. Shelton Steelworks were on both sides of the canal but closed in 1978. The rolling mills had a roof over the canal, closed in 1995, and the last load was taken by *Shad* to the National Waterways Museum, Ellesmere Port. Etruria was the site of the old Wedgwood works and it was used for the 1986 National Garden Festival. It was the setting for H. G. Wells' *The Cone.*

The China Garden, built for the festival at the Festival Basin, remains in good order. The Festival Park has a Waterworld pool, flumes, wave pool, rapids area, aqua assault course, cinema, bowling, ski slope and Quasar. Brightly muralled abutments carry the A53 across the canal.

The Caldon Canal climbs away from Etruria Junction. The Summit Lock at Etruria had a toll house and a wooden shed over the lock until settlement reduced the headroom too much, door tops being at waist level. One of the locks has a fall of 3.99m, one of the deepest narrow locks around. Next to the second of the three Etruria Locks is the Etruscan Bone & Flint Mill Museum, including the last-surviving steam potter's mill, with the 1820s beam engine *Princess*. The former Etruscan gasworks were said to have the largest gas holder in Europe. Further across is the centre of **Hanley** (from the Old English *hean-lea*, high glade) with the City Museum & Art Gallery containing the world's largest collection of ceramics and a Second World War Spitfire, designed by locally born Reginald Mitchell. A less happy resident of Hanley was a man prevented from marrying his sweetheart by both sets of parents. He

Etruscan Bone & Flint Mill at Etruria. In spring the lock is surrounded by blossom.

A pair of bottle kilns in the centre of Stoke-on-Trent, restored in 2008 by a housing estate developer.

Television for the nestbox with all modern conveniences?

sold everything and went to live in a cave for the next 70 years until he died.

Trees in blossom soften the industrial atmosphere of the three locks of the flight in the spring. Below them is a cemetery, opposite which two restored bottle kilns stand in a new housing estate.

The Manchester to Euston line crosses over so close to Cockshute Lock that the lock arms are cranked yet recesses have had to be cut in the plate girder bridge to take them. Near here were British Railways' last working stables, used until the 1960s.

The new Stoke Lock feeds into a concrete canyon, which hides much of the locality from view, beginning by threading through the roundabout where the A52 crosses the A500 and past the station, the university, the line of the Roman road from Chesterton to Rocester, Stoke Basin, the junction with the derelict Newcastle Branch, a former clay pit, a museum and the A5007. The River Trent is crossed for the first time. In 1961 the canal bank collapsed into the Trent. **Stoke-on-Trent** comes from the Old English *stoc*, dairy farm and Trent is British Celtic for flooding river.

In 1487 the Battle of Stoke resulted in the defeat of impostor Lambert Simnel, who was then put to work as a palace turnspit, the last battle of the Wars of the Roses, bringing an end to the Lancastrians and, perhaps, medieval England. A hard-fought battle, it involved more combatants than Bosworth.

A Sikh temple is built on the site of the Kerr Stuart Locomotive Works where L. T. C. Rolt worked. This pound would have been one he saw each day. There is a Twyfords sanitary ware factory opposite and a wagon works before the A50 crosses and the A500 moves away.

Stoke City Britannia football stadium has been built on colliery waste by the canal, the latest football development in a city that produced Stanley Matthews, probably the greatest British footballer of all time.

The water takes on a skin of bright British Waterways green, which seems very ready to form large bubbles when it rains, not the healthiest of environments.

After the complexity of the city centre comes a dead straight 900m run with the railway hiding colliery spoil on the left bank and a line of pylons fronting open ground with long views on the right bank. At Trentham, recent housing leads to more mature but smart properties backing onto the canal. At Trentham Lock the canal returns to Staffordshire from the City of Stoke-on-Trent. Across the railway, the Wedgwood Visitor Centre, opened in 2000, has demonstrations, an art gallery, a restaurant with food served on fine Wedgwood china and one of the largest china factories in the world, moved here from the city centre in 1940. Visitors can make their own pots and have them sent on after firing.

Barlaston has the Plume of Feathers and then the plumes of greater reedmace plus herons and Canada geese. A golf course stands back on the right bank as the canal twists through tree-lined land, which is largely free of buildings. Meaford power station used steam locomotives until 1970.

The four Meaford Locks used to be a staircase of three. They are now well spaced out. The A34, which has been running parallel to the canal, now follows next to the bank to make its presence conspicuously felt.

Few towns are so clearly welcoming of canal traffic as **Stone**. There is obvious pride that this is where the canal began. A tree at the bottom of a canalside garden has a nestbox mounted with a television aerial, a juxtaposition that suggests they give their wildlife every possible amenity. After the first lock, the Manchester to Stafford railway crosses. On the left bank is a very large-scale miniature railway and there are three dry docks under canopies on cast-iron columns. This is where Rolt's *Cressy* was broken up. One lock has a horse tunnel with a rope-guide roller at its entrance. Downstream is a bank

Economy sized model railway beside the canal at Stone.

The lock at Stone with its shortened balance beam to clear the road bridge.

of ivy with a large wooden statue of a woman among the greenery. A recent fire beacon basket stands by the main mooring area. A friction crane on the town wharf, currently just a base, is being restored. Stone takes its name from the cairns on the graves of two 7th-century Mercian princes, killed by their pagan father, King Wulfhere, for practising Christianity. Stone was the home of Admiral Jervis and the site of a brassworks. Most of the shops are 18–19th-century. The Crown Inn was designed in 1779 by Henry Holland for mail coaches. Just above the A520 crossing are a restaurant and the Star, the latter with its only entrance on the towpath at the bottom lock.

Stone slalom course and Stafford & Stone Canoe Club's premises are across the field to the right. Five-times world slalom champion Richard Fox was a member in his heyday and the club has produced more British team members than any other.

These days the A51 leaves the A34 at Aston-by-Stone to cross the canal and follow it to Rugeley. Aston Lock is close to Aston's church and a moated hall.

From Aston Lock the line is almost totally rural to Rugeley, other than the noise from the railway and A51 which follow the left bank. The right bank of the Trent valley is gently rising farmland on a ridge that is just high enough to hide Stafford.

The canal slips quietly through the open farmland, not even noticing Burston, until past Sandon Lock, after which the 20ha Sandon Park stands on the ridge on the left. The 13–15th-century parish church in the grounds contains the 1603 family tree of the Erdeswicks. The 19th-century Jacobean Sandon Hall is owned by the Earl of Harrowby and notable features of the estate are a Gothic shrine to Spencer Perceval, assassinated in the House of Commons, and a conspicuous column rising above the trees to commemorate Pitt the Younger. Opposite is the hamlet of Salt and, beyond it, Hopton Heath, scene of a 1643 Civil War battle.

After crossing Gayton Brook, the canal passes under the A518 near Weston Hall and skirts round **Weston**, a village with a factory fronted by a mock-up of an angler seated by the water, dressed in yellow and complete with Thermos flask. Weston Lock is well south of the village.

Where Amerton Brook enters, a track leads up to a lane to Hixon. In 1968, a 120t electric transformer was being transported up this lane on a heavy-load trailer to English Electric's depot on the former airfield when it was hit on the level crossing by a 120km/h Manchester to Euston express with the loss of 11 lives on the train and dozens of injuries. The time sequence from the train-approach warning being given to the arrival of the train was not adequate for a 162t rig travelling at 3km/h to clear the crossing. All large loads have since had to ring for permission to cross all level crossings.

On the other side of the river, Ingestre Hall was the Earl of Shrewsbury's Jacobean building, damaged by fire in 1882 and restored as an arts centre. The front was by Nash, the landscaping by Capability Brown and the adjacent St Mary's church of 1676 by Sir Christopher Wren. This is where Edward VII spent his holidays.

After Hoo Mill Lock, powerlines cross and there is a garden centre and tea rooms before Great Haywood Junction, where the Staffordshire & Worcestershire Canal leaves. The Trent & Mersey Canal now forms part of the Black Country Ring to Fradley.

Behind Haywood Lock and a restaurant at Great Haywood, the River Sow joins the River Trent.

To the right of the canal lies Shugborough Park, restored as a 19th-century working estate. An 1805 farm is the Staffordshire County Museum with farm machinery, working cornmill, demonstrations of farming methods and historic breeds of livestock. There are restored estate interiors, a kitchen, butlers' pantry, brewhouse, laundry, coach house, shops, domestic life exhibits, costumes, toys, crafts, steam locomotives and a café. The white colonnaded mansion was founded in 1693, greatly extended about 1750 and altered by Samuel Wyatt in 1790–1806.

It was bought by William Anson from the Bishops of Lichfield in 1624 and since then has been the home of the Ansons, who evicted a village in order to expand their estate. William's great grandson, George, became First Lord of the Admiralty in 1751 and repaired the neglect of the Royal Navy. His stay in Canton, during his four-year round-the-world voyage, resulted in the Chinese House. Other art treasures include 18th-century French and English china, French furniture, tapestry and paintings, silver, Vassalli rococo plasterwork and eight beautiful neoclassical monuments, including an 18th-century monument by James Athenian Stuart. It is the earliest neo-Grecian building in the country. It was partly occupied by photographer Patrick Anson, the 5th Earl of Lichfield, a descendent of the original owners. Victorian terraces, rose gardens and garden and woodland walks are backed by the greater expanses of Cannock Chase. The hall of Shugborough, which means meeting place of witches and hobgoblins, is set in 3.6km² of park.

The iron bridge carried a drive to Shugborough Hall with its pines, beeches and rhododendrons. The Essex Bridge was a late 16th-century packhorse structure of sandstone. At 1.5m wide externally, it was the longest in the country. Fourteen of the 42 arches remain and it is now 94m long with a triangular cutwater at each pier. It was built as a hunting access bridge for the Earl of Essex to Cannock Chase, 50km² of heath Area of Outstanding Natural Beauty containing fallow deer, cowberries, bilberries, conifer and heather moor in an outlier of the moors of the north of England.

At Little Hayward there is a railway junction where, in 1986, a London-bound express on the West Coast Main Line had its path crossed by a northbound train with a resulting crash.

Close by is a moat and Colwich Lock, with a Victorian cottage and the remains of a roller on the bridge parapet. Muddy fields around ancient farm buildings and the church tower peering down through a horse chestnut tree give the impression that construction of the canal has been the only sign of change in recent centuries.

At Wolseley Bridge, Midland Crafts have a display of onyx and other mineral specimens, jewellery and a geological garden. Wolseley Garden Park comprises 18ha of formal gardens opposite Bishton Hall.

Dredgings on the bank contain freshwater mussel shells, showing that the canal is or has been more muddy than polluted.

An aqueduct carries the canal over the River Trent and this is the first time since Stoke that the river can be seen from the water although the two have run side-by-side for 30km. Brindley's Bank takes its name from the canal's engineer. The Bloody Steps date from 1839 when Christina Collins was carried up to the Talbot Inn. She had been raped and murdered by the drunken Pickfords boat crew, with whom she had been a passenger; two were hung and one transported for 14 years for the crime, the basis of the *Inspector Morse* episode *The Wench is Dead*. The sandstone steps, since replaced, were said to ooze blood on the anniversary and ghostly sounds and apparitions are still experienced. Another **Rugeley** miscreant was Dr William Palmer, publicly executed in Stafford in 1855 as a multiple poisoner.

The B5013 crosses, to be followed a little later by two freight railway bridges, the first of stone with massive concrete buttresses after it partly collapsed into the canal in the 1920s.

The A513 follows the right bank of the canal while the left is screened by a line of poplars, not enough to hide the power station, where two of the cooling towers

Bridge protection cut by towlines at Wolseley Bridge.

A TV aerial and a metal gate are probably the only changes to this view at Colwich since the canal was built.

Spode House has given its name to fine porcelain.

Constricted space in the Armitage Narrows.

The Armitage factory and the Lady Hatherton, formerly the inspection boat of the Staffordshire & Worcestershire Canal Company's directors.

Silver birches line the reach through Fradley Wood.

The Swan at Fradley Junction faces the end of the Coventry Canal, always a hive of activity in the summer.

are white and two red in an attempt to break up their visual impact. Next to it was Lea Hall Colliery, closed in 1991, with reported reserves of 200 million tonnes of coal.

At the Ash Tree the canal passes Spode House, the name of which became synonymous with fine porcelain. Until 1972 the Armitage Narrows were a tunnel, one of the first two to have towpaths going through, but the sandstone became unstable as the result of colliery work at Lea Hall and the roof was removed and replaced with a concrete-beam bridge carrying the A513 at such an oblique angle that it is, effectively, a tunnel again. This new roof can be jacked up in the event of further subsidence but this has not been needed.

After passing the Plum Pudding Inn and Spode Cottage, the canal winds round the back of the church in Armitage with its Saxon font, its 1690 tower and the rest rebuilt in the 19th Century in Saxon- and Norman-style. Inside is a loud 200-year-old organ from Lichfield cathedral.

If Spode has given its name to fine porcelain, Armitage has managed no better than toilet bowls but there is a steady demand for Armitage Ware, as shown by the large fleet of their lorries parked next to the canal.

The West Coast Main Line crosses for a last time and heads away towards the south-east. By the Crown Inn is a garden with some brickwork that includes a recess. This takes a lifebelt neatly and another is filled by a model of a human skeleton.

After Handsacre, the canal becomes rural except for the various overhead powerlines but the appearance is of heathland quality vegetation, slightly stunted oaks interspersed with gorse and birch. Alders increase before the A515 crossing, after which the right bank has Scots pines interspersed with rhododendrons while the left bank has birches.

Near Wood End Lock, the canal reaches its most southerly point, turning sharply north-east to continue its journey. To the south can be seen the spire of Lichfield cathedral. To the north is the open farmland of the Trent valley. A stream runs northwards and it is the valley of this that causes the pronounced corner in the canal's route. A former airfield is seen through the trees to the east.

Shade House and Fradley Middle Locks drop the canal to Fradley Junction, the second-busiest point on the inland waterways, where the Coventry Canal joins opposite the Swan, an inn built to serve the boat trade. Now with a field of caravans behind, it finds plenty of land-based custom.

The Trent & Mersey Canal becomes part of the East Midlands Circuit, the flight continuing with Fradley Junction, Keeper's, Hunt's and Common Locks. Several

Flower gardens are tended in odd corners on the Fradley Flight.

For the only time before its last lock the Trent & Mersey Canal makes brief use of the River Trent itself.

The church at Wychenor, approached by a rather makeshift looking bridge.

Beside the A38 at Wychenor.

locks have corners that are not easily accessible which have been turned into small gardens of flowers or shrubs by local residents. The crane is an import from Horninglow Basin.

Bagnall Lock stands alone. The A513 makes a final pass over the canal at Alrewas Lock. Between the two somebody has been doing some interesting topiary with a wavy-topped-willow-lattice fence and a circular lined hole through it. Perhaps this harks back to the village's fame for basket weaving. Next to the lock is the Old Boat with a traction engine in the garden.

The village has black-and-white thatched buildings, a working flour mill and a 13–14th-century stone church with a pinnacled tower. Lady Godiva reputedly used a former Saxon church here.

The Trent is braided here and the canal passes through another lock and then uses river channels for over a kilometre. Alrewas takes its name from the alder washes used for growing the trees that provided the materials for basket making.

The main channel leaves over a large diagonal weir. A bridge across from the church at Wychnor gives a narrow channel for powered craft, often with a significant flow.

The lock cut at Wychnor collects driftwood. There are the remains of a towrope roller on the bridge parapet. The lane can be busy as it leads to a country club. It is not busy with people rushing to claim the flitch of bacon that still hangs in Wychnor Hall for any couple able to prove their marriage is completely harmonious, a custom probably started by John of Gaunt in 1347. A red-brick Georgian farmhouse was the Old Flitch of Bacon coaching inn.

From Wychnor Lock the canal joins a major traffic corridor, opened up by the Romans with their Ryknild Street. This has now become the A38, a dual carriageway forming the major link between the West and East Midlands. The canal follows the A38 for 15km to Egginton, often right alongside but otherwise just far enough away for the roar of traffic to die back. Gorse and blackberry bushes line the canal with cow parsley, white deadnettles and other flowers adding colour.

Powerlines cross where the canal begins to edge away from the road. Across the fields, a 16th-century battlemented church tower locates **Barton-under-Needwood** at the foot of Needwood Forest. This was a royal forest with its only access from Barton Turn. These days Barton Lock is followed by a junction of the B5016 with the A38, one slip road looping over the canal.

A relative newcomer to the transport corridor is the Birmingham to Derby railway, which follows to Findern. Barton was chosen as the site of the Central Rivers depot for servicing Virgin's West Coast Main Line Pendolino tilting trains.

Confusingly, Tatenhill Lock is accompanied by the Barton Turns public house and Barton Turn Marina. This area is dominated by gravel working for concrete, resulting in the 16ha Branston Water Park with windsurfers, model boats, anglers, picnic area, children's play area, wild flower meadow and wildlife. A bench, like another to follow in Burton upon Trent, is carved with a selection of water birds and the information that this is the Trent & Mersey Canal through the National Forest. Presumably this includes the ash tree on the near bank and the willow and two hawthorn bushes on the other side.

The Bridge Inn at Branston would seem the ideal place to stop for a ploughman's and pickle but the menu seems to major on variations of curried chicken. On the other side of the canal there are extensive views to the ridge of Tatenhill Common, where a motorcycle trials route winds its way up and down the hillside.

Canada geese have taken over Branston Lock. The A38 makes its only crossing of the canal and then there is a notice over the canal advising that **Burton upon Trent** is the largest town in the National Forest. The trees were felled from a conservation area between Branston and Burton for Centrum 100, a massive distribution site with large warehouse, offices and parking for 114 lorries for Bass and Carling. On the other side of the canal at Shobnall, Marston's have their own brewery.

Burton upon Trent is dominated by a large tower with the Bass name round the top. Perhaps it also had a tower when it was the Old English *burh-tun*, a fortified village. This is Britain's brewing capital, often marked by the odours of brewing, an activity thought to have been

Topiary in the making at Alrewas.

A carved bench faces the canal near Burton upon Trent.

187

started by monks in the 13th Century, using water which was very clear because of its high gypsum content. The first brewery was established in 1708 and there were a score at one time. Burton's most popular product is India Pale Ale, initially made for export only, until a boatload sank and was released onto the home market by the underwriters. In the town, the Bass Museum includes England's oldest microbrewery, an Edwardian bar, shire horses, a 1905 Robey steam engine, a steam lorry, a bottle car and an N-gauge model of Burton upon Trent in 1921. There is also a Heritage Brewing Museum.

A conspicuous, riveted, pipe bridge across the canal advertises Marston's, who used to occupy the adjacent Shobnall Basin, now a marina. The Bond End Canal provided a link to the River Trent although there was a bar across at the Trent & Mersey end. It also used to flush the town's main sewer.

A5132 and B5008 cross, the latter with a picnic site by the Green Dragon. The 585-berth Mercia Marina is being built in a former trout angling lake.

The Stoke-on-Trent to Derby railway crosses, having taken a more direct route than the canal. Between here and the river is a large power-station complex with powerlines heading off over the canal.

The Canal Turn public house is premature. First comes Findern, a village marked by its black church spire. An apprentice here was Jedekiah Strutt, whose contribution to society was his invention of the ribbed-stocking frame.

As the canal turns from north-east to south-east it passes under the Birmingham to Derby railway and swaps its company for the current line of the A50. The road mostly stays far enough away not to be intrusive. A freight railway line follows, with coal trains to Weston-on-Trent.

The canal mural beside the canal at Horninglow Basin.

The bridge with the odd parapet line at Barrow upon Trent.

The toll wharf at Swarkestone.

A cottage near Dallow Lane Lock was considered a suitable site for a waterway activity centre.

At **Horninglow** Basin, a large canal mural decorates the side of the embankment of the A38, now back alongside.

Swans nest on the bank and herons fish as the canal approaches Stretton. There is still a large works to come and a large sewage works with the tower of Bladon Castle between them on the other side of the river. Building has been taking place at Stretton. The Mill House pub turns its back on the sewage works.

The canal moves from Staffordshire to what is soon to become much more rural Derbyshire as it crosses the first of the two channels of the River Dove. The canal uses a dozen low brick arches, at 73m the longest aqueduct on the canal, while the A38 uses a classic style bridge despite the road's being of almost motorway standard. A pillbox protects the aqueduct. Near the Dove is a dovecote. Mussel shells are found on the bank.

At the Every Arms, the canal parts company with the A38 and the route becomes more peaceful. There is a small works at the approach to **Willington**, where the

After the marina of the mid 70s and the Bubble Inn at Stenson Lock, which has a 3.8m fall and is the first wide-beam lock at this end of the canal, there are occasional extensive views southwards across the Trent Valley.

The 18th-century stone-and-brick Arleston House farmhouse is followed by the Ragley Boat Stop public house. A minor road crossing the canal to Barrow upon Trent has a very odd parapet line although it probably looks fine to road users.

The following bridge crosses from Swarkestone Lows to Swarkestone and the Lowes Bridge, a five-arch structure that crosses the Trent, continuing across the meadows for a further 1.2km. From the 18th Century, part remains from the 13th Century, having been built by two sisters of the Harpur family whose fiancés were drowned trying to ford back through the flooded Trent after being called across to a meeting of barons. This is where Prince Charlie's Jacobite army of 1745 abandoned their attempt on London and began their retreat to Scotland.

A narrows with an old wharf crane and a group of

traditional canal buildings enabled control to be kept of boats emerging from the Derby Canal. The toll house was to catch traffic passing from this to the Swarkestone Branch or Shunt, which dropped through four locks to the River Trent. The deep Swarkestone Lock is at the end of the Derby Canal, closed in 1964, the remains being used for moorings.

The freight railway crosses at **Chellaston** and long views open up to the south. Another of the Willington powerlines marks where a further former railway bridge crosses, now carrying the Derby Cycle Route, just a little less rural than in the days when *diuraby* was from the Old Scandinavian for deer, giving the future city its name.

A Ukranian farming settlement was established near Weston Cliffe. A church with a 14th-century spire stands near a reach where increasing beds of greater reedmace join the willows and alders to Weston Lock at Weston-on-Trent. This is another power station which remains quite well hidden. Donington Park motor racing circuit across the valley is also hidden but audible when in use. Aircraft climb over it out of East Midlands Airport.

The freight railway crosses again and departs and a last set of powerlines cross. Up from Aston Lock, around which lakes are dotted, is Aston-on-Trent, at the centre of which is a 12th-century stone church with its noted Victorian stained glass. Sweeping round the back of the village and crossing the canal is the new line of the A50.

Red-brick waterways cottages flank the grey stone Georgian Shardlow Hall. Shardlow, with the former A6 crossing the canal in the centre, is a canal jewel and one of the first inland ports, dating from 1797. After Shardlow Lock is the Clock Warehouse, a four-storey building of 1780 with a barge entry point, now converted to a popular family public house. Shardlow Heritage Centre is in the Salt Warehouse and there are maltings and the Malt Shovel of 1799. Everywhere there are warehouses. In 1839, 140 boatmen at the port called for the canal to be closed on Sundays for religious reasons. The canal company took up the call and tried unsuccessfully to get Parliament to apply this to all canals and railways. The village claims the largest number of public houses in the country for its population size.

Moored among the many wharves are large numbers of narrowboats, likely to leave a film of spilt fuel on the water surface. The Lady in Grey restaurant is approached through a hole in a wall but the New Inn at Great

Salt Warehouse at Shardlow, now a heritage centre building.

Wilne is less coy. Chapel Farm Marina is by Derwent Mouth Lock. This final lock is overshadowed by a large 19th-century horse chestnut that British Waterways wanted to fell because it is damaging the structure of the lock but a petition against this and a subsequent Tree Preservation Order have saved it.

There is a four-way deep navigation junction where the canal finally rejoins the River Trent at Wilden Ferry. The River Derwent was abandoned for freight after the Derby Canal was opened but the Trent is open for navigation both ways at this point.

Wharf buildings and moorings in Shardlow.

Derwent Mouth Lock with its horse chestnut tree.

Distance
150km from the Preston Brook Branch to Derwent Mouth

Navigation Authority
British Waterways

Canal Society
Trent & Mersey Canal Society Ltd
www.trentandmersey.btinternet.co.uk

OS 1:50,000 Sheets
117 Chester & Wrexham
118 Stoke-on-Trent & Macclesfield
127 Stafford & Telford
128 Derby & Burton upon Trent
129 Nottingham & Loughborough

35 Macclesfield Canal

Dramatic views and advanced landscaping

The Macclesfield Canal was one of the later canals to be built, being opened in 1831. It was designed to act as a shortcut around part of the Trent & Mersey and Bridge-water Canals. Thomas Telford selected the line and used bold embankments and cuttings to produce a shorter journey time, rather than using the more economical construction method of following contours. This produced a canal with dramatic views, helped by the proximity of the Peak District, whereas the Trent & Mersey had kept to more even and lower ground further to the west. Notwithstanding the extensive cut and fill, engineer William Crosley managed to cut a couple of kilometres off Telford's line. He produced landscaping that was advanced for its time and often the surrounding land blends up or down to the canal edge in a way not achieved by most of our motorways.

The attractiveness of the canal and its importance as part of the Cheshire Ring make it an ideal cruising waterway. Powered craft keep the water free of weeds but leave it muddy.

Marple Junction is where the Macclesfield Canal leaves the Peak Forest Canal at the top of the **Marple** flight of locks.

To enthusiasts of canal architecture, the most attractive spot on the canal comes immediately, with stone buildings adjacent to a stop lock, which was probably never used because of good relationships between the canal companies, particularly over water supply. The whole ensemble is framed at each end by a pair of immaculate roving bridges, of which the canal has several excellent examples.

Bridges down the canal are generally elliptical in section, the abutments curving in at the bottoms and the grooves of towropes cutting the corners on the towpath side.

The canal heads south through housing estates that terminate at Hawk Green. Across the road is the imposing brick block of the Goyt mill, now with small industrial units. A hoist still projects over the canal, a reminder of when boats carried the goods.

The canal quickly breaks into open country although it is just possible to glimpse the spread of spires, gas holders and tower blocks of the south of Manchester through gaps in the hedge on the right. The noise of jets using Manchester Airport can be heard.

At High Lane, the canal passes the backs of more houses.

The High Lane Arm is the home of North Cheshire Cruising Club. It has corrugated-iron boathouses, some of which are considered to be heritage.

Just before leaving High Lane, the Bull's Head is passed on the right while misleading signs on the other bank are meant to indicate that the Dog & Partridge offers food 200m up the road the other way, although the arrows point along the canal.

An aqueduct carries the canal over the railway on its first high embankment and gives views through to Disley, a castle-like structure in Lyme Country Park and the Peak District National Park, which runs parallel to the canal for much of its length and comes within 600m of the canal at one point.

Higher Poynton was a mining district for over 200 years and now suffers from subsidence. Canal banks and bridges have to be raised regularly. There is still some evidence of coal tips. A short arm by bridge 15 used to be a mine connection.

Once again, the canal is on high embankment with fine views to the west. Occasionally it passes through wooded cuttings. Alders predominate. Snowstorms of rosebay willowherb and thistledown are encountered in the autumn. Much of the countryside is used for cattle grazing and, although the canal is never far from towns or villages, its atmosphere is mainly rural. The people met are noticeably friendly.

At Wood Lanes there is a marina. Desirable residences are passed from time to time, each in rural settings. One at Booth Green seems to surpass the others. Helicopters sitting on the back lawn just put the cherries on the cake.

Arrival at **Bollington** is marked by the Clarence mill, now accommodation, business premises and shops. This time it is of stone but still has the hoist over the canal. An aqueduct takes the canal over the river before a

The broad sweep of the canal at Higher Poynton.

The canal runs at a high level with a backdrop of the Peaks at Higher Poynton.

A fine roving bridge at Mode Hill.

timberyard. Strangely, Bollington lies on the River Dean while Macclesfield is on the River Bollin. The Adelphi mill houses commercial premises.

High-level passage gives views across the rooftops to the right while peaks such as Nab Head, White Nancy and Tower Hill dominate to the east. Bollington was once supported by its cotton mills and another stands silent on the right of the canal, the four turrets on top of its tower mocking its quietness.

Mode Hill has another roving bridge. The bridge at Tytherington was restored by the Macclesfield Canal Society, including an invalid chair ramp.

Astra Zaneca' drugs factory flanks the canal at Higher **Hurdsfield**, its fences guarded by closed-circuit television. The towers and spires of Macclesfield are visible beyond, with fine industrial dignity built upon an 18th-century silk industry and now relying on textiles and pharmaceuticals.

The canal runs above most of the town. At the A537 bridge there are various shops and the Puss in Boots Hotel on the left of the canal with Peak Forest Carriers across the road. Just down the hill is the Bridgewater Arms, a strange choice of name as the sign indicates that reference is being made to the Bridgewater Canal which lies away on the other side of Altrincham.

Hovis mill stands beside the canal as high-class apartments, its moorings now used by Macclesfield Marine Centre and Freedom Cruisers. A heavy retaining wall flanks the hill on the left, its base surrounded by heather. Bridge 41 is much higher and wider than other bridges, without a constriction of the canal, but it is built in similar style and probably took a tramway across. Bridge 43 is another roving bridge and between the two there are views across the orange rooftops of a newer part of **Macclesfield**.

Along the canal there are many abutments of former swing bridges but one survives at Lyme Green.

Approaching Oakgrove, tier upon tier of trees rise up in front, the top being surmounted by the transmission tower on Croker Hill, directly ahead.

Beyond Sutton Reservoir, chevrons on the bank indicate the presence of the A523 if its traffic noise has not already done so but it soon goes away again, leaving the rest of the journey noise-free.

At Oakgrove, the Fool's Nook provides food and beer. A low lifting bridge crosses the canal. The hillside on the left is wooded for a while. A large house by the canal has its opening windows plain while the others are leaded, an effect that looks incongruous.

The canal has been lined with reedmace sporadically but as the ground falls away to the right, taking the canal with it, the rushes close in and narrow the otherwise wide canal for the first time. The canal was designed so that all the locks come together in a flight of 12 at Bosley. They have mitre-top gates because of the depth of the locks. Side pounds exist at all except the top lock, which now has a garden instead. Both these and the locks themselves are built on embankment in places. Taken with the rapid fall of the ground and the

adjacent disused railway embankment, there is some dramatic topography. Overlooking it all is the Cloud, a high hill with the remains of ancient earthworks beyond the bottom of the flight. Part of the craggy appearance is from a quarry that provided stone for the canal and then the railway.

Having reached the level of its lower pound, the canal might be expected to run in cutting for a while but, instead, it leaves the lock flight on a high embankment and a single-arch stone aqueduct carries it over the River Dane. Parallel to the aqueduct is a long viaduct carrying the railway.

Milestones are met before and after the railway passes over the canal. An extreme example of Telford's no-nonsense line comes in the form of a 2km dead straight length to the east of **Congleton**. Unfortunately, the same thinking was not employed at bridge 75, where the concrete edging of a flyover does not line up with the stonework of the original of the three bridges sharing this number 75 and forms a chicane for narrowboats. The third bridge carries the railway.

At the other side of a cattle

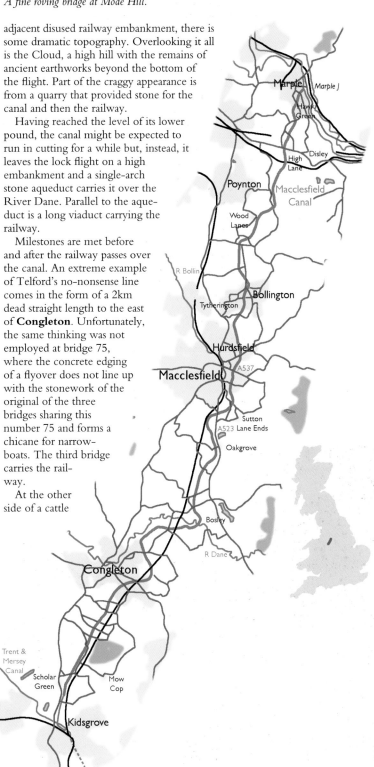

Bosley bottom lock.

An aqueduct to the south of Congleton has white railings rather than the usual stone sidewalls, giving a view onto cars below whilst crossing. A mill on the right has been converted to flats and faces a modern, concrete water tower on the other side of the canal.

Arrowhead and yellow water lilies begin to make their appearance for the first time in the vicinity of a golf course.

Mow Cop dominates the eastern skyline. Its summit is capped by a folly in the form of a ruined castle dating from 1750, inside which a 14-hour prayer meeting was held in 1807, resulting in the birth of Primitive Methodism.

Ramsdell Hall, standing back from the left bank of the canal, overlooks a magnificent spread of lawn, a meadow-sized putting green. Looking back on the other side of the canal, it is possible to see Little Moreton Hall just over a kilometre away. Built between 1559 and 1580, it is surrounded by a moat and is probably the best example of Elizabethan half-timbering in existence. It has a fine collection of pewter and oak furniture, is opened to the public by the National Trust and can be approached by the footpath leading from one of the canal bridges.

Access across the canal at Scholar Green is on a swing bridge and this is soon followed by a narrow swing foot-bridge.

The final lock at Hall Green is supposed to be a stop lock but low water levels in the Trent & Mersey have resulted in a drop of some 600mm across the lock. Here it joins the Hall Green Branch.

Kidsgrove appears on the hill ahead. At Red Bull, an aqueduct carries the Hall Green Branch over the Trent & Mersey, one of only three such canal flyovers in Britain. A sharp left turn by Red Bull Marina brings the Hall Green Branch parallel to the Trent & Mersey, which it joins at Hardings Wood Junction after another 600m, during which time the Trent & Mersey has climbed through two more locks to meet it. Many boats have cruised back and forth along this stretch of the Trent & Mersey, trying to find the entrance to the Macclesfield Canal on the north side.

The Trent & Mersey is stained a rusty, orange colour by ochre from the ironstone strata but the Macclesfield is kept clean as water flow is in a southerly direction.

On the other side of the Trent & Mersey is the Tavern. Just to the south lie the Harecastle tunnels.

meal mill the Railway public house offers the landlady's home-made pies. The Queen's Head is keen to promote its food, too, and is approached from bridge 76 which, like the bridge beyond, is a fine roving bridge.

Distance
42km from Marple Junction to Hall Green

Navigation Authority
British Waterways

Canal Society
Macclesfield Canal Society http://adkins-family.org.uk/macclesfieldcanal/mcs

OS 1:50,000 Sheets
109 Manchester
118 Stoke-on-Trent & Macclesfield

The Cloud rises above the Dane aqueduct.

36 Caldon Canal

Although a branch of the Trent & Mersey Canal, the main business of the Caldon Canal was self-contained when it was constructed in 1778. Its purpose was to bring limestone from the quarries at Caldon Low to the industrial area at Etruria, site of Josiah Wedgewood and Thomas Bentley's factory producing Jasper Ware, red-and-black figure porcelain vases and cameos. The siting is where the Etruria marl meets the coal measures of the North Staffordshire coalfield, the marl and coal being used 1:6 in the production of the porcelain. It was the canal that established the commercial importance of **Stoke-on-Trent**.

The area has been turned into the 3ha Etruria Industrial Museum with a waterside public house, museum, steam beam engine from about 1850 and the Etruscan Bone & Flint Mill with its iron-banded brick chimney, erected in 1857. Also included are the canal maintenance buildings and dry dock at the junction with the Trent & Mersey, just above Etruria Top Lock.

Beyond a cricket pitch, an anti-aircraft gun guards a parking area for several army lorries before brown-brick-clad tower blocks. Brick bottle kilns remind of pottery factories, now being replaced by housing estates. Ivy House lift bridge, at the far end of this reach, carries cars.

Between Northwood and Abbey Hulton is a landslip area.

At Sneyd Green the canal turns sharply right just by the Foxley public house and the blocked-off end of the 800m Foxley arm, disused since 1934.

Beyond a brick bridge arch heavily reinforced with cast iron at Milton, the canal winds behind a line of houses.

This long pound ends at Norton-in-the-Moors, with lock 4, Engine Lock, taking its name from a steam-powered beam engine that pumped water from a nearby mine. The conurbations of Stoke-on-Trent fall back, too, and open country is met for the first time.

The River Trent passes under the canal at Norton

Brindley's statue of 1990 faces the Etruria works.

The Bedford Street staircase at Etruria.

New housing arises around bottle kilns in Hanley.

The Caldon Canal has one of the most dramatic changes of scenery over its length of any of the British canals. The start is past new housing, fronted by a 1990 statue of Brindley. A gently arched bridge has X-braced parapets, perhaps inspired by Horseley Ironworks bridges

The first two of the 17 narrow locks come as a staircase at Bedford Street. Like the following locks, they have large side weirs and have split footbridges across their lower ends, supported by decorative ironwork.

The water is largely clear of flotsam, other than some arrowhead, and has no debris in it. As the canal curves round towards Planet Lock alongside a new hospital it passes a bank planted with birch trees. Bracken lines the other bank.

The line cuts through the ornamental **Hanley** Park with its ornately decorated canal bridges, bandstand and other public facilities before emerging to give the first views of the moorlands, sweeping round to the transmission mast at **Fenton**.

Green but there is also a feeder connection to allow it to deliver water from Knypersley Reservoir.

The canal now begins to blend into the landscape in a most agreeable fashion as fields slope gently down to water level rather than being cut into by the canal. This is particularly so at Stockton Brook, where irregular hilly fields dominated by clumps of deciduous trees tumble down to meet the canal as it climbs its final flight of five locks, accompanied by sculptures showing features of local industry and life. These five begin with Waterworks Lock, which takes its name from the Victorian waterworks alongside, and follow with Fens Lock, Railway Lock where the line crosses

193

Above Fens Lock at Stockton Brook.

over, Road Lock by the Sportsman and Top Lock, which has its large side weir discharging beneath a set of old canal buildings. The canal is now 148m above sea level and takes off with a broad straight pound instead of the narrow twisting route so far.

This is interrupted at **Endon** by a circular island in the canal, the base of a swing bridge for a former light railway. The main line used to continue through what is now Endon Basin but was diverted to the south in 1841 to accommodate the railway and feeder from Rudyard Reservoir. Another feeder from the south is supplied by Stanley Pool.

Endon, dominated by a creeper-covered house on a hilltop, obtains fame for two days in late May each year as the only village outside Derbyshire where well dressing is practised.

The canal now begins to follow the right bank of the Endon Brook valley. With the wind following, the sun casting a muted yellow light over one shoulder and deep blue storm clouds gathering harmlessly over the moors ahead, the boater is able to enjoy canal travel at its best.

The occasional rock outcrop, brick bridges being progressively replaced by stone and cows feeding in woods at the base of a steep bank, lead up to Hazlehurst Junction where the 4km Leek branch leaves on the right. This route to the Capital of the Moors was built in 1801, partly to bring water from Rudyard Reservoir.

A wrought-iron footbridge here is matched by decorative wrought-iron railings around the first of the deep side ponds to the three locks, opposite the whitewashed lock keeper's cottage.

Although it was never officially abandoned and remained passable to Hazlehurst, the canal was restored from near dereliction by the Caldon Canal Society with the assistance of Stoke City Council, Staffordshire County Council and British Waterways and reopened in 1974.

The main line drops through three locks and then passes under the Leek Branch. Hazlehurst Aqueduct was constructed in 1841 from brick rather than the usual stone although its appearance has not been improved by an anaemic coat of whitewash or, rather, pale greenwash.

Beyond the Hollybush at Denford, the canal swings southwards into the valley of the River Churnet and some of the finest scenery in the Midlands, deep wooded valleys with overhanging crags.

At **Cheddleton** comes the first of several unusual side weirs with the towpath taken across the face of the weir while a narrow footbridge adjacent enables the man leading the tow horse to keep his own feet dry.

Cheddleton is renowned for its mill complex with buildings dating back as far as the 13th Century. Its pride is a magnificent pair of undershot mill wheels. One of these flint mills was probably designed by James Brindley.

The banned tunnel at Froghall.

The main line drops away below the Leek branch at Denford.

The height difference approaching Hazlehurst Aqueduct.

The matched water wheel pair at Cheddleton Flint Mill.

The canal passes under a wooden building and the A520.

Restored in 1967 and open to the public on weekend afternoons, the mills were used to grind flint from the English Channel before sending it up the canal to the Potteries. Other exhibits include a Robens steam

The Churnet section flows placidly through Consall Wood.

Froghall's substantial limekilns.

Trip boat moored at Froghall Wharf.

engine, a haystack boiler, a model Newcomen engine, a section of plateway and lime and ochre kilns. On the hill overlooking the complex are the remains of a 14th-century church.

Beyond a wooden building across the canal, two locks drop past a restaurant and a farm-gate factory before the canal reaches the station for the restored Churnet Valley Railway, which follows the canal, and the Staffordshire Way footpath comes alongside.

Leaving the Boat Inn, the canal cuts below a hillside with the scars of motorcycle scrambling engraved upon it. The alders give way to Scots pines in Consall Wood as the canal approaches Oak Meadow Ford Lock, where the navigation takes to the River Churnet. If the level at the lock is more than 150mm above normal, all boats are forbidden from passing.

Normally dark and slow, the river winds down to Consallforge, once an extensive water-powered ironworks and recently restored limekilns, where the canal breaks away from the river past the Black Lion and into woodland that is almost oppressive in its solitude. There is barely room for the railway to squeeze in between the canal and river and a wooden footway takes the towpath under the railway bridge. The railway platform was rebuilt in 2007, cantilevered out low over a particularly narrow section of canal, and the railway waiting room is cantilevered out even further.

Unguarded weirs discharge surplus water down steep falls under the railway into the river.

The final lock, Flint Mill, is in red stone, as is the adjacent water-driven mill, which grinds sand as a constituent of pottery glaze.

Cherry Eye Bridge, with its pointed, stone arch, marks the start of Rueglow Wood, with its picnic area and mooring. This reach has been re-lined in concrete, its idyllic scenery only deteriorating as pipes begin to emerge in the wood and concrete planked and metal spiked fences appear around Thomas Bolton's copperworks.

At Froghall, the canal turns sharply into the low and irregular 69m tunnel, listed by British Waterways as being by far the shortest British canal tunnel from which unpowered craft are specifically banned in all circumstances. The reasons are far from clear.

Froghall Wharf, the terminal basin, has a canal craft centre, horse-drawn narrowboat trips and a picnic area. It has ruined limekilns and, when constructed, was connected by horse-drawn tramway to the quarries at Caldon Low.

In 1811, the canal was extended southwards to Uttoxeter. The Uttoxeter Canal, as the new part of the Caldon Canal was sometimes called, was built as a blocking move to prevent a broad-beam canal being built to rival the Trent & Mersey. It ran for a further 21km with 17 locks. In 1846 the North Staffordshire Railway bought the Trent & Mersey Canal, of which this was a branch, and converted most of the canal extension to a railway which has, in turn, been closed. However, there are now restoration plans. The first lock on the Uttoxeter Canal at Froghall has been restored, giving access to a lower basin.

For 2km at **Alton** the bed has water, albeit shallow, putrid and blocked by the occasional fallen branch or dry section. Although it skirts the southern boundary of Alton Towers, the pleasure park remains hidden and nothing more than the screams of punters is heard. On the other hand, the castle, with its mixture of Scottish and French Gothic styling, is the dominant building, high on its crag on the south side of the valley. Water appears in the canal just beyond a small sewage works but conditions are such that boating is not a serious proposition at the moment.

At **Rocester** the giant JCB factory straddles the line with buildings and ornamental lakes. The last water is at Combridge.

Future prospects at Alton.

Distance
41km from Etruria to Combridge

Navigation Authority
British Waterways

Canal Society
Caldon & Uttoxeter Canals Trust
www.cuct.org.uk

OS 1:50,000 Sheets
118 Stoke-on-Trent
& Macclesfield
119 Buxton
& Matlock
128 Derby &
Burton upon Trent

The restored basin below the first lock on the Uttoxeter Canal, to continue beyond the far bridge.

37 St Helens Canal

Section of Standish Street railway bridge erected by the canal.

There is some debate as to which was the first canal of the modern era. There is no doubt about which was completed first, the St Helens Canal or Sankey Brook Navigation being opened in 1757, a couple of years ahead of the Bridgewater Canal, but the case is put that the St Helens was not a true canal but the making navigable of the Sankey Brook, although almost all of it was actually in new cut. Initially, the canal ran between St Helens and the tidal Sankey Brook at Warrington, the section to Widnes being added later, and had ten locks including a two-rise staircase. Its purpose was to carry coal from the St Helens coalfield, destined for the Cheshire saltfields.

Closure came in 1948 and today it exhibits the full range of states of repair although there are hopes of reopening it. Some sections of the canal are in water while others have disappeared without trace. Its terminus is in **St Helens**, emerging next to Tesco's store. The town is synonymous with Pilkingtons, whose glassworks surround the head of the canal. Broken glass is a regular feature along the banks, some of which have been restored to act as public walkways. The World of Glass museum places it all in context.

Disused railway tracks have to be crossed at one

stoppage, then across a road and slightly downhill past Kentucky Fried Chicken's premises. A section of the Standish Street railway bridge ironwork has been set up beside the canal as an example of industrial craftsmanship.

The Old Double Locks are the oldest, working staircase in Britain. They have been restored but lack of navigable water means that they are deteriorating again with disuse. Overlooking them are the Burgy Banks, mountains of debris from the glassmaking industry. Ravenhead Greenway is the title given to an open amenity area next to the canal although the canal often fails to match it. From the Old Double Locks the canal is just a narrow rocky stream surrounded by high vegetation. A six-stepped weir in place of the New Double Locks drops the water down in front of the A58, access on foot being up the hill and then keeping to the left of the stream, thus avoiding the area fenced-off for the large sewage works.

There were branches at Gerrards Bridge, Boardman's Bridge and Blackbrook, only the Boardman's Bridge branch not being largely intact. From Blackbrook, the canal lay on the left side of the Sankey Brook and just the occasional pond remains in water. Wharf appears in

The only remaining cone house, part of the World of Glass.

The Old Double Locks, restored but deteriorating again owing to disuse. Beyond are the Burgy Banks.

The world's first passenger-carrying railway crosses the line of the world's first industrial canal.

The canal comes back into water beyond the railway.

Bradley Swing Bridge across the canal at Wargrave.

local road names in some parts of **Newton-le-Willows** as a reminder.

Earlestown had a railway wagon works near the canal although its pride is the Sankey Viaduct, which carried the world's first passenger railway over both the world's first modern canal and the river. The first railway viaduct of any size, it was based on canal aqueduct design and has nine semi-circular arches 15m x 21m high with each pier resting on 200 timber piles 6–9m long.

At Wargrave the canal comes back into good condition and has been landscaped as part of the Sankey Valley Park. The first of the attractive swing footbridges crosses. The setting is pleasant with banks of oak trees and bluebells in the spring despite the obvious nearness of houses, pleasant enough for the kingfisher to be seen.

The infilled Hey Lock is passed near the former Ruston Diesels engineering works and Vulcan Foundry, which made locomotives, reputedly sending a steam locomotive to India every week for a century until closed in the 1990s. The nearby Vulcan Village has matched houses, as are likely to appear on the levelled locomotive works site. The railway alongside was a branch to Warrington off the Liverpool & Manchester Railway but subsequently became part of the Liverpool to London main line when the Grand Junction Railway was opened from Birmingham.

As the Merseyside/Cheshire boundary is crossed, the canal dries up. In 1974, British Waterways decided to use the canal bed south of the boundary for tipping rubbish, thus creating a long section down to Bewsey which needs redigging.

Locks were built above and below Winwick Quay and the M62. The first of these, Winwick Lock, is being re-excavated. West of the canal are Burtonwood services and a disused airfield with the motorway being carried economically down the centre of the main runway. The airfield was the U. S. Air Force's largest during the Second World War. Beyond the M62, the last surviving drydock is opposite the canal maintenance yard. Hulme Lock cottage had its lower storey below ground level and is being partially restored by the canal society and Warrington Borough Council.

The Sankey Brook occupies a section of the canal deepened to reduce flooding. A dry section of canal leads through a series of recent housing developments on the outskirts of **Warrington** (Old English village at the weir). Almost continuous water then begins and the park has been extensively landscaped. Lack of use of the water permits duckweed and water fern to flourish later in the year.

Also disused on the right bank is Bewsey Old Hall, a building originating in the 13th Century, which has seen fires, murders, abductions and visits by royalty and is due to be turned into apartments. There is also a ghost story about the White Lady of Bewsey and a white rabbit relating to it. It was this tale which is believed to have inspired Lewis Carroll's *Alice's Adventures in Wonderland*.

The canal moves down towards another railway viaduct. This has rounded arches except for a square hole over the canal.

A culvert carries the canal below the A57 Sankey Way dual carriageway. The last of the benches installed is passed near a bandstand. The canal approaches Sankey Bridges and a low road bridge next to a timber and builders' merchants on a former boatyard. The hand-cranked swing bridge, used every time the low bascule bridge upstream was out of use, now stands partially restored near a former lead works. Coastal sailing flats used to be built here and there was a time when this reach was a sea of masts.

A low railway bridge crosses and brings the railway alongside for most of the remaining journey. The canal bends right. Formerly, the canal went straight on to join

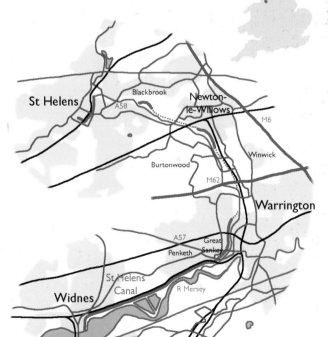

Old swing bridge at Sankey Bridges, a view formerly filled with ship masts.

Empty coal train leaves Fiddler's Ferry power station with the pipeline alongside.

Final lock and Spike Island moorings on a misty day.

the tidal Sankey Brook near what is now a sewage works.

The canal passes under the first of several more wooden swing footbridges by a large chemical works, forming the southern edge of Great Sankey.

At Penketh there are extensive views to the left over the River Mersey and across to the prominent water tower on the Daresbury Laboratory.

A wall across the canal, acting as a level control weir, marks the start of the section that acts as the Fiddler's Ferry Yacht Haven. A swing footbridge crossing the top of the lock allows access from the adjacent River Mersey. Part of a slipway remains, used for launching concrete boats during the First World War The sudden appearance of all these moored yachts with their high masts comes as a shock after the general disuse upstream.

The major landmark is the massive Fiddler's Ferry power station with its eight cooling towers. Large-diameter pipes from it cross the canal in a bund, carrying waste to lagoons between the canal and river. With the large pipes and high fence on one side and coal trains feeding into the merry go round unloading system on the other bank, the situation can be quite claustrophobic.

Carterhouse Swing Bridge, signal box and chemical works by the canal at Moss Bank.

Distance
21km from St Helens
to the River Mersey

Canal Society
Sankey Canal
Restoration Society
www.scars.org.uk

OS 1:50,000 Sheet
108 Liverpool

A high concrete bridge across from the power station is followed by two bands of vegetation across the canal, although it is possible to force a passage through both. A concrete wall crosses with a gap and the next section is then much reduced in width, overgrown with rushes and barely passable because of a very thick skin of slurry on the surface. Vegetation is starting to grow on the surface, surprising, as it is acidic power station waste ash.

The canal is severed again to let a stream flow across at low level. A concrete wall is parallel to and about a metre out from the left bank and just below the surface, hidden by the scum.

The world's first canal, dock and rail freight transfer complex.

After this blockage the canal returns to its wide state again with just greater reedmace and bulrushes at the edges (with a surprising number of balls resting among them) and a surface covered with duckweed.

Spoil heaps line the right bank and chemical works are not far away. Racks of exposure samples face the river in what has been a centre for the chemical industry since the first alkali works opened in 1847, coal, sand and Triassic salts being readily to hand. An abnormally high-tapered chimney is something of a landmark.

When the canal was extended, its western end at **Widnes** (Old English broad headland) was a resort noted for its fresh air, clean water and fine views. Much has changed.

The end of the West Bank Yacht Club's moorings are marked by a barge across the canal.

Almost opposite a brick air-raid shelter are the Swan inn and the interpretation centre and ranger huts for Spike Island, a final area of restoration that takes in the site of the world's first canal, dock and rail freight transfer complex. The Catalyst museum covers the chemical industry. One of the locks has been restored to permit use by the yacht club. Below high tide it is difficult to reach the river by any other route because of the deep mud around the lock. The flow is quite fast as it sweeps past the remains of timber sailing flats, beached to resist erosion. There are a children's playground and a red-stone church at West Bank.

Currents flow fast through the Runcorn Gap under the Runcorn Bridge. When opened in 1961, the bridge was the third-longest bowstring girder bridge in the world and is still the longest steel bridge arch in Europe. With a main span of 330m and a clearance of 23m, it replaced a transporter bridge of 1905, the design being chosen as a suspension bridge could have oscillated in some wind conditions because of the proximity of the railway bridge immediately downstream. The road bridge was widened in 1975–7.

One of its 76m side spans crosses the Manchester Ship Canal, Britain's largest and (until the Ribble Link) newest canal, the antithesis of the St Helens Canal. The lock gates into the Manchester Ship Canal from the Mersey are silted up and unusable.

Progressively higher are lines of spare dock gates, a swing bridge, Runcorn capped by its church at Halton and airliners heading towards Speke to land at John Lennon Airport.

38 Bridgewater Canal

The Bridgewater Canal was one of the engineering wonders of the age, the first Industrial Revolution canal to be built away from the course of an existing river, beginning the canal building era. To this day it remains the largest British water project to have been financed by one person. The Duke of Bridgewater turned his attention to business after being jilted and employed John Gilbert and James Brindley to build the canal. The canal was begun in 1759 and reached Castlefield by 1765. A wide canal, it followed the 27m contour and had no locks, despite some long straights, except at Runcorn, to which it was later extended to join the Mersey estuary, opening throughout in 1776.

It was bought by the Manchester Ship Canal Company in 1885, the builders of the biggest modern canal, which superseded it, although it continued to carry commercial trade until the early 1970s. Today it forms part of the Cheshire Ring from Castlefield Junction to Preston Brook.

The canal starts by the Pack Horse, fed by the River Medlock, emerging from under a low bridge. This is the Castlefield Urban Heritage Park, Britain's first such. Close by is the Arndale Centre, Europe's largest covered shopping area, which was devastated by an IRA bomb, and Manchester Central, the North West's newest and largest conference and exhibition centre, located in the former Central Railway station. The **Manchester** Hilton includes apartments above the 25th floor. Manchester takes its name from the British Celtic Mamucium after the breast-shaped hill on which it was built, that hill now being lost among a forest of buildings.

The canal immediately passes under the A56, which follows a similar line as far as Preston Brook. Here the canal is home to gudgeon, perch and roach.

On the right side, beyond the bridge, is the congregational chapel that was refurbished as a studio for Peter Waterman, who has had more pop number ones than the Beatles or Elvis. The traditional-looking footbridge crossing to it is, in fact, recent. Beyond it on the right, crossing the front of Grocers' Warehouse, are a pair of drawbridges. The warehouse has a waterwheel operated crane that was used to lift coal boxes 14m up a shaft in the red limestone cliff to Castle Street.

Castle Quay, formerly Middle Warehouse, has been restored behind a drawbridge and a dock. One of those to take residence was writer and broadcaster Mike Harding, whose interests range from rambling to folk music, preferably with a north of England flavour.

Merchants' Warehouse of 1800, on the right, was partly destroyed by fire but has been restored as a wine bar and restaurant. Next to it, the Rochdale Canal joins at Castlefield Junction.

The highly distinctive state of the art Merchants' footbridge has crossed the Bridgewater Canal since 1995.

The chapel turned recording studio with new footbridge and old drawbridge.

Curved in both plan and elevation, it is supported by a single arch, like a bowstring girder but leaning upwards and towards the inside of the curve at 60°. Its structural integrity depends on torsion, a concept more commonly used in aircraft design than in bridges. It is 67m long. The more traditional kind of bridging is seen at its best in the Castlefield Viaducts, massive Victorian girder structures that carry everything from the line to Liverpool to the Manchester Metro, which follows the canal to Timperley. To build the viaducts involved evicting hundreds of slum dwellers from Alport Town without compensation.

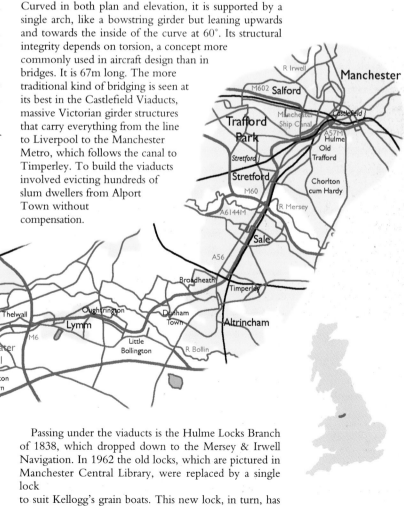

Passing under the viaducts is the Hulme Locks Branch of 1838, which dropped down to the Mersey & Irwell Navigation. In 1962 the old locks, which are pictured in Manchester Central Library, were replaced by a single lock to suit Kellogg's grain boats. This new lock, in turn, has been superseded by Pomona Lock, to suit a local trip

Castle Quay with its drawbridge.

boat. Branches run into several of the factories, the towpath still being taken over the entrances in a series of arches.

In 1850 St George's church was Manchester's finest ecclesiastical building.

The railway crosses, followed by Cornbrook bridge, where the stonework is protected from horse towropes by gouged wooden panelling. A south-easterly wind brings a range of aromas from Duckworth's essence distilling factory close by.

Pomona Docks, on the Manchester Ship Canal, were for coastal and perishable traffic, now linked to the Bridgewater Canal by the Pomona Lock. The docks were opened in 1894, the third largest in the country, amazing for a location so far inland. No 3 dock was a Ro-Ro off terminal, with Colgate's Wharf opposite. The land between the two canals was previously the pleasure gardens. Pomona Palace had political meetings with Disraeli speaking to 28,000 people in one hall, not to mention concerts and shows. Blondin wheeled a man in a barrow on a high wire and the botanical gardens were opened in 1831, John Dalton selecting it as the cleanest suburb in Manchester after a tour in which he wiped leaves with his handkerchief to check for pollu-

tion. In 1857 it was the site of an Art Treasures exhibition with an even larger Crystal Palace exhibition in 1887. It was all cleared to build the docks, a process repeated when the docks became an Enterprise Zone for redevelopment in 1981. Now they are better known as **Salford** Quays with such amenities as the Lowry and the Imperial War Museum.

One of the more dramatic buildings has a chain and hook on each side, each link of the chain being several metres high, the chains being freestanding and leaning in towards the building. The adjacent arched bridge crosses the Manchester Ship Canal. Beyond Throstle Nest bridge, which must be the most clearly named bridge on the canal system, is another dramatic building, several storeys high, with its end entirely given over to a bright mural featuring 90 years of businesses from the **Trafford Park** industrial estate, the workshop of the world, and some of its more notable products. This refreshingly meaningful mural took Walter Kershaw nearly five years to paint and has been left alone by those who usually like painting on walls. The huge Trafford Park industrial complex was built on land owned by the de Trafford family since the time of Canute and sold in 1896. Activity in the complex peaked during the Second World War, when 75,000 people were employed, producing Lancaster bombers, Rolls Royce aero engines, the PLUTO pipeline, penicillin, DDT and dyestuffs for camouflage. The *Daily Telegraph* was produced here. There is an oil refinery and a derelict power station, oil tanks, Massey Ferguson tractors, GEC, factories being knocked down and new works being built. Kellogg's were the last commercial users of the canal with grain being delivered to

The Trafford Park mural covering ninety years of activity in the industrial park.

their wharf until 1974. The steel dumb barge *Bigmere* was one of the last used for this purpose and now houses an exhibition of historical canal routes in National Waterways Museum, Ellesmere Port. Two brick warehouses had 600mm thick walls for storing cotton, now used for chilled food.

To football fans, Trafford Park means the home of Manchester United Football Club, the only team to have won the Cup and League double twice. It is one of the spiritual homes of the game, the stadium reaching almost to the bank of the canal. The Old Trafford cricket ground lies a little to the south-east.

The former Trafford Park power station was served by the Trafford Road coal wharves, from where coal was sucked onto conveyors through pneumatic tubes. Trafford Edible Oil Refiners, formerly Kraft, issue some unpleasant smells and use water from the canal for cooling. There may be tropical fish in the canal here. Two freight railway lines cross the canal after a container terminal to serve the engineering works.

Stretford Junction or Waters Meeting brings in the Stretford & Leigh Branch, the first section of the canal to be built, and the route follows that original line briefly, passing playing fields after being crossed by another freight line and the Piccadilly to Liverpool railway. The 1761 terminus lay between **Stretford** gasworks and the A56.

An inlet at Longford has mooring posts, crane mountings and a dry dock. Boats were slipped sideways here because of the relative narrowness of the canal.

Stretford Wharf, by the cemetery, had steps up out of the canal so that any horse that fell in could be recovered. There were stables close by at the packet changeover point, connected by the lengthsman's cottage. His most important task was to watch Barfoot Bridge over the River Mersey, hence Watch House, which gives its name to the boat cruising club that occupies the building today. The water is generally opaque. The canal company built a weir upstream to pass floodwater away from Barfoot Bridge.

The Watch House packet changeover point, lengthsman's cottage and stables.

The valley contains Sale Water Park and Chorlton Water Park, with activities in gravel borrow pits for the M60, which now crosses, Junction 7 being just west of the canal.

Manchester University Boat Club are on the corner by the Bridge Inn. They travel at speed, looking the opposite way, and their blades sweep almost the whole width of the canal.

Dane Road manure wharf has now become a picnic area. A highway depot precedes **Sale** warehouse and wharf. By the Upper Crust is the first of several venerable but well maintained cranes with piles of stop planks of substantial size for this wide canal. There is also a coal wharf here, followed by Roebuck Lane wharf.

The 19th-century St Paul's church stands prominently by the canal, perhaps too conspicuously as it was hit by an incendiary bomb in 1940, destroying the roof and leaving scars that still show on some of the pews. The cemetery which follows was fashionable and its occupants include Richard Pankhurst and James Joule.

Trafford Rowing Club adds to the activity and what was Walton Park grain store is now a sports centre with bowling greens, tennis courts and model trains. There is also an adventure playground close by.

The Manchester to Chester railway leaves the line of the canal after Timperley station. Once the railway was opened in 1849, the packet boats stopped here and discharged their passengers so that they could travel on to Manchester by the much quicker trains. The line was electrified in 1930. A freight line and oblique bridge that previously carried another line both cross.

A sluice drains excess water to Timperley Brook. Rotalac Plastics are sited at a former manure wharf for Manchester night soil. There are more horse steps out of the canal. While the Old Packet House is 18th-century, the Navigation was rebuilt in 1937. The canal was widened at Broadheath Wharves to allow grain boats to enter a warehouse. A cotton warehouse, now used for making prams, nursery furniture and toys, had enclosed loading. There was a coal wharf opposite MFI and B&Q.

The Roman road from Manchester to Chester didn't bother with the A56's detour to Altrincham. It crossed the canal's line somewhere near the 1897 factory of Linotype, with its elaborate clocktower, a factory that has played a vital role in the development of printing. Broadheath developed after 1884 because of the rail transport for its heavy engineering products and the canal for coal supplies. These days it has more to do with storage, distribution and DIY.

Oldfield Quay provides modern moorings for pleasure craft, adjacent to the canal.

The Bay Malton at Seamon's Moss bridge takes its name from a racehorse. A dovecote is for pigeons which are permitted a more casual life.

The move from built-up area to open country is dramatic for its suddenness. There are now extensive views over the mosses, peat bogs formed after the Ice Age and now used for farming. A depressing feature, though, is the angling place numbers at regular intervals along the towpath, well over a thousand of them, covering many kilometres.

Dunham Town had a night soil wharf near Dunham School bridge. Both the bridges near the village are well notched by towropes. Bluebells add a carpet of azure in the spring.

The one square kilometre Dunham Park has 150 fallow deer. It is a Site of Special Scienic Interest with 180 species of beetle, old trees not being felled nor fallen timber removed. An obelisk, dating from 1714, terminates one of the tree-lined avenues in the park. Dunham Massey Hall is a Georgian rebuilding of a Tudor house, the family home of the Booths and then Earls of Stamford. It has a superb collection of 18th-century furniture, Hugenot silver and portraits of the Booth and Grey families, including Lady Jane Grey, our briefest reigning monarch, on the throne for just nine days. Thirty rooms are open to the public, including the great hall, vast kitchen, library, laundry, 18th-century stables (containing a restaurant) and Elizabethan cornmill, converted to a sawmill in the 19th Century and working most Wednesdays. Around the park are an orangery, well house, slaughter house and Edwardian water gardens. A break in the trees and railings replacing a section of the hedge give a clear view northwards from the hall to the canal. Whiteoaks Wood contains an obelisk said to mark the grave of a favourite racehorse that was shod with gold although it actually formed part of the landscaping on Dunham Park in the mid 1700s.

Railings on the aqueduct over the B5160 give clear views down onto cars as they screech to a halt where the road narrows to a single track with little warning.

The Bridgewater Canal crosses the River Bollin on its highest embankment, 10m above the river, not so much an aqueduct as a concrete tower block with a flat arch cut into the bottom and a rapid passing below. In 1971 a breach occurred and the water level had dropped 360mm in Manchester before the stop planks could be placed. Meanwhile, the canal water had carved a 30m gorge in the embankment, carrying sand, gravel and rocks into the river. A new concrete and steel trough for the canal was opened two years after and the Bridgewater Canal Trust was set up to finance proper maintenance of the canal.

Views are extensive in all directions. Prominent

beside the River Bollin is the five-storey Victorian Bollington corn mill, restored as a residential building.

At the river, the canal moves from Greater Manchester into Cheshire and passes Little Bollington, the diminution being added after local government reorganisation in 1974 to avoid confusion with the Bollington on the River Dane. Bollington coal wharf and warehouse were sited here.

Three sets of powerlines pass over in quick succession. Ye Olde No 3 across the fields to the right is a 17th-century coaching inn, the third stop on the road from Liverpool to London. It has several ghosts, mostly heard rather than seen.

Along the canal banks are a couple of marinas, a caravan park, the Admiral Benbow, Agden wharf and a house that has been converted from the Old Boathouse Inn of 1779, with stables for the packet boat horses. Burford Lane warehouse and wharf, formerly used for grain, have been taken over by the Vincent Owners' Club for manufacturing motorbike spare parts.

Cotton manufacturer George Charnley Dewhurst paid for the 1870 Gothic church seen beyond the woods, across the rich dark soil of the fields at Oughtrington. The three-storey cottages by the bridge were open on the top floor, being used for cutting fustian on long tables to form velvet.

Bream, carp, chub, perch and roach frequent the canal.

Lymm had stables and a warehouse with a canopy that was used for packet boat goods and passengers until 1924. It still has a notice of 1884, warning boats not to moor there. More recent additions are a dovecote and, on the opposite bank, modern houses with verandas above the towpath.

Whitbarrow Aqueduct had an overflow to the Slitten Brook. A Victorian park is sited in the valley which had a slitting mill that flattened iron bars into strips for making into nails, tools and barrel hoops, being used for woollen manufacture after 1800. There was also a corn mill. Assorted watering holes include the Golden Fleece, Saddlers wine bar and Lymm Conservative Club.

A cutting in wet sandstone has a selection of ferns and liverworts and curtains of dark ivy everywhere.

A well kept park with swings leads down to the canal behind a winding hole.

The A56 crosses again at Ditchfield Wharf, where there is another crane with stop planks and an iron bar to protect the bridge from towlines. More noisily, the M6 also crosses as it rises onto Thelwall Viaduct to cross the Manchester Ship Canal. The viaduct has been dualled, the first viaduct having suffered badly from salt corrosion. Fifteen tonnes of salt is applied to the structure per year to keep it free of ice. The replacement cost of the viaduct was £10-million or £27,000 per ton of salt applied.

Thelwall Grange, built in quieter Victorian times, is now a nursing home. Further along the hillside is Massey Hall, now a school but previously the house of the owner of Rylands Wire Works. Thelwall was founded in 923 by Edward the Elder, with a camp to guard the Mersey from the invading Danes.

The top part of a green marine navigation buoy at Cliff Lane Bridge Wharf draws attention to S&A Marine in a former tannery owned by the Co-operative Wholesale Society. Soon after, the A56 crosses on a heavy concrete bowstring bridge.

A notable building in Grappenhall is the 16th-century St Wilfred's church by the canal, some parts of which date back to the 12th Century. In 1874 it was enlarged and restored. At the gate there are stocks. Inside there is a Norman red sandstone oblong font, a 13th-century dugout chest made from a tree trunk and a 1275 effigy of Sir William Boydell. Some of the stained glass is very old and at one time it was said to have more stained glass than any other church in Cheshire. There is a 180-year-old

sundial and a stone cat above the west window. Lewis Carroll's father preached here at times and would have been accompanied by his family, giving credence to the suggestion that the cat was to become the Cheshire Cat in *Alice's Adventures in Wonderland*.

There is a sports club with playing fields on the left.

The canal effectively acts as the southern boundary of Warrington as far as Lumb Brook underbridge at Stockton Heath, which has fine stonework by Brindley. Views from the canal are extensive again.

Stockton Heath warehouse and wharf have semi-circular steps that served as the interchange between fast packet boats from Manchester and Runcorn and stage coaches for Warrington or the south, the London Bridge taking its name from the A49 London Road which crosses, the Roman road from Middlewich. It was home to the *Duchess-Countess* packet boat, which took five hours to Manchester and carried a blade on the front to cut any towlines of other boats that were not dropped in time. A model of the boat is to be found in the National Waterways Museum, Ellesmere Port. Stables for the packet boat horses were in the house of the bank rider, now occupied by Thorn Marina. The dovecote is a recent addition.

The Delamere Way footpath follows the canal from its Stockton Heath end towards Frodsham, passing the Fox Covert cemetery and older Baptist one beyond (where Oliver Cromwell once worshipped) and leaving across the golf course. On the other side of the canal is a cricket pitch. Hill Cliffe Reservoir contains 45,000m^3 of water for Warrington. Hillfoot Farms' 17th-century barn has an arched end and round holes for pitching hay or straw down from the loft. Water from Appleton Reservoir is supplied to Warrington at a rate of 2,200m^3/day through a pipe carried on a wooden frame at Hough's Bridge, a crossing named after a local family.

Another local family of great significance at Higher Walton were the Greenalls, noted for their brewing. Two generations of the family rebuilt the village, the Victorian and Edwardian houses now mostly having been sold to tenants. The Gothic revival church of 1885 was built by Paley and Austin for Sir Gilbert Greenall and he had Walton Hall built in 1836, these days with a children's playground, pitch and putt course, outdoor chess, museum, children's zoo and gardens laid out by the first Lady Daresbury. The canal runs in a broad sandstone cutting.

The A56 crosses for the last time at Chester Road Wharf. Once again, there are extensive views. There used to be a manure wharf before the next bridge. This reach has curled pondweed which shelters minnow, perch, roach and tench.

At Moore, there is a memorial to Ken, who lived rough on this stretch of canal until his death in 1984 and for whom flowers have subsequently been put out.

The broad sandstone cutting past Walton Hall.

Moorfield bridge has another stop plank crane and winding hole, on a section of canal above two railway lines and with views across the Mersey to Fiddler's Ferry power station. Up the hill, beyond the potato fields, is the tower of All Saints church at Daresbury, the church where Lewis Carroll's father was the vicar, something clearly indicated by his characters in the stained-glass windows. Court House in Daresbury has become a Lewis Carroll centre. Many boat crews were not prepared to come up to the church, however, so the vicar converted a mission boat at Preston Brook for their use.

A more conspicuous tower, these days, is the white one of Daresbury Nuclear Physics Laboratory, housing the world's largest tandem Van de Graaff generator. The laboratory also had the first synchron radiation source for research, cooled by water drawn from the canal.

George Gleave's bridge is one of the most attractive on the canal, steeply humped and with its corners protected by wooden posts to prevent rope wear.

There is a drain sluice down towards the Keckwick Brook before powerlines come alongside for the last kilometre to Waters Meeting. Despite the increasing noise from the traffic on the M56, which crosses next to the canal fork with the A56 joining the M56 at Junction 11 alongside, the area is quiet enough for a heron to feel comfortable. On the far side of the railway, at Red Brow, there is a white cottage with gargoyles.

Ahead, the short Preston Brook Branch connects through Preston Brook tunnel with the Trent & Mersey Canal. The plan was that transhipment should be done at Middlewich but, by one of those unfortunate little errors, the tunnel was just too narrow for Mersey flats to get through so it all had to be done here, to the considerable financial advantage of the Duke of Bridgewater. Two kilometres of wharves, warehouses, offices, houses and stables were set up from the tunnel onto the main line

A stop plank crane with the Daresbury Nuclear Physics Laboratory tower behind.

Stitts warehouse, the last remaining of the canal transhipment facilities at Preston Brook. It is now home of Pyranha, Britain's largest canoe manufacturers. The water tower is visible on the hilltop.

towards Runcorn, forming one of the busiest inland ports in the country, later becoming a canal/railway interchange with the Warrington to Crewe line passing under the end of the line to Runcorn.

One of the few remaining canal buildings is Stitts warehouse, occupied by Pyranha, one of the world's leading canoe manufacturers and source of much original thinking in canoe design as well as being behind such landmarks as the descent of Everest and the majority of Richard Fox's world slalom championship successes. At one time, Norton warehouse covered the canal arm here but now it is open.

A marina for 300 boats was built opposite in 1974 and the towpath diverted across the canal and round the back of what is now Pyranha's site for security reasons. From here the canal forms the northern boundary of **Runcorn**'s housing.

The Warrington to Chester railway passes over and tunnels under a hill topped by a massive 1892 water tower, which is a key feature in Liverpool's water supply from Lake Vyrnwy. More powerlines follow the canal as it skirts Windmill Hill, on which there is no longer a mill.

A canalside notice invites water users to visit the remains of Norton Priory, an Augustinian centre from 1134 until the Dissolution in 1536. The museum has the most extensive display on medieval monastic life in Britain and features the largest excavation of its kind in the country. There are a church, 12th-century undercroft, chapter house, dormitory and cloisters displayed, woodland gardens of a 16th-century house, summerhouses, Victorian rock garden and stream glade with azaleas, Georgian walled garden, shop and refreshments. It took a five-year legal battle to buy land from the owner of the priory to complete the canal.

Runcorn has a network of dedicated roads for single-deck buses and one crosses the canal from the direction of Castlefield, where the remains of the castle are to be found on top of the hill. The Barge, beside the canal, has been attacked by the graffiti merchants.

An industrial estate lies between the canal and the Manchester Ship Canal, hidden by the embankment of the A533, which runs alongside and then crosses. Watering holes come quickly, the Royal Naval Association, the Grapes Inn, the Navigation and the Egerton Arms. There were also various commercial premises, Astmoor wharf and tannery, Highfield tannery, Halton Road gasworks and wharf and the Bridgewater foundry.

The Sprinch was a loop containing Runcorn dockyard, left after a bend was straightened in 1890. When a minor road was dualled to form the new A533 a further section of the loop was lost, leaving two spurs. Victoria Dockyard was the main one on the canal, building and repairing the Bridgewater Company's extensive fleet. Bridgewater Motor Boat Club, formed in 1951, the oldest on the canal, use the crane every Sunday to remove the gate from the large drydock.

The end of the present section of canal is reached past the new Brindley Theatre & Arts Centre, the Hotel Campanile and the Waterloo Hotel. Beyond Waterloo Bridge the A533 viaduct leads up onto the Runcorn Bridge, blocking the canal.

There were two flights of locks down to the Manchester Ship Canal. The New Line flight of ten double locks was closed in the 1960s and has been built over, the lock gates being reused on the River Avon. The Old Line of five double locks has been preserved. When the second Mersey crossing is built, it is hoped that the flight will be restored, creating a second Cheshire Ring via the River Weaver.

Among the rubble beyond the old locks stands Bridgewater House, owned by the duke during the canal's construction. It is surrounded by the campus of Halton College and new apartments.

Distance
45km from Castlefield to Runcorn

Navigation Authority
Manchester Ship Canal Company

Canal Society
Bridgewater Canal Trust, Runcorn Locks Restoration Society www.runcornlocks. org.uk

OS 1:50,000 Sheets
*108 Liverpool
109 Manchester*

39 Leeds & Liverpool Canal, Leigh Branch

A link that is important rather than pretty

The Leeds & Liverpool Canal was the last of the three trans-Pennine canals. The Leigh Branch was opened in 1820. It runs south-east across Greater Manchester from the main line at Ince-in-Makerfield to join the Stretford & Leigh Branch of the Bridgewater Canal, thus avoiding craft having to cross the Mersey estuary to travel south.

The junction with the main line at **Ince-in-Makerfield** is just above **Wigan** Pier in the valley of the River Douglas. Wigan power station was demolished in 1989 but had been supplied along the branch until 1972 by short boats. These craft were characterised by ornate baroque scrolls, flowers and birds on the bows and stems and by square wooden chimneys, which were prone to catching fire.

While an important link, there is no way that this is a pretty canal but it is improving. Bracken grows from between the blackened stone blocks, red clover enhances the mown grass along the towpath and sycamore and poplar trees line the first reach. After the first road bridge there is a large patch of arrowhead.

Most of the buildings around are recent houses.

The two Poolstock Locks are the only locks on the canal and their size determined the 18.9m x 4.3m of the distinct variety of wide beam barges used on the main line and on the Bridgewater Canal. Its wooden construction was based on Mersey and Weaver river barges with crew accommodation in fore and aft cabins below deck. Its maximum load of 45t was large and this contributed to the prosperity of the canal, even resisting railway competition at first. Steam engines were used from 1890, especially the V-shaped twin units of the Leeds & Liverpool Canal Carrying Company in Wigan, which worked in pairs as motor and butty, the last one being launched in 1936 and the design finally being withdrawn in the 1950s.

One lock was built with a wooden chamber to deal with subsidence problems. Both have penstocks that open sideways instead of vertically as would usually be the case and the lower lock has windlasses with chains to pull the balance beams shut instead of relying on leg power, which would normally be enough. It also has a footbridge with baffle plates to resist wheeled traffic.

The stone roving bridge with bulldozer just visible on the Wigan Alps beyond.

Poolstock Lock Number 2 with Poolstock church beyond.

Dover Lock Inn with no longer a lock in sight.

One of the unusual windlasses on the lower Poolstock Lock

The Plank Lane lifting bridge in front of the Britannia Hotel.

Beyond the Eckersley Arms is a blackened but very substantial church. A young wood on the other bank includes many birch trees.

A dismantled bridge formerly carried a railway line, which bypassed Wigan.

The route has been badly affected by repeated subsidence. The banks have frequently been raised and the canal is deep and is largely edged with sheet piling, concrete planks or in situ concrete with frequent additions. Views from the banks and not infrequently from the water can be extensive. The area was one of the earliest centres of the Industrial Revolution because of good coal seams at the surface. Deep mining replaced it. Scotsman's and Pearson's Flashes are the first two subsidence lakes, to be followed by a number of others. They are being filled with refuse and material from the Wigan Alps, the pet name for the colliery spoil tips.

A stone roving bridge, which carried the towpath from right to left sides, would no longer bear a horse as the deck has been removed and replaced with a foot-bridge span.

Two bridges carry important railway lines but this does not prevent the kingfisher from occupying the intervening reach, emerging from the reedmace to streak away in a blur of azure and orange.

Abram begins suddenly with houses right beside the canal but then draws back to leave the hillside on the left as grazing for horses, overlooked by another blackened but substantial stone church. To the right there are long views across the valley of the Glaze Brook. The canal runs on embankment, at times, with low ground on both sides. The Dover Lock Inn draws attention to a series of bank undulations that were once Dover Top Lock and Dover Low Lock, these two being eliminated when subsidence made them unnecessary.

Other redundant structures include Edge Green Basin, seen as a blocked-off inlet on the left, and a railway line that once crossed close by. Plank Lane Bridge was a swing bridge carrying a busy road and opened mechanically

Stop board crane in Leigh, ready to address subsidence.

by a bridge keeper, until it was replaced by a troublesome, electrically operated, lifting bridge.

Next to the bridge is the Britannia Hotel. Bickershaw Colliery, the final working mine in the area, has now gone without trace.

A significant quantity of spoil has been tipped into the largest of the subsidence lakes, the Flash, but it still has plenty of room for Leigh & Lowton Sailing Club, not to mention the water birds for which it is a popular venue in the winter. The Flash is visible from the canal. There are extensive views over low land at its east end towards Pennington.

The canal takes a rather odd line on its approach to **Leigh**, a series of wiggles that are so slight it is not clear why a straight line was not chosen.

A new road crosses on the line of a former railway. New housing stands beside the canal.

Beside the canal is a brick warehouse that has been restored as the Waterside Inn.

The Leigh Branch meets its Bridgewater Canal equivalent on the far side of the A572 bridge in Leigh. Leigh comes from the Old English word *leah*, glade, although that meaning seems rather lost these days.

Distance
12km from the main line to the Bridgewater Canal

Navigation Authority
British Waterways

OS 1:50,000 Sheets
108 Liverpool
109 Manchester

Pennington Flash at the former Bickershaw Colliery site.

Derelict warehouse in Leigh, restored as the Waterside Inn.

205

40 Bridgewater Canal, Stretford & Leigh Branch

The scheme that launched the canal building era

Derelict canalside mill in Leigh.

The Bridgewater Canal was the first of the modern canals to be constructed entirely independently of rivers. The first section was opened in 1761 and it began the era of canal building. A contour canal with no locks, it was a broad canal with many features that would not have seemed out of place on much later canals and which served as a model for many canals that were to follow.

It was constructed to bring coal from the mines of the 3rd Duke of Bridgewater at Worsley to the centre of Manchester. Construction of the new canal immediately halved the price of coal in Manchester, made a fortune for the duke and made the name of James Brindley who, along with James Gilbert, engineered the canal. Scheduled passenger services began in 1776.

The Stretford & Leigh Branch was subsequently extended by the Leigh Branch of the Leeds & Liverpool Canal, which took it on from Leigh.

The end of the Bridgewater is immediately east of the A572 bridge in **Leigh**. The limit of the Bridgewater Canal is marked by a stop board suspended from a manually operated iron and timber crane, one of a number that were installed to protect the 50km of water on one level. Other than the occasional shopping trolley or floating armchair, the water is generally clear of debris.

Leigh was a coalmining town and also has a number of mills, with prominent chimneys, dating from Edwardian times. After crossing the line of a disused railway, the canal passes Butt's Basin with its wharf and canal office. Most craft are narrowboats but a concrete boat is moored among them.

A new brick bridge is built in a style in keeping with the original brick-arched bridges, followed by a school on the left. The housing is a mixture of modern estates backing onto the canal and traditional canal terraces with roads meeting the canal at right angles. Occasionally an immaculately kept garden or some other gem turns up unexpectedly, just when the canal seems to be settling in for a poorer reach. Perhaps it is the results of the efforts of Wigan Groundwork or of the City of Salford that have generally resulted in these oases of inspiration.

A bridge gives one of a number of reminders of the 6km/h speed limit with another stop board crane close by. Glimpses of more open country appear with lower land leading on to Chat Moss, away to the south, and views northwards to the Pennine and a prominent aerial.

Another school follows. The Leigh Manufacturing Company Ltd of 1909 is one of the mills that has been built beside the canal in grand brick style but now lacks its former glory with windows holed and large entrances bricked up.

The A580 East Lancs Road crosses and the canal is into country which seems derelict except for a couple of pipe bridges passing over.

Astley Green is a colliery village, dominated by its pithead gear, the disused mine being selected as the location of the National Mining Museum. A red-brick Victorian church stands to the south side of the canal with the waterside Old Boathouse public house built opposite.

Spoil tips stand off to the south in an industrial wasteland and high banks show signs of subsidence. Much of the canal has concrete edging although planking is sometimes used instead. Picnic tables are erected at intervals in a derelict area with a couple of bridges that have lost their decks and most of their abutments now having Bailey bridge spans thrown across the gaps.

The former course of a railway closes in, originally coming alongside at Boothstown basin, a dock with soil tipped in its entrance and several sunken barges lying within.

The canal forms a southerly loop as it follows the

Magnificent Packet House and passenger landing steps in the basin at Worsley.

Another stop board at Worsley.

One of the bridges across the approach to the mines.

contours, young birches and stands of gorse growing along the banks. At intervals a blackened but delicate church spire stabs the sky above the housing. Up the hill from the one at **Worsley** is Old Hall, a 16th-century manor house once owned by the Duke of Bridgewater. Exten-sively remodelled in the 19th Century, it now has restaurants and hosts Jacobean banquets. Briefly back to the current day, as the canal is overlain by Junction 13 of the M60, the canal passes unexpectedly into one of the jewels of the British waterways network, the basin at Worsley and its surroundings. The canal goes into cutting, passing below a stop board crane and heavily timbered house and under a brick arch into the basin. The basin is breathtaking. At its head stands the magnificent, ornate, timbered Packet House and former passenger boat landing steps, with a packet boat once again moored below. Scott's church of 1848 contains a rich collection of Duke of Bridgewater monuments.

From the north-east corner of the basin leads a channel that divides while ducks rest on any surface, oblivious of the Casserole restaurant above. The channels pass under the A572: the one on the left through rough hewn arches festooned with ivy, the one on the right through brick arches with inter-esting holes going off at angles, the two routes rejoining in a pond surrounded by a rock cliff. In the corner, covered by a wire grill, is the top of a double-track inclined plane that operated between 1797 and 1822, protected at the top by locks. The inclined plane served the mines which were worked from the 14th Century onwards and in which the Duke of Bridgewater cut 74km of canal tunnel. Despite the fact that this was the first of the industrial canals, over two centuries ago, those tunnels still account for 52 per cent of the canal tunnelling in Britain and exceed the total for the canal tunnels in the rest of the world. A narrow barge lies in the water outside the entrance, showing the small size of the Starvationer craft that were used. The mines are now considered too dangerous to enter but water is pumped out and the iron ore in it colours the canal orange for a considerable distance. There are hopes that some of the mined area can be made safe for public access.

The southern end of the canal basin is flanked by wooded lawns and backed by a post office, village store and the Bridgewater Hotel. It leads on to Worsley dry-docks (now used for working on narrowboats), Worsley warehouse (now fully restored for alternative use) and various traditional housing around the Worsley coke ovens site. In the midst of it all, the Sea Cadets have their T. S. Ilex. There are many craft moored, including a narrowboat with paddles at the stern instead of a screw.

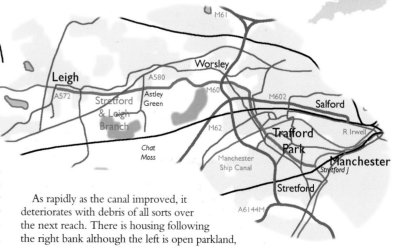

As rapidly as the canal improved, it deteriorates with debris of all sorts over the next reach. There is housing following the right bank although the left is open parkland,

Worsley Wharf and drydocks.

The entrance to the coalmines and a Starvationer.

once crossed by a railway passing close to the canal at Monton. The Bargee inn and restaurant is situated at this point. A busy road follows the left bank until the M602 crosses both. The reach passes under the Liverpool to Manchester railway, after passing mills now being used by the Great Universal mail order business. The canal bends round a hospital at Patricroft and another main road follows the right bank from the Wellington through to Barton upon Irwell. By the crossroads in Patricroft, the bank has been landscaped. Amenities on the other side include the Packet House. The Dutton Arms is situated beyond a canal basin and there is an interesting set of warehouses.

Several tower blocks are passed before the run down to the Barton Swing Aqueduct, another wonder of the canal system. Originally a 183m x 12m clearance stone aqueduct crossed the River Irwell in three arches; it lasted from 1761 to 1893 when the Manchester Ship Canal was built. The original plan was to lock down and up again, subse-

quently considered as a boat lift to clear the Manchester Ship Canal before the present solution was developed by Sir Edward Leader-Williams. Situated with the pivot point on an island in the middle of the Manchester Ship Canal and a tall brick tower at the west end, it has control huts on the two ends of the Bridgewater Canal which operate stop gates. The girder swing aqueduct has a 72m x 5.8m x 2.1m deep tank which is swung full to save time draining. With 23m clearance over the Manchester Ship Canal, it weighs 1,600t and is opened half an hour before a ship is due on the major waterway in order to give the ship time to stop in case of a malfunction, any collision having potentially catastrophic consequences.

Perhaps it is not surprising that the authorities do not rush to move this vast structure to let a small boat across. However, the walking route is not easy down a steep bank, through a narrow gate which leads directly onto the main road, along the footway of the adjacent swing road bridge with an inevitable crosswind and finally along a narrow footpath hemmed in between a fence and a hedge before a significant drop from sandstone blocks with rounded edges. Passed on the walk is the richly decorated Catholic church of 1867 by Pugin.

The character of the canal now changes as it passes into the huge **Trafford Park** industrial complex. Built on land owned by the de Trafford family since the time of Canute, it was sold in 1896.

The branch passes a building selling pies and sandwiches before arriving at Stretford Junction or Water's Meeting and the main line of the Bridgewater Canal, part of the Cheshire Ring, with another Liverpool to **Manchester** railway line crossing on the far side.

At this point the canal is rather hemmed in with high security fences and the nearest road access point is in **Stretford**.

Distance
17km from Leigh to the main line

Navigation Authority
Manchester Ship Canal Company

OS 1:50,000 Sheet
109 Manchester

The massive Barton swing aqueduct carries the Bridgewater Canal across the Manchester Ship Canal.

41 Rochdale Canal

The Salt Warehouse at Sowerby Bridge.

Todmorden. A picnic table faces the Salt Warehouse with boat works. The 12th Halifax Sea Scouts are located around it and a canal museum is incorporated.

In front of the William IV is the second lock and then the recent Tuel Lane Tunnel. The 100m long tunnel replaces a section filled in to carry the A58 and a road off it. Reopened in 1996, this is one of the most complicated examples of canal restoration work. The tunnel curves at the end and is subject to considerable turbulence from the lock. It may not be entered without the lock keeper's permission. Locks on the canal were set at a 2.7–3.0m rise to allow interchange of gates. The Tuel

During the era of canal mania it took just an hour in 1791 to raise £60,000 in subscriptions to build the first trans-Pennine canal. Calderdale was the centre of the British wool industry in the 17th and 18th Centuries and completion of the Rochdale Canal in 1804 provided a much improved link with Manchester. Surveyed by John Rennie, it was built by William Jessop, who didn't like tunnelling, preferring to add extra locks to climb over obstacles. This is the only trans-Pennine canal without a summit tunnel, offering splendid scenery, particularly on the West Yorkshire side of the summit, but producing a large number of locks: 36 to the summit and 56 down to Manchester. It was built as a wide beam canal, passing its last working boat in 1937. Apart from the Cheshire Ring section, it was abandoned in 1952 but remained as a water supply feeder, requiring structures to be maintained. Since restoration, there have been several serious breaches.

The Bolton Brow Gallery is at the canal basin at the head of the Calder & Hebble Navigation in Sowerby Bridge, featured in the BBC's True Tilda series, and the Moorings bar is in an old canal building. The Rochdale Canal leaves the Calder & Hebble Navigation just east of the basin at the Kirkham Turn, named after Ralph Kirkham, the founder of the Rochdale Canal Society.

The first lock is Albert Wood Lock, named after the canal's main trader. On the south side of the lock is the

River Calder, which the canal is to follow to

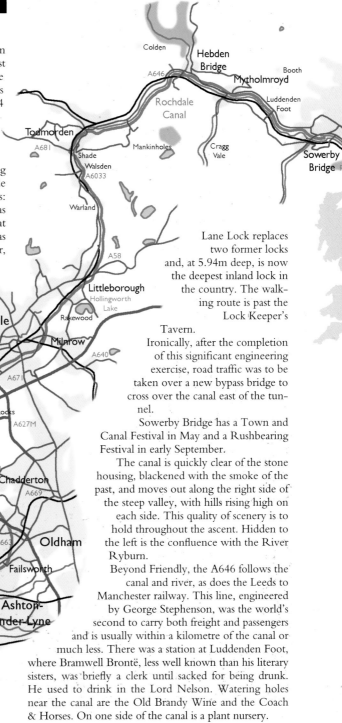

Lane Lock replaces two former locks and, at 5.94m deep, is now the deepest inland lock in the country. The walking route is past the Lock Keeper's Tavern.

Ironically, after the completion of this significant engineering exercise, road traffic was to be taken over a new bypass bridge to cross over the canal east of the tunnel.

Sowerby Bridge has a Town and Canal Festival in May and a Rushbearing Festival in early September.

The canal is quickly clear of the stone housing, blackened with the smoke of the past, and moves out along the right side of the steep valley, with hills rising high on each side. This quality of scenery is to hold throughout the ascent. Hidden to the left is the confluence with the River Ryburn.

Beyond Friendly, the A646 follows the canal and river, as does the Leeds to Manchester railway. This line, engineered by George Stephenson, was the world's second to carry both freight and passengers and is usually within a kilometre of the canal or much less. There was a station at Luddenden Foot, where Bramwell Brontë, less well known than his literary sisters, was briefly a clerk until sacked for being drunk. He used to drink in the Lord Nelson. Watering holes near the canal are the Old Brandy Wine and the Coach & Horses. On one side of the canal is a plant nursery.

Tuel Lane Tunnel at Sowerby Bridge.

Washing hangs between terraces at Hebden Bridge.

At Charlestown near where the Pennine Way crosses the canal.

Mytholmroyd is sited where Cragg Brook joins the River Calder. In the 18th Century it had coin counterfeiters. It was the birthplace of former Poet Laureate Ted Hughes and is renowned as the venue for the world dock pudding championships. Beyond the White Lion, the A646 crosses, to cross back over the recent Falling Royd tunnel, a structure made of corrugated arches on a curved line through which daylight is not visible; it is surprisingly dark inside for its large size.

One of the stone mills houses Walkley's Clogs, together with a Cloggers & Coiners Museum and a selection of shops in Victorian style. In **Hebden Bridge** itself, a town developed on the manufacture of fustian, there is a Childhood Reflections museum with dolls, toys and miniatures, including working Meccano models. Hebden Bridge Vintage Week, in early August, follows the town's arts festival. A small but open marina is sited in front of the Railway and Hebden Lodge Hotel. A neighbouring house has a prominent slogan about freeing the weed, apparently not owned by a gardening enthusiast.

Beyond the Little Theatre, a four-arch aqueduct carries the canal over the River Calder, just downstream of its confluence with Hebden Water, which carves a deep valley down from the moors and passes under a stone arch bridge of 1510, replacing a medieval timber bridge. Mill buildings after the aqueduct have become the Hebble End Works with small businesses, particularly refurbishment related.

Colden Water is the other confluence of the Calder in the town, near the Stubbing Wharf canalside public house.

Surprisingly, a large enough piece of waste land has been found in the town for construction of a 5,000m^3 balancing pond to serve the canal.

The railway line cuts obliquely across the canal in the middle of a long stretch of hillside covered by deciduous trees, particularly oaks. The Woodman is an appropriate public house near Rawden Mill Lock.

Winding its way down through the woods and crossing at Callis Lock is the Pennine Way, the senior member of our family of long distance footpaths, both in length and age. The surroundings are slightly less attractive here with a large sewage works squeezed into a narrow site between the canal and the river.

Running along the side of the valley at Luddenden Foot.

Canal activity at Mytholmroyd.

Rhododendrons and dense greenery at Eastwood, hardly typical Pennine scenery.

A landmark high on the southern rim of the valley is Stoodley Pike Monument, built to commemorate the defeat of Napoleon at Waterloo. Near the canal, a black rockface is populated by ravens. Almost as black is a burnt-out mill, charred scaffolding spilling from one end.

Approaching Lob Mill Lock there is a notice telling boaters to sound their horns to warn children canoeing. A Scout narrowboat is moored nearby. As on the rest of the canal, there are no swans but plenty of domestic and Canada geese. A picnic area is sited with a backdrop of a large railway viaduct.

Todmorden was a cotton town. The former Fielden cotton mills were one of the world's largest and the company even had their own fleet of ships. In August 1842 the millworkers were involved in fierce rioting.

Honest John Fielden became MP for Oldham and promoted the 10 Hours Act of 1847 for women and children, by which time his own workers already had an eight hour day. The magnificent town hall and Unitarian church were paid for by Fielden, of whom there is a statue in the town. There are also sculpted figures to represent Yorkshire and Lancashire industry when the county boundary was here.

These days the disruptive behaviour is on the canal by a minority who take exception to the passage of boats. Streams of profanity have been delivered to women and men alike on boats, maggots have been fired and concrete blocks have even been dropped to foul lock gates, it is reported.

A policeman claimed to have had his activity disrupted in November 1980 when he was abducted into a UFO.

Mayroyd moorings.

The burnt-out mill at Castle Street.

Moorings alongside an old mill at Eastwood.

Trees give way to moorland near Lumbutts.

A local whose flying activities were better documented was Nobel Prize winner Sir Geoffrey Wilkinson, who helped develop Concorde. Another Nobel Prize winner from the town was Sir John Cockcroft who split the atom. Todmorden carnival takes place in May and its agricultural show is in June.

The A6033 replaces the A646 and follows the canal to Littleborough. At its first crossing, the towpath is taken through a horse tunnel.

The north bank of the canal is dominated by the Great Wall of Todmorden, a 12m curved wall of four million blue engineering bricks that support the railway embankment. At Shade, the railway crosses the canal on Gauxholme Viaduct, the last remaining in use of three arched viaducts on the line, part masonry viaduct with castellated turrets at the ends and more hidden girders added in 1906, which now carry the live load. Beyond the A681 the railway crosses back.

Steadily, the trees are left behind and the landscape takes on more of a moorland character, despite being in a valley.

All the locks had unusual horse towrope guide prongs and more of them have survived undamaged on the higher locks. From time to time overflow weirs required tow horses to ford sections of towpath while those on foot used narrow bridges alongside.

Walsden, the valley of Welshmen, has textile mills and dye works. What look like early square, stone railway sleepers edge the canal.

Travis Mill Lock lifts the canal past the Cross Keys. A succession of further locks continue the process. Near Bottomley Lock, Stephenson gave up the contest and took the railway into a 2.6km tunnel, the longest railway tunnel in the world when he built it in 1840. In 1984 a train of petrol tank wagons burned in it for several days and the damage caused took six months to repair.

A backpumping scheme has been installed at Warland Lower Lock to assist water levels at the summit. Warland Upper Lock, in front of the Bird i'th Hand, is the boundary between West Yorkshire and Greater Manchester.

The summit level is 183m above sea level and those who get this far are entitled to purchase brass summit plaques at Longlees Lock, in aid of the canal society. Canal society members carrying out restoration work in 1974 included Dr Harold Shipman, Britain's greatest mass murderer, and his wife, Primrose. On the road, Todmorden Turnpike tollhouse at Steanor Bottom Bar still lists the toll charges. An inclined trackway ran down to the wharf from a quarry and brickworks. The Summit public house has a backdrop of a high stone quarry face at the foot of Blackstone Edge, on which were built Warland and Light Hazzles Reservoirs as canal feeders.

Part of the Great Wall of Todmorden and a disused mill.

Gauxholme Viaduct on the world's second-oldest passenger and freight railway line. They don't build them like that any more.

Wild scenery makes the summit pound seem remote.

The first pound beyond the summit, below the distinctive Cowberry Hill.

The infant River Roch crossing the railway tunnel mouth.

The canal crosses the narrowest part of the Pennines, where the anticline is steeper on the west side.

From Summit, the canal follows the infant River Roch as far as Smithy Bridge. As the railway emerges from its tunnel, the river is carried over it in a snaking iron trough.

At Sladen Lock, old mill buildings are used by an engineering textile company. Directly above Bent House Lock is a spur with the remains of a crane.

The A58 returns to cross the canal again at Durn Lock, having taken a route over Blackstone Edge, a line close to that used by the Romans for a 5.5m road which featured a central braking ridge. Near the canal, the Coach House Heritage Centre is housed in a late 18th-century Grade II listed building.

By the Waterside Inn, the B6225 is carried over.

Littleborough used to be a textile town. Nowadays it has progressed to PVC and chemical plant lines the canal opposite Stubley Hall.

A wharf at Smithy Bridge has pieces of broken glass, a feature of much of the rest of the canal. To the east of Smithy Bridge is Hollingworth Lake, dug as a canal feeder and now the centrepiece of a country park.

There can be little doubt that the eastern half of the canal is the more attractive. As the canal crosses Greater Manchester the scenery becomes less spectacular, the stone buildings giving way to brooding red brick. There are sections where greater reedmace has taken hold.

After the Tophams Tavern, for a while there are extensive views to the north towards Rochdale and then to the south past Milnrow.

The Oldham to Rochdale railway crosses over between the A640 and A664. There is a busy road crossing by the Lord Nelson. The Greater Manchester Fire Service Museum includes a Victorian street, a fire station and a 1940 Blitz scene. Derelict brick mills between Moss Upper and Lower Locks have been ruined, perhaps since that time, but their impact is softened by a substantial wall of pale yellow roses along the front of one of the old mill buildings.

More open countryside below Littleborough, heading west on the Greater Manchester side of the summit.

The loop round the Irk valley at Chadderton.

Beyond the A671 is the end of the Rochdale Branch which ran northwards for a kilometre towards the centre of the town. **Rochdale** – named after the River Roch, in turn named after the British Celtic *rached*, river by the forest – was a cotton town and, like Todmorden, its mill-workers were involved in fierce rioting in August 1842. Two years later 28 weavers set up the original Co-op store in Toad Lane (t' owd lane), a scheme which worked where others had failed because it bought in bulk, sold at market prices and shared the profits in proportion to purchases. By 1915, three million people were claiming their dividends from the Co-operative Wholesale Society. The Rochdale Pioneers Museum is located in the original store. The town hall is Victorian Gothic with fine stained-glass windows, paintings, carvings, ceramics and a hammer-beam roof; it is one of the finest in the country. A similar accolade was extended to singer Gracie Fields, who hailed from the town.

A section is filled with reedmace and the final reach before a large roundabout has been occupied by large terrapins.

The canal crosses the roundabout at the end of the A627M and along the toe of the embankment but the route for walkers is less than obvious.

The canal is taken through the three Blue Pits Locks and past the Blue Pits Inn and Bridge Inn. Maden Fold was the end of the 2km Heywood Branch. The closely following Leeds to Manchester railway has a depot here, with track maintenance equipment.

The M62 Trans-Pennine Motorway is taken over a bridge that provided access to a farm and now takes the canal and a floating towpath, to the farmer's annoyance. There were a series of problems in the area, including breaches and attempted breaches of the canal embankment, and all the balance beams were sawn off the lock by the M62. Eventually, the farmer was found to be responsible and jailed in 2008.

As the traffic roar falls away there is not a building in sight, just golfers up on a ridge to the right. Powerlines follow the canal to Chadderton and the six Laneside or Slattocks Locks are crossed by the A664 and the railway again. A vast Tesco building sits on top of the ridge to the left as the railway crosses back on a cast-iron bridge for the last time and the canal makes a loop round the River Irk valley. To the west is **Middleton**, with the Bobby Charlton International Soccer School.

Chadderton is the haunt of the Broadway Boggart, identified on misty nights by flickering blue lights in the sky and crackling and screeching noises.

Small lilies join the canal plant life. The canal heads past a power station, perhaps accounting for some of the Broadway Boggart's trademarks.

From the Boat & Horses to the Morrisons super-market,

The rebuilt section of canal alongside the M60.

the canal was obliterated by construction of the M60 and a new concrete-lined canal channel with distinctive decoration has been built to one side with a tunnel under the motorway and a separate bridge over.

The Oldham to Manchester railway crosses over at **Failsworth**, where the canal has been rebuilt past a mill and a new Tesco store.

Six kilometres of the canal through Newton Heath and Miles Platting were capped with concrete to prevent suicides but the navigation has now been fully restored. The New Crown Inn, Navigation, Brown Cow, Jersey

The gable ends of Ducie Street Warehouse are barely visible among all the new building as the Ashton Canal joins.

A brick mill at Miles Platting starts the build-up to central Manchester.

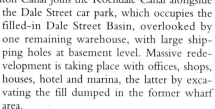

Manchester has plenty of fine brick architecture. To the rear the modern Rodwell Tower block is visible.

Lily and Ancoats Hotel come at intervals among the housing, together with a large Morrisons supermarket. The Huddersfield to Manchester railway crosses.

Ancoat's Lane Lock was once nominated as one of the worst eyesores in the North West of England.

Brownsfield Lock is at the head of the pound that brings in the Ashton Canal and the Cheshire Ring at Ducie Street Junction.

Ducie Street Basin had a massive warehouse for storing high-class salt from Northwich but it was burned down a few years ago. Still present is the impressive Ducie Street Warehouse, a railway warehouse built in 1867 with wrought and cast iron to reduce fire risk.

The Ashton Canal joins the Rochdale Canal alongside the Dale Street car park, which occupies the filled-in Dale Street Basin, overlooked by one remaining warehouse, with large shipping holes at basement level. Massive redevelopment is taking place with offices, shops, houses, hotel and marina, the latter by excavating the fill dumped in the former wharf area.

The last few locks are known as the Rochdale Nine, particularly by those who have not passed the previous 83. Dale Street Lock leads to the Piccadilly Undercroft, a forest of concrete columns supporting the 18-storey Rodwell Tower office block above Piccadilly Lock. The concrete walkway is always lit but shady characters hang about in the gloom.

The other blocks around include County Hall and the glass-and-concrete Piccadilly station, which began life as Store Street and became London Road in 1846. In 1881 it had a major rebuild with four spans of 24–30m arches on lattice girders supported on cast-iron columns. Its 12,000m^2 undercroft is an abandoned warehouse with heavy brick arches and cast-iron columns, the spaces between the columns being considered ideal as a tram interchange for the Greater Manchester Metrolink. The possibility of fire risk or of a tram striking one of the

supporting columns was suggested and the terminus is now encased in a 1.2m-thick concrete box with more concrete round the nearest columns. The redesign cost an extra £5 million and used 10,000m^3 of concrete.

This is one of the finest Victorian city centres anywhere, **Manchester** having been built on textile wealth that developed from the skills of Flemish weavers in the 14th Century. Although the final section of canal through Manchester is largely below road level, the whole area has been restored and sees a lot of city centre activity.

Chorlton Street lock keeper's cottage is built over the canal. Italianate buildings are the old Minshull Street magistrates' court. Manchester Metropolitan University is in an 1895–1912 French Renaissance-style listed building. The North West Film Archive has material from 1896 onwards. The National Museum of Labour History is in the former Mechanics' Institute building, the first meeting place of the Trades Union Congress in 1868, with displays on the co-operative movement, trade unions, the Labour party, the women's movement and the world's largest collection of trade union banners. Central House is a listed warehouse of 1884.

After David Street Lock, the A34 crosses close to the River Medlock. Pipes formerly carried steam along the towpath from Bloom Street power station for the world's first district heating scheme, an environmentally sound concept taken up enthusiastically abroad but abandoned here. The power station was built in 1902 to power trams, the most advanced in Britain. It was hoped to convert it to burn domestic refuse.

A moored restaurant barge that blocks half the canal and a relatively low bridge across the canal to the restaurant have caused an ongoing row between canal interests and the local authorities.

Manchester Art Gallery has one of the country's best collections of pre-Raphaelite art. After Oxford Road Lock comes the line of the Manchester & Salford Junction Canal, built to link the Rochdale Canal with the River Irwell, its tunnel being used as an air raid shelter during

A fusion of ancient and modern arts near the Hacienda.

struction of the 200-arch 19th-century railway viaduct, which now carries the Manchester to Northwich and Metrolink lines by the canal. The Dukes 92 public house takes its name and number from Duke's Lock, the last on the canal, and is located in the stables of the Merchants Warehouse of 1827, the oldest surviving great warehouse. The lock was built by the Duke of Bridgewater with short balance beams, opened by chains because of the closeness of the bridge. The lock attendant's house of 1890 was built by the Manchester Ship Canal Company.

The Rochdale Canal joins the Bridgewater Canal as canals lead off in all directions and a network of railway and footbridges provide a hive of transport activity.

the Second World War. The first part was reopened in 1996, passing under a bridge with wrought-iron musical motifs incorporated. It opens into a basin below the glass edifice of the Bridgewater Concert Hall, opened in 1997 as the new home of the Hallé Orchestra. Beyond is the Manchester Central exhibition centre in the former Central station railway terminus, a station with a roof span exceeded only by that of St Pancras in London. Built in 1880, the 64m-arch rises 27m to provide an extensive unobstructed exhibition hall.

Over the wall from the canal is the Palace Theatre, built in 1871 and one of the premier theatres in the north of England.

Tib Lock has the Canal Bar alongside with plenty of the customary broken glass and the A5103 passing over a tunnel. The Manchester Sound derived from the Hacienda Night Club. What at first glance appear to be four wharf cranes on the side of a modern building metamorphose into four griffin-like sculptures with chains in their mouths.

Deansgate Tunnel was cut as a canal arm in a cave into the red sandstone, the original end of the Bridgewater Canal having been in tunnel to Castlefield.

Many Roman remains have been found around the area and there would probably have been more if the fort of 79 had not been largely destroyed by the con-

Bridgewater Hall on its opening weekend with the Manchester Central exhibition centre behind.

No-compromise Victorian engineering is evident in successive railway viaducts straddling the canal.

Distance
26km from the Calder & Hebble Navigation to the Bridgewater Canal

Navigation Authority
Rochdale Canal Company

OS 1:50,000 Sheets
103 Blackburn & Burnley
104 Leeds & Bradford
109 Manchester

42 Ashton Canal

A canal with many important connections

Despite its connections, the Ashton Canal, designed by Thomas Brown, was built as an isolated narrow canal from Ashton-under-Lyne to Manchester. It was opened to Ancoats in 1796, extended to Ducie Street in 1799 and connected to the Rochdale Canal the following year. Initially it was used to take fresh vegetables from the market gardens of **Dukinfield** into Manchester and later it served many small coal mines in the area. Its connections increased its importance considerably and its position as part of the Cheshire Ring has done more than its environment to encourage its restoration. Despite its background of the Pennines, it has a high proportion of derelict buildings along its route. Glass on walls, barbed wire, spiked railings, burglar alarms, razor wire and video cameras all emphasise the depressed nature of the area.

Commercial traffic ceased in 1957 with the last pleasure boat forcing a passage in 1961. Restoration began in 1968 and 600 volunteers removed 500 lorry-loads of junk in one weekend. The high spot came in March 1971. Operation Ashton saw 1,000 volunteers in action one weekend and full restoration was then promised, also preserving the Cheshire Ring, including the lower end of the Rochdale Canal. There is now considerable use by narrowboats.

The connection with the Huddersfield Narrow Canal is at its final lock in **Ashton-under-Lyne**. The canal dives into a tunnel. Walkers have to divert to the A627

through an Asda car park with the route unmarked but this gives a chance to see red lines painted at the exit. There is a warning that trolley wheels will lock here if an attempt is made to take them any further, technology which would be welcomed elsewhere.

The Tameside Sea Cadet Corps building is passed. The notice on the wall of a derelict mill, blaming closure of this second less-important trans-Pennine route on a bureaucratic blunder, has been removed.

Ashton-under-Lyne was a small market town and, indeed, still has a thriving outdoor market most days. With local coal and good transport by canal and, subsequently, railway, it became an important cotton centre. It saw the start of the Chartist action in 1842 when strikers turned workers out of the mills.

The Peak Forest Canal also brought Macclesfield Canal traffic – coal and lime for industry, building and farming – approaching across an aqueduct over the River Tame and under a slender, stone, towpath bridge for the Ashton Canal at Dukinfield Junction. The aqueduct was built by the promoters of the Ashton Canal in readiness for the Peak Forest Canal. The adjacent area of benches is Weavers' Rest, which takes its name from the suicides of cotton workers in the 1860s cotton famine and the depression of the 1930s. Although the Ashton Canal is now entirely in Greater Manchester, before 1974 it was in Lancashire. The far bank of the River Tame was Cheshire. Lancashire police paid a shilling for reporting the finding of a body while Cheshire police paid ten shillings. Thus, corpses were frequently floated across the aqueduct to the Cheshire side before being 'discovered'.

Opposite the junction was built the New Ashton Warehouse in 1834, to serve the growing number of mills on the west side of Ashton-under-Lyne. All but the ground floor was destroyed in a fire in 1972. The remains have been rebuilt as the Portland Basin heritage centre, a museum with free entry and toilets. A prize possession outside is the 7.3m-diameter breast shot waterwheel of advanced design, which was used to power the cast-iron hoists. A jagged white post is the remains of a crane.

Moving westwards, the canal has an overflow weir to the River Tame. It passes Junction Mill, which has a massive chimney with courses of blue engineering bricks and a band of lattice brickwork. Next to it are the premises of Barcrest, who make fruit machines.

A footbridge with a handrail missing carries the towpath over the entrance to the former Princess Dock, constructed for easier transhipment between railway and canal although this worked to the railway's advantage. Brookside Sidings now occupy the dock site for 1.2km, flanked by the Huddersfield to Manchester railway line. This is joined just before Guide Bridge by the line from Glossop. The Glossop to Piccadilly route is being considered as a possible future extension to the Greater Manchester Metrolink tram network.

Oxford Mills, opposite the sidings, were built from 1845–51 by the Mason family for spinning. They were supplied with housing and community facilities, including sports ground, swimming and washing baths, institute and library, plus smoking and chess rooms, an advanced concept for the time.

Birch Mill and those which follow have decorative coal and ash holes, serviced from the canal. A modern unit has a recent canalside sculpture on its wall, made up from jagged fragments of metal flooring.

Guide Bridge takes its name from the bridge over the canal and its signpost to Ashton, Manchester and Stockport.

The Peak Forest Canal joins at Dukinfield Junction.

Since it acquired its name, the bridge has been widened to an 85m tunnel although the original bridge is still obvious from water level. Guide Bridge Mill on the right was a late Victorian spinning mill.

The next bridge formerly carried a freight railway and is followed by a bridge that has been subject to mining subsidence, a fact underlined by the colliery spoil heaps on the north side of the canal. Coal from Ashton Moss Colliery was tipped from wagons into boats at the Shunt. Not all the coal that fell in the canal was accidental and diving in the canal for coal during the hard times of the 1920s resulted in the canal being deepened at this point.

The Ashton Packet Boat Company Ltd basin has narrowboat restoration and horse-drawn canal trips. To the west is another freight line and, over that, a road bridge with some intricate brickwork under its high, skewed arch. The gap in the buildings has been used for the M60 Manchester Outer Ring Road, near Audenshaw Reservoirs.

Housing begins on the right at the line of an embankment leading to a former railway bridge site. A winding hole follows.

Robertson's factory is on both sides of the canal. Established in 1896 and famous for its golly symbol, it uses water from the canal for cooling. The warm water attracts ducks. Disappointingly, the smells are not of jam, marmalade or mincemeat but are extremely pungent.

Park House is some of the most attractive accommodation on the canal, sheltered housing established in 1969 by Air Vice Marshal Johnnie Johnson. Well tended gardens, trees and ducks add interest. Behind the new housing on the other bank is Fairfield, a 200-year-old Moravian Settlement village.

A modern factory on the right, with a line of long parallel drainpipes down the front, produces an unusual architectural effect, almost a caricature of Doric pillars.

Fairfield Junction brings the stump of the Hollinwood Branch. This climbed away through Droylsden to Oldham. It was abandoned in 1932 but the first 160m is now being restored as a marina. Another section remains in the Daisy Nook Country Park. There was also a Fairbottom Branch. This has disappeared.

Now begins the series of 18 locks descending to Piccadilly. The locks are in good condition although the use of scaffolding-style safety rails owes more to economy than to beauty. Many of the locks have overflow channels and about half of these are on the verge of being shootable, subject to sufficient water flow and boats of the right dimensions and materials.

The first two locks were doubled in 1830 but only the left one of each pair has been retained in use. Between

Fairfield Lock, lock 16, one of the ones with open spillways.

them is a packet boathouse of 1833 with services to Ashton, Stockport, Stalybridge, Hyde, Marple and Manchester. The services were regular and efficient and Ashton to Manchester took only two and a half hours, including working through the locks. The building is now part of a water adventure centre. By the second lock is a school.

An open reach passes under two low, swing bridges, the second lower than the first, before the Yew Tree at Openshaw.

Locks 16 to 8 form the Clayton Flight over the next 1.6km, several with distinctive characters. Lock 16, Fairfield Lock, is near the Friendship. Lock 15 has allotments, pigeon-racing lofts, pigeons and even the odd lapwing tumbling in the sky as it trills out its song. From lock 14, Edge Lane Lock, to lock 13, Clayton Top Lock, the left side is dominated by the Acre Works of Eva Brothers Ltd, who undertake hand forging with heavy hammers. The latter lock is overlooked by the Strawberry Duck public house. Between locks 11 and 10 is the bridge carrying the towpath at Clayton Junction over the Stockport Branch. This ran for 8km, serving much industry. It was not used from the 1930s and was closed in 1960. Now it is no more than a stump pointing to a waste area, the highest aspirations of which are to become a motorcycle training area.

Standing back from lock 9 and the Bridge Inn is the church of St Cross with St Paul. It was built in Polychromate style in 1866 by William Butterfield, one of three Grade I listed churches in **Manchester**. Between here and lock 8, Clayton Lock, the left side is occupied by the Clayton Aniline Company which manufactures dyes with water from the canal. Established in 1876, it also used the canal for bulk handling and was unique in opening a new branch for the premises with a swing bridge to carry the towpath. The area is industrial but a kestrel may be seen hovering over this unlikely environment.

Working through the top lock at Fairfield Junction, one of the two double locks. Beyond, members of a canoeing class prepare to go afloat at the Water Activity Centre, which does much for disadvantaged youngsters.

Manchester City football stadium, seen from the canal.

The Medlock Aqueduct hides beneath a solid railway bridge.

Housing in the Picadilly Village development.

A large girder bridge once carried a freight railway but now has only bushes on it. The new A6010 bridge crosses near where the Clayton Branch served a complex of chemical works until it was closed in 1953. Before it on the right is the National Velodrome while it is followed on the right by the Indoor Tennis Centre and on the left by the National Squash Centre and the stadium of Manchester City, the Manchester football club supported by Mancunians rather than the one supported by the rest of the world. The concentration of sports facilities is a Commonwealth Games legacy.

Locks 7 to 4 form the Beswick Flight over 500m. The first has a lock keeper's cottage of 1865. At this point the Medlock Valley Way footpath joins the towpath and faces the turn at the west end of the lock. This is very constricted. Between here and the next lock, Bradford Lock, the Bradford Branch on the right served the Bradford Colliery. Both have now gone. The Beswick Locks are the last two of the flight. Then an overflow weir disappears

Distance
11km from the Huddersfield Narrow Canal to Ducie Street Junction

Navigation Authority
British Waterways

OS 1:50,000 Sheet
109 Manchester

under a concrete-plank fence to join the River Medlock, which is underground at this point. The river resurfaces at the Medlock Aqueduct, an arrangement that is made more complicated by a large railway bridge crossing over the top and a solid brick wall reaching from the railway bridge to the towpath on the right.

Two large gas holders follow, opposite the site of the former Beswick Sanitary Works. Gas is also made on the left, this time liquid carbon dioxide for the drinks industry.

At Ancoats, the canal enters a canyon of brick cotton mills and warehouses up to seven storeys high, a windy and dismal area despite the neat grass alongside the towpath.

The final three Ancoats Locks come in 200m. There were several small branches, the largest being the 400m Islington Branch. This left Ancoats Junction on the right, running past the heavily fortified British Waterways office, to supply coal, salt and sand wharves. An arm opposite served the Manure Wharf, which supplied night soil and manure from the corporation stables to the Beswick Sanitary Works and farmers further along the canal.

Piccadilly Village was opened in 1991. This is a red-brick mini docklands scheme with housing located around a couple of short branches. A roving bridge, a lifting bridge and a crane add interest.

At the far end of the development is Store Street Aqueduct. It leaked in 1964 and was closed for several years until repaired. It has been equipped with illuminated bollards.

Another crane on the right fronts one more area of restoration, Whittle's Croft Wharf, the original canal terminus.

In the early 1800s, the canal was extended back beyond Jutland Street, the steepest street in Manchester, sketched as Junction Street by L. S. Lowry in 1929.

Ducie Street Basin had a massive warehouse for storing high-quality salt from Northwich but it was burned down a few years ago. Still present is the impressive London Warehouse, a railway warehouse built in 1867 with wrought- and cast-iron to reduce fire risk.

The final bridge over the canal now has a high building over it, just part of the extensive redevelopment in this area.

The Ashton Canal joins the Rochdale Canal at Ducie Street Junction, alongside the Dale Street car park. This occupies the filled-in Dale Street Basin, overlooked by one remaining warehouse. To the west, a lock continues the Cheshire Ring through the centre of Manchester.

The massive brick London Warehouse. The Rochdale Canal lies to the right under new offices in complementary styling.

43 Peak Forest Canal

The Peak Forest Canal must have the most inviting name of any of our canals. Anything it lacks in the way of forest these days is compensated for by peaks, particularly on the upper section.

Designed by Benjamin Outram, it was opened in 1800 to service the limestone quarries at Doveholes. Although the upper end of the canal is one of the highest parts of popular. The towpath is much used by walkers, who appreciate the fine views.

Footbridges take the form of drawbridges at first. They are used by the local kids as seesaws. Later, they give way to swing bridges.

The rubbish tip and sewage works before Furness Vale are followed by a marina. As in the marina at **New Mills**,

Just one part of the Buxworth canal terminal complex.

the British canal system still in operation, it did not reach all the way. In 1799, the 10km horse-drawn Peak Forest Tramway was opened to connect the two. A vast complex of wharfs was constructed at Buxworth or Bugsworth, as it was called before the inhabitants changed the name. The wharves are located around the Navigation Inn. The tramway was closed in 1926. The area with its limekilns was handling 600t of limestone and lime daily in the 1880s.

The Whaley Bridge Arm provided a connection with the Cromford & High Peak Railway and thence to the River Trent.

The canal follows the Goyt valley, clinging to the left side and offering spectacular views out over it through not infrequent breaks in the dry-stone walls. The canal is

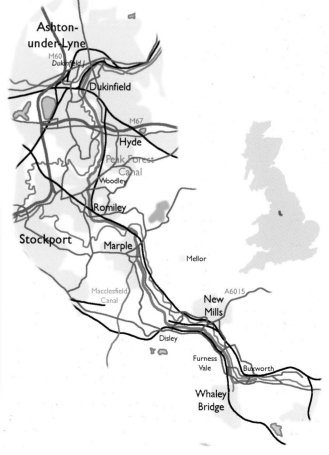

Typical canal bridge near Furness Vale.

219

owners spruce up their vessels, watched, invariably, by their black-and-white sheepdogs with folded-down ear tips.

The bridge carrying the A6015 at New Mills is heavily shored. Its courses of stonework show a marked dip above the canal. Road vehicles are reduced to single line working in an effort to keep the two forms of traffic apart.

New Mills has had textile printing, engraving, a canal foundry and engineering industries. The largest mill is a sweet factory and the smells emanating are likely to produce a craving for them. The canal user should plan ahead by carrying a suitable stock.

Fencing off an old stone quarry has produced a field of the kind that only goats can fully appreciate. The resident family have trodden paths in places that gravity should have placed out of reach. A copse follows, banks of wood sorrel sloping down to waterside marsh marigolds. These are followed by neatly mown lawns each side of the canal at Disley. Here are some of the best views of all. It is easy to understand why this has been the scene of two breaches, the first in 1940 and the other in 1973, sweeping away several cruisers after the bank collapsed, owing to a prolonged spell of torrential rain.

A lifting access bridge is accompanied by a fixed footbridge at a high enough level for canal traffic to move under.

The Ordnance Survey have placed a trig point just 200m from the canal at Marpleridge. Oak trees and yellow-flowered gorse bushes cover fields to the west of the canal. Breaks in an ivy-covered wall and a stand of pine trees allow a last look across the Goyt valley to Mellor Moor and beyond. This is excellent touring country.

By this stage the high land on the west is already subsiding and the canal leads out onto a peninsula, where it is joined by the Macclesfield Canal emerging under a roving bridge to become part of the Cheshire Ring. The Macclesfield Canal did much to improve the prosperity of the Peak Forest Canal.

While the upper Peak Forest Canal has always been kept open, the lower section was abandoned in 1944 and reopened only in 1974.

Conveniently, all 16 locks on the Peak Forest Canal are in a single flight at **Marple**, a flight which some consider to be the most agreeable in England. It was not completed until 1804, a tramway being used up and down the hill for the first four years of the canal's life.

For nearly a century from 1840, narrowboats were built and repaired at the top of the flight. The drydock these days acts as a sunken flower garden. Industry is now limited to a floating restaurant barge.

A road separates the locks from the onlooking row of houses on the left. On the right, long side ponds run out along the contours with modern houses between each one. After the fourth lock, Possett Bridge is a particularly fine structure, carrying the road across. It has a horse-shoe-shaped arch on the left for tow horses and this leads out onto a paved area, surrounded by flowerbeds. There

A sweep of the canal with fine views at Disley.

The curious old Possett Bridge at Marple.

The Marple aqueduct drops to the River Goyt way below.

are interesting buildings all around the flight, from a converted mill to the quaint Brabyns No. 1 Cottage. The flight descends further through tree-covered fields, overlooking Brabyns Park. Three-quarters of the way down, the railway tunnels under the flight, seeming not to be deep enough to clear the bottom of a lock, especially as some locks are close contenders for the title of deepest narrow canal lock. The locks are also unusual in having stone-arched bridges across their lower ends. Uphill walls are curved so that a boater, pushing the balance beam to open the gate, walks round from the side of the lock onto the bridge. The balance beams then prevent the bridge being used easily with the bottom gates open.

Almost immediately comes the magnificent Marple aqueduct, its massive proportions only overshadowed by the nearby railway viaduct. The three arches of the aqueduct, designed with circular voids through the piers below the trough, consumed 600m^3 of masonry in their construction. Repaired after a partial collapse, it is now a listed monument. Water level is well over 30m from the River Goyt below. There is no parapet on the east side, allowing the boater to look down onto the tops of mature trees. The waterway is narrow, narrow enough for the local lads to jump across.

Rose Hill cutting is equally narrow, with a high retaining wall on the right. Its unusual shape results from the fact that it began life as a 100m long tunnel, not being opened out until 1820.

Old factories along the canal at Hyde.

At this point the canal is about to leave the River Goyt, which lies just to the left.

The canal dives into the 282m Hyde Bank Tunnel, a squat structure with the towpath running over the top and a slight kink in the middle, to emerge among wooded banks which are covered with celandines, bluebells and red campion in early summer. Arrowhead makes its appearance in the water.

Beyond **Romiley**, a high railway bridge of almost tunnel proportions crosses the canal. The first half has significantly smaller dimensions than the far end and the step must have caused the arches to ring with curses from southbound boats many times over the years.

A restaurant is passed before reaching the final tunnel, the 161m Butterhouse Green tunnel at Woodley. Like the railway bridge, the towpath is taken through this one.

Little by little, the amenity value of the canal drops as the Tame valley is reached. Debris on the banks increases, as do flotsam and clumps of floating vegetation, in surprisingly large quantities for a frequently used navigation.

Naylor's abrasive mill is the first of an increasing number of mills, many derelict. Some provide an atmosphere of decaying charm while others could benefit from the attentions of a bulldozer, an idea that is receiving some positive application around **Hyde**.

Bridges often have the corners of their abutments protected by lengths of railway line, the flanges grooved by tow ropes and the webs eaten through by rust.

Captain Clark's Bridge, at Hyde, is a fine, stone roving bridge and is paired with another roving bridge with cast-iron deck to take the towpath across the canal. After passing a mill with more security fencing than many of our military establishments, the slightly realigned canal is crossed by the M67 on a bridge that is unusual in having the luxury of lighting underneath in the daytime.

The surroundings in **Dukinfield** are far from beautiful but attempts are being made to improve the land around the sewage works. A rubbish tip has been landscaped and a railway bridge is worthy of comment. Rebuilt in 1978, its shallow concrete arches follow the shape of the original, the cast-iron facades of which have been incorporated into the new structure. Even the three-colour paint scheme has been retained.

The final bridge at Dukinfield Junction is also a flat arch with long, gentle approach ramps carrying the towpath for the Ashton Canal, which the Peak Forest Canal joins.

A cast-iron roving bridge at Hyde.

Distance
23km from Buxworth to Dukinfield Junction

Navigation Authority
British Waterways

Canal Society
Marple Locks Heritage Society www. marplelocks.org.uk

OS 1:50,000 Sheets
*109 Manchester
110 Sheffield & Huddersfield*

44 Huddersfield Narrow Canal

Britain's highest canal and longest tunnel

The Huddersfield Narrow Canal was designed by Benjamin Outram as the most southerly and shortest of the three canal links across the Pennines. Begun in 1794, it was opened in 1811. It was not a great success commercially.

Joining the eastern end of the Ashton Canal in **Ashton-under-Lyne** at its first of 74 locks, it had a direct connection to the Cheshire Ring, allowing access to Lancashire, Merseyside and the Black Country.

The closure came in 1944 but it was retained as a canal water supply. Most of the canal is exceptionally pleasant.

The canal harbours healthy specimens of leech. It crosses the River Tame, the valley of which it is to follow to Dobcross. The aqueduct lacks rails although the towpath crosses immediately to the right, giving a feeling of security. Below, the Tame resembles cold, milkless tea, as it drops over a small weir.

Dilapidated mills flank the canal. The occasional expensive car among the debris shows that appearances can be deceptive.

Stalybridge has seen extensive restoration in 2000. Tesco take up a favourite position as one of the retail outlets beside the canal.

The canal leads off past a conveyor belt that fed a power station but is now truncated high above the

Distinctive British Waterways logo mooring pins.

The second lock after the power station.

The canal crosses the River Tame with virtually no protection.

Climbing up into the Pennines.

The canal passes under Saddleworth Viaduct at Uppermill.

Wool Road Transhipment Warehouse.

towpath. From here, the views of the Pennines ahead improve steadily and built-up areas become progressively smaller.

Metalwork is kept well painted and grassed areas are mown and often set out with tables as picnic areas. Other councils could learn from what has been done in Greater Manchester.

Crags line the steep Tame valley. The canal turns to the right and comes to the 200m long Scout Tunnel. Just before the tunnel, as at a number of other points along the canal, a roller is set at the edge of the towpath to assist with pulling out the drain plug.

Climbing from **Mossley** to Greenfield, a number of large mills are passed. Most are derelict but some are active and are fronted by landscaped lawns, rhododendron bushes and flowerbeds leading down to the canal.

The Roaches Inn has its outside area decorated with scenes from *Rubadub's Garden*, explained in a children's tale about a dog, bought from the bar.

Approaching the prominent radio aerial above Grasscroft, the canal crosses back over the river on another small aqueduct. The placid cricket ground nearby on the right has a panoramic backdrop of crags and scars leading up to Saddleworth Moor, the Peak District National Park and an obelisk above Tunstead to watch if the game gets tedious.

The lock island at Grasscroft has been taken over by the owner of one of the adjacent houses and has been converted to flowerbeds, the lock being a riot of colour in the summer.

A treehouse has been built by the canal. After Saddleworth Rangers Rugby League Football Club ground, the athletic activity continues with a running track and a sports field the other side of the A669. After the A670, the canal flanks a landscaped area at Uppermill. A grassed area and car park are surrounded by the Granby Arms, the Waggon and a museum and art gallery.

Bilberries join the bracken along the stonewall-lining to the canal, accompanied by large clumps of heather.

Slabs of stone roof Wool Road Transhipment Warehouse, which projects over the canal. Artwork is displayed under the bridge as the A670 comes back across the canal for the last time.

So far, the locks have been surprisingly well spaced. From Dobcross to Diggle they come as a flight, climbing up onto more exposed moorland above the treeline, passing a mill which manufactures wooden pallets. From the top lock the water level is 196m, the highest canal level in Britain. As the canal approaches Diggle, set in a basin of hills, the railway closes alongside and suddenly the tunnel mouth appears ahead – an

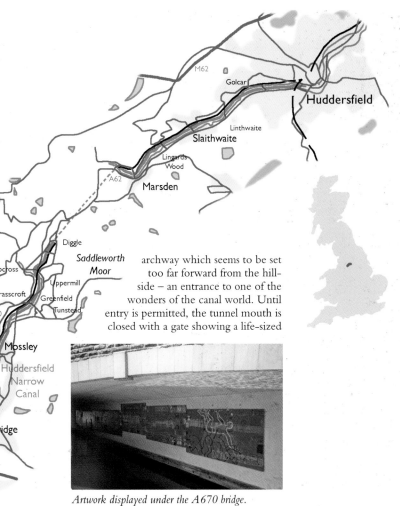

archway which seems to be set too far forward from the hillside – an entrance to one of the wonders of the canal world. Until entry is permitted, the tunnel mouth is closed with a gate showing a life-sized

Artwork displayed under the A670 bridge.

The mouth of the mighty Standedge tunnel, which seems longer than it needs to be. The roller is one of a number to remove drain plugs from the canal bed.

narrowboat bow with leggers on each side. This is Standedge tunnel. At 4.95km, this is the longest canal tunnel of them all, burrowing under the Pennines at depths up to 180m. Opened in 1811, it took 17 years to build at a cost of £160,000. It is unlined and rocks come in a spectrum of colours. As well as acting as a water supply for the canal, the tunnel is a drainage channel for the railway. It has four larger caverns to allow boats to pass and connecting adits to the adjacent railway tunnels, two of which are now disused. In the days of steam trains, smoke would blast from these adits, adding further drama to a passage through the tunnel. There is no towpath and walkers face having to use the A62 or of finding their own way over Standedge and the Pennine Way with little help from signposting.

The tunnels all emerge at **Marsden**, the canal tunnel still being lower than the other three and projecting a little further forward. A highly unusual layout takes the

Narrowboats emerging from the tunnel at Marsden.

The British Waterways depot in the Colne valley at Marsden, now a visitor centre.

young River Colne, the valley of which the canal and railway are to follow, on an aqueduct over the active railway line, over the end of the canal tunnel and down a cascade of steps to disappear underground. It re-emerges beyond the disused railway lines.

A pair of old cottages at the tunnelmouth have been restored as a canal visitor centre and as a base for the country ranger service.

Information boards have been erected beside the canal. More entertaining than informative is the revival in the 1990s of the Celtic Imbolc cross-quarter festival, in early February, with fireworks, drummers, live music, theatre and a torch-lit procession near the canal.

The treeline comes as high as the canal on the east side of the Pennines and heather, gorse and bracken line the canal. The start of the long flight down soon arrives with the Railway facing its namesake across one of the first locks. Also crossing here is the stiff 20km Colne Valley Circular Walk, which re-crosses at Golcar.

As the canal drops towards Lingards Wood, it passes between a pair of feeder reservoirs, with another on the left, approaching Slaithwaite. On the right, the River

The canal runs right through the centre of Slaithwaite, once again.

Colne is never far away and sometimes exposes bare crags, which enhance the scenery, normally steep hillsides with farms, weavers' cottages and, increasingly, mills. These are all in darkened stone.

Slaithwaite, marked by its railway viaduct and by the aerials above it on Pole Moor, is steeped in history. Slaithwaite Manor is a fine Elizabethan house, dating from the 1560s. The nearby St James' church was built in 1789 to replace one that was subject to flooding by the River Colne. Its shape is based on the Puritan ideal that it was easier to be devout in a barn than in a temple. Its name was twisted into Sanjimis, a four-day festival of feasting, dancing bears and Waffen Fuffen bands, dialect for waifs and strays playing in the contemporary equivalents of jug bands. The present funeral parlour was the old Slaithwaite Free School, a fine building. The canal passes Empire Brewing, public toilets, the Shoulder of Mutton and the Commercial before moving behind the Globe Worsted Company and a pallet factory.

From 1825, Slaithwaite was a spa based on mineral springs in the bed of the River Colne. It modelled itself on Harrogate. Slaithwaite docks, on the canal, filled in 1956, handled grain and coal barge traffic. The village

Emerging from the new tunnel by the university.

was a hive of industry. Smuggling was rife in the 19th Century too; the usual procedure was to place things in the canal by day and then hook them out by night. One gang, caught in the act by the king's men, pretended to be drunk and claimed to be trying to rake the moon's reflection out of the canal, thus acquiring the title of Slawit Moonrakers. The canal has been re-excavated through the centre of the town and now has pride of place once more.

The route becomes rural again. It is progressively more overlooked by houses on the A62 at Linthwaite and Milnsbridge. Old mill buildings have been fitted with incongruous balconies and now serve as residential buildings, accompanied by a tall chimney.

The canal swings right on a small stone aqueduct across the River Colne. It passes a chemical works.

By the Four Horseshoes in Milnsbridge, the B6111 bridge has been widened.

The Luddites were active in Milnsbridge the year after the canal was opened. They shot a mill owner from Marsden, broke into a number of mills and smashed up cropping frames, which they considered would put them out of employment.

Entering **Huddersfield**, the canal swings back across the river again. A railway crosses both on a high aqueduct.

After the A62 crosses and another lock, the canal enters a modern 600m tunnel. Walkers have a much longer route, signposted at first.

The canal emerges from under a series of steel buttresses and heads towards the final lock by the modern buildings of the university. The Huddersfield Broad Canal, otherwise known as the Sir John Ramsden Canal, continued south from Aspley Basin to here and then joined the River Colne.

While the Narrow Canal had long locks, the Broad Canal had short ones and so goods had to be transhipped or boats had to be short and narrow if they were to work through. The former option was usual. The Huddersfield Broad Canal gives access to the Calder & Hebble Navigation.

Distance
32km from the Ashton Canal to the Huddersfield Broad Canal

Navigation Authority
British Waterways

Canal Society
Huddersfield Canal Society www. huddersfieldcanal.com

OS 1:50,000 Sheets
109 Manchester
110 Sheffield & Huddersfield

225

45 Calder & Hebble Navigation

Linking industry and improving the River Calder

Working canal sculpture at Sowerby Bridge.

The construction of the navigation parallel to the River Calder and, in places, using its course was completed in 1770. This was one of the canals making a major contribution to the Industrial Revolution, linking the Aire & Calder Navigation with the Rochdale Canal and providing a route from industrial Yorkshire over the Pennines. At no point is it far from conurbations yet it is surrounded by far more open country than the map might suggest.

Its western terminus is the basin at **Sowerby Bridge**, deep in the valley of the River Calder. Hillsides rise steeply to north and south. Roads, some still cobbled, are equally steep as they climb away from the grimy industrial buildings. The canal basin is just one place where new life has come to the old buildings in various forms.

Just beyond the canal basin, the Rochdale Canal joins from the right, opposite the first of the Navigation public houses.

Old gasworks are matched by new propane tanks among the darkened stonework. Assorted industrial smells permeate the air at times all along the canal yet trees gradually replace the buildings. The open hills rise behind. Railways and roads intertwine around the navigation in a three-dimensional tangle. The railway crosses for the first time at Copley, the viaduct passing over both canal and river.

Along the length of the navigation, signposting is

A viaduct carries the railway over for the first time at Copley.

clear. The first warning comes at Salterhebble, where the remains of the Halifax Branch joins. This branch allowed **Halifax**, founded on the cloth industry, to be built back from the main line of the canal. One of the nearest features is the Perpendicular St John's church, with its pinnacles, parapets, gargoyles and fine woodwork.

Looking east from the Sowerby Bridge basin.

The end of the top pound at Salterhebble.

Looking towards Elland Park Wood with lock keepers' cottage boarded up.

A canal warehouse converted to a private dwelling at Elland.

Canalside mills near Rastrick.

Salterhebble brings the end of the top pound, with a flight of three locks that replace a former staircase, wasteful of water, and the whiff of an adjacent sewage works. When the water authorities were created in 1974, the new Yorkshire Water Authority acquired this large works which was the most neglected major works in the country. At that time, the effluent going into the Calder was more polluted than when it arrived at the works.

A stream passes under the navigation before the third lock. Beyond the lock is a modern, electrically operated guillotine gate. There is a small horseshoe-shaped tunnel under the road towards the Calder & Hebble public house.

The next lock, Long Lee, brings the dark River Calder alongside for the first time. At **Elland**, square, tapered chimneys with iron banding add a further dimension to the industrial architecture. A warehouse, that had large end doors which opened to admit a narrowboat, has become a private dwelling. Opposite, the Barge & Barrel has a welcoming notice for boat people.

Beyond the town, Elland Park Wood, on its hillside, faces the power station with its cooling towers. Rusty notices warn boaters not to drive mooring pins into the towpath because of the presence of high voltage cables.

Sadly, lock cottages are mostly boarded up. There are

39 locks, of which 12 are flood locks. These may be left open when the River Calder is not in spate. Lock approach walls are often quite high. The locks themselves are unusual in having open panels to allow excess water to weir over them and by being operated by a bar of wood used as a crowbar rather than the conventional windlass. Substantial rope guide pins mark the corners of locks.

The Colliers Arms is conveniently close to the navigation while a sign for the Rawson Arms points vaguely at the sky in the direction of a wooded bank.

Beyond Cromwell Bottom, the navigation passes between lakes which look man-made and the waterski jump on one is an indication that they are well used. The Grove Inn stands facing the navigation across one of the lakes in a pleasantly rural setting. This disappears at the next corner, just as the navigation crosses Red Beck at surface level. It passes between **Brighouse** and Rastrick, names evocative of coal mining and brass bands. The flowerbeds

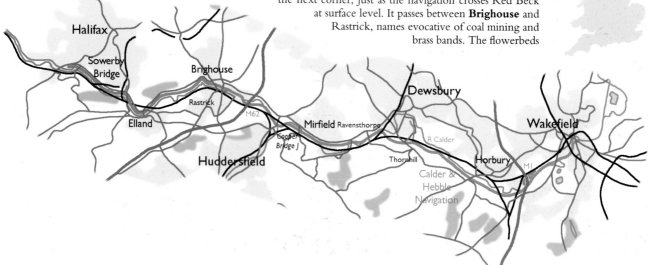

Double roving bridge at Brighouse, each end having straight and curved ramps.

Brighouse Basin ends the first section of cut.

couple of quaintly named works sit adjacent to each other, the mill of Squire A. Radcliffe and Sons Ltd and the premises of the British Bung Manufacturing Company.

A confusing part of the navigation is by the Swan at Shepley Bridge Lock, where two sets of lock gates face the boater. The left set guard a barge dry dock, the navigation running right to rejoin the river opposite the Ship. One more brief canal section comes before Ravensthorpe, with the Thornhill power station. A warehouse, opposite, displays contemporary gantries and a wharf crane.

The Perseverance lies in a bleak derelict area in company with the large hall of Yorkshire Transformers. By the Savile Hotel the surroundings are getting a little more interesting. The **Dewsbury** Branch leads to the Savile Town Basin, serving this old industrial and heavy woollen manufacturing town.

The Figure of Three Locks at Healey now lack the third lock.

A prominent and striking church in **Horbury** was finished in 1791, built and paid for by its architect, John Carr. A subsequent vicar, Sabine Baring-Gould, composed *Onward Christian Soldiers* for his younger congregation members to sing while walking to and from church.

Below Broad Cut Top Lock is the last of the Navigation public houses.

The M1 crosses while the navigation is in a river section, avoiding the need for a second bridge.

Another canal leg cuts off a large meander in the river and heads through a recent industrial estate on the outskirts of **Wakefield**.

A railway bridge over the river is continued as a 99-arch viaduct, the full expanse of which is not seen from the river. It is possible to see the 15th-century Perpendicular cathedral, an uprated parish church.

At the British Waterways depot, the final cut runs past a timberyard and some larger moored boats that are used as houseboats. The final lock, Fall Ing Lock, takes the navigation down to the River Calder for the last time.

The River Calder sections are of a larger scale and lack a towpath or, sometimes, any path at all.

of the Black Swan come down to the waterside, a foretaste of the landscaping that has gone into the canal basin at Brighouse. An astonishing double roving bridge at the start of the basin seems to offer towing horse lines excessive options, especially as the towpath does not continue beyond the basin. Indeed, the towpath is present only on the sections of cut but not alongside the natural river. Signposting for walkers is poor.

Two locks take the canal from the basin out onto the River Calder for the first time.

Just before the M62, the canal cuts right, avoiding two weirs. The country opens out again and the giant television mast on Emley Moor is seen for the first time. This is a replacement for one that blew down soon after first being erected.

After rejoining the river, the supposed burial place of Robin Hood is passed at the top of the wooded hill on the left. Although it might not be a particularly handy spot for Sherwood Forest, this is not as unlikely as it might at first seem as the grounds are those of the 16th- and 17th-century Kirklees Hall, formerly those of the Cistercian Kirklees Priory, of which his cousin was prioress.

The Huddersfield Broad Canal on the right once linked through to Huddersfield and the south of Manchester.

A steel roving bridge has a banked bend for the horses. It drops from the deck level, underneath and up the far side to deck level again immediately adjacent to the bridge abutment.

A canal leg runs past a wharf complete with original wharf crane. A river leg ensues before the route returns to the canal at **Mirfield**.

Barges built at Mirfield for the navigation have been based upon Humber keels, resulting in the short wide beam locks. Lack of space has resulted in some spectacular sideways launchings.

A bridge over the canal connects two parts of a brewery. There is an adjoining Navigation public house. A

The railway bridge in Wakefield and the long viaduct.

Distance
35km from Sowerby
Bridge to Fall Ing

**Navigation
Authority**
British Waterways

OS 1:50,000 Sheets
104 Leeds & Bradford
110 Sheffield
& Huddersfield

46 Leeds & Liverpool Canal

Work started on the Leeds & Liverpool Canal in 1770. By 1777 it had been completed from Leeds to Skipton and from Wigan to Liverpool. It then became delayed by technical problems and expenditure on the Napoleonic Wars and it was not until 1816 that John Longbotham managed to complete the longest single canal in the country and the one that took the longest time to build. Once built, it was one of the most prosperous, especially from 1820 to 1850, promoting work in the mill towns and carrying coal, limestone, cement, machinery, wool, cotton, groceries, beer and spirits. Liverpool corporation had helped with the financing. Most traffic was at the two ends rather than across the Pennines but the finances were sufficiently healthy to support the whole canal and it has been the only transpennine canal to remain open throughout. From the 1870s it lost long-haul traffic to the railways, the decline being exacerbated by water supply problems, but it was the 13-week freeze in 1963 that finished much of the local traffic.

It is perhaps England's finest canal for scenery and variety. As it is almost the most northerly in England and on the extremity of the canal network, it is much less frequently visited than popular routes to Llangollen and Oxford. Although it is a broad canal it has short locks that keep out some of the longer narrowboats. The broad beam reduced the impact of the railways. Two of the canal's wide boats are on show in the National Waterways Museum, Ellesmere Port.

The fastest commercial crossing was in 52 hours, the towpath has been run in 35 hrs 5 mins and it has been cycled in a day.

The canal leaves the Aire & Calder Navigation at River Lock in **Leeds**, an area where otters are present.

Locks and bridges are mostly of dark millstone grit. The bridge arch is often picked out in white and the centre point of the channel is marked, usually offset from the keystone at the centre of the arch, giving a lopsided look. The legendary dour Yorkshire countenance seems to be lacking. There are bream and roach in the canal but fewer on this side of the Pennines. Only the swans are more aggressive than usual on this canal, although mallards are much more sociable.

Overlooking the junction is the Hilton tower block, fine warehouses, Granary Wharf with its Victorian shopping arcade, a 19th-century wharf crane and the large Leeds station. The elevated railway hides the cathedral

Britain's longest single canal

River Lock in Leeds, at the junction with the Aire & Calder Navigation. Bridgewater Place towers above everything else.

and museum. The indoor market is claimed to be Europe's largest.

The canal follows the River Aire closely all the way to Gargrave, initially running over coal measures. The ladenses, British Celtic for a violent river, gave the city its name.

The second lock is beside the canal office. Beyond it is the Tower Works with distinctive campaniles.

The Leeds to Bradford railway crosses. By St Ann's Ing Lock is the former Leeds Forge, which made railway chassis as well as boilers for the Royal Navy. Adjacent are the Leeds & Thirsk Railway roundhouses site and the Castleton flax mill.

A feature of the canal is that locks are frequently grouped into flights or staircases and this begins after the A58 crossing at the end of the A58M. Alongside the Oddy Two Rise Locks staircase murals are mounted on a wall that is surprisingly free of graffiti. It is interesting how the graffiti artists seem to respect the work of other wall painters, here and elsewhere. Beyond the wall is Armley prison and Yorkshire TV's studio, the world's largest on one floor, used for filming their *Emmerdale* series.

The Leeds to Harrogate railway crosses after Spring Garden Lock. Armley Mills museum follows in what was the world's largest woollen mill. It is now a museum of the industrial past with steam locomotives, static engines, water wheels, textiles, a mill cottage and a 1920 cinema.

There are wooden rope guide rollers on a bridge. The Leeds to Shipley railway crosses on an oblique girder bridge and follows the canal to Shipley.

Already the city is starting to pull back and allow a green corridor. A steep-wooded bank on the left leads to Upper Armley and its golf course. Burley is more obviously on the other side of the valley beyond a large electricity substation, from which powerlines lead alongside the canal to Shipley. Picnic tables are set back behind alder trees. There are blackbirds, long-tailed tits, robins, wrens and pigeons.

By the B6157, 1,100 Leeds Metropolitan University students have their halls of residence in the former Kirkstall brewery on the edge of Bramley. Headingley Rugby Union Football Club is nearby.

Across fields of rhubarb stand the remains of Kirkstall Abbey, dark and roofless but very complete and with a good west portal. It was built in 1152–82 in transitional Norman style by a breakaway group of Cistercians from Fountains Abbey, who rejected Barnoldswick because of

The canal office by Office Lock.

Tower Works with their campaniles beyond Office Lock.

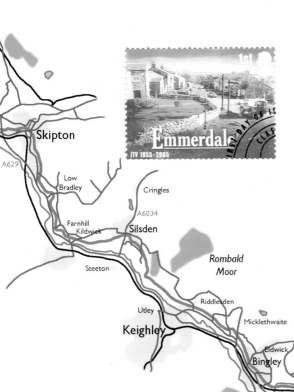

the abbey and many more structures. They are now set in bluebell woods with kingfishers on the canal. Many of the locks on the canal have open overflow channels past them like miniature dam spillways.

Newlay Three Rise Locks are another staircase. The minor road to **Horsforth** crosses the canal and then the river on an attractive cast-iron bridge of 1819, the Micklefield Estate toll bridge, similar to the Scarborough Spa Bridge. It is 3m wide with two 900mm footways and spans 25m with rhomboid openings.

The canal and river make a large loop round a sewage works. The scenery around Rodley Wharf and the Rodley Barge public house is surprisingly attractive, considering Rodley became synonymous with heavy railway lifting cranes from 1820, the works now derelict. On the hill is an observatory.

itsclimate. It has the ghost of a 12th-century abbot, who met an unfortunate end. It also has a folk museum in a gatehouse, with three streets of 18–19th-century cottages, workshops, shops with period furnishings, 1760 costumes, toys, domestic items and items from the abbey excavations. Kirkstall Abbey crypt was painted by Turner in 1824.

There is a Jacobean hall at Hawksworth.

Forge Three Rise Locks are a longer staircase, named after the 17th-century Kirkstall Forge, from which the thump of the drop hammer still resounds. It now makes road vehicle axles for GKN. Quarries provided stone for

Mills before Oddy Two Rise Locks.

The canal has some fifty low wooden swing bridges. Canada geese mock as the first is met. There are redwings in the hawthorns.

After the A6120 crosses, the Railway warns that the tracks are returning and will cross beyond Calverley. *A Yorkshire Tragedy* was written in 1608, possibly by William Shakespeare, about Walter Calverley, who had murdered his two sons and was pressed to death in 1605. This had been the seat of the Calverleys for centuries.

Herons, moorhens, fieldfares and jays might be seen along this stretch and past Lodge Wood.

At Apperley Bridge, where a former boathouse built the transatlantic rowing boat for Chay Blythe and John Ridgeway in 1966, the A658 passes over and the canal goes from Leeds to Bradford near Barratt's shoe factory. After the Dobson Two Rise Locks, the Leeds to Shipley railway crosses and dives into Thackley Tunnel under the Nosegay. Those still in the open air have the odours of the large Esholt sewage works, which had its own

Murals beside Oddy Two Rise Locks.

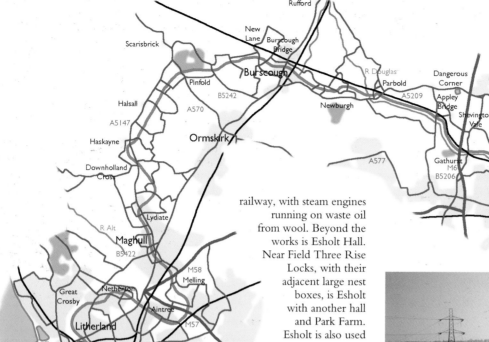

Hirst Lock is sited in Hirst Wood, with its beeches, oaks, hazels and pussy willows. This venue attracts blue and great tits, mistle thrushes and treecreepers. The canal crosses to the opposite bank of the river on the seven-arched Dowley Gap Aqueduct as the river doubles back in an S-bend. The wooded cutting high above the river leads past a canalside tree house to the Dowley Gap Two Rise Locks, an area with snowdrops in the spring. Below is Cottingley where, in 1917, two young girls claimed they played

railway, with steam engines running on waste oil from wool. Beyond the works is Esholt Hall. Near Field Three Rise Locks, with their adjacent large nest boxes, is Esholt with another hall and Park Farm. Esholt is also used for filming *Emmerdale*.

At the far end of Brick Wood the railway emerges from its tunnel. Alders provide some screening of **Baildon**. A mast rises on the skyline in Shipley.

The Ilkley to Shipley railway crosses over. A spur is the end of the former Bradford Canal. It closed in 1922 and was so polluted that the water burned with a blue flame. *Room at the Top* was filmed here. An aqueduct takes the canal over the river. There is a bridge with historical murals painted underneath, sadly this time not respected by the graffiti sprayers. The Noble Comb public house and a high chimney are among the landmarks. In Shipley, the Alhambra had its dance floor replaced with timbers from the short boat Cedric. Shipley Wharf has been restored. Canopies still project over the water and a bus stop sign shows it is a Metro Waterbus stop for Bingley. This textile and engineering town has 16–17th-century mills, a battlemented Salvation Army citadel and Windmill Manor.

Saltaire was a model village, built from 1850 by Sir Titus Salt with 850 good terraced houses for the workers in his mohair and alpaca mill, now a World Heritage Site. The 4ha site has a school, institute and almshouses modelled on Italian villas but no public house or pawnshop. Salt's Mill of 1853, the Palace of Industry, is a six-storey, 22m high building, 166m x 15m, and the upper floor was thought to be the largest room in Europe. The Italian-style buildings include a chimney disguised as a campanile, modelled on Santa Maria Glorioso in Venice. An adjacent single-storey, 100m weaving shed had 1,200 looms and there was a 64m x 34m combing shed. These days there is the largest display of David Hockney works. An adjacent United Reformed church of 1859 is a particularly splendid building. The Victoria Hall has a reed organ and harmonium collection.

On the other side of the river, the Shipley Glen Tramway dates from 1895. The park is said to have been one of the most beautiful in the world and was one of the world's first amusement parks until the tow on a toboggan ride broke in 1900 with fatal results. Four lions, near the grammar school, are ones that were rejected from London's Trafalgar Square as being too small.

Dobson Two Rise Locks at Apperley Bridge give extenstive views southeast.

Junction with the former Bradford Canal.

Shipley Wharf. Note the bus stop sign.

Kirkstall Abbey.

Salt's Mill at Saltaire.

At the foot of the Three Rise Locks is the Damart factory, the original manufacturer of thermal underwear, aimed at the elderly rather than the outdoor market. Pipes take water from the River Nidd over the canal to supply Bradford.

Then comes another remarkable set of structures: the Bingley Five Rise Locks. These are the country's highest staircase, lifting the canal 18m. At the top the Five Rise Lock Café & Store replaces early stables. Ash trees attract grey wagtails, chaffinches, crows and greenfinches. There are goosanders and little grebes here.

At Micklethwaite there is a goose farm and ugly goslings build their nest outside the fence on the bank of the canal.

Houses alongside the B6265, which follows the canal, hide East Riddlesden Hall. This 17th-century merchant's house is in a small walled garden with beeches and a 17th-century 37m oak-framed medieval barn, one of the finest barns in the north of England, near a monastic fishpond. Built in the Civil War by passionate Royalists, the house has a host of ghosts, Yorkshire oak furniture and panelling, wonderful plasterwork, embroidery and pewter. The last three Murgatroyds, the owners, ended up in York Debtors' Prison in Charles II's reign. Finances have been helped by using it to film Wuthering Heights. A wooden warehouse has a flyboat semaphore message signal on the corner. Across the canal is the Marquis of Granby public house.

One of the Cottingley fairy photos.

Saltaire's United Reformed church beside the canal.

The Bingley Five Rise Locks, Britain's highest lock staircase.

with fairies by the waterside. They borrowed a camera, asked to be showed how it worked and duly came back with pictures, which they did again in 1920. Experts could find no evidence of forgery and Sir Arthur Conan Doyle said he could see movement in these pictures, which baffled the nation. In 1983 one of them admitted it had been done with cutouts on hat pins but the other died still claiming there was no trickery.

There were Ice Age moraines in the valley between Shipley and Keighley. The river has cut through one at **Bingley** to form the Aire Gap. Space is restricted and the canal has been moved to allow for the current line of the A650 across the South Bog Viaduct. It is a series of arches over peat up to 10m deep, the largest arch being over the twin bores of the Nidd Aqueduct, which supplies Bradford.

Bingley was a medieval wool and market town, chartered in 1212 but named after the Old English man Bynna, Bynnings *leah* being his people's glade. There are an ancient market cross, Georgian market hall and Bronze Age round barrows. The 16th-century Holy Trinity church, with its massive spire, has an early Norman fort and fragments of a Saxon cross.

Wooded quarry at Lower Holden.

Keighley is a manufacturing town. The Cliffe Castle Museum is in a Victorian mansion that was owned by worsted manufacturer Henry Butterfield. It features the history of Airedale, natural history and geology, craftsmen's workshops, reception rooms with French furniture and art treasures and extensive gardens with a conservatory and aviary. Keighley was another Old English man's glade, this time Cyhha, and was home of poet Gordon Bottomley.

Utley was home to groundbreaking kayak expedition leader Mike Jones.

Ground was also broken in 1952. This length of canal was always prone to leaks but a breach caused sufficient damage to the golf course for it to require four policemen to control the crowds of spectators.

On the other side of the canal are oakwoods with occasional quarries. Coltsfoot, celandines and dog's mercury are found around Low Wood Nature Reserve & Scout Activity Centre and then rhododendrons. From Alder Carr Wood the scenery opens out with extensive views to Steeton Moor to the south-west and Rombald Moor to the north, moorland indicative of what is to come. The piping curlews add to the mood. At one point there are old limekilns.

In Silsden, Cowling Bridge carries the A6034 over the canal next to the Bridge public house and the canal moves from Bradford to North Yorkshire. There are stone warehouses with canopies over the canal. The Co-op had their own barge to bring provisions to the shop. Rather incongruous is a large polyethylene boathouse to allow work on narrowboats out of the weather. Trout are also found in the water now.

Parson's Bridge, beyond the White Lion in Kildwick, connects two parts of the cemetery of St Andrew's, also called the Lang Kirk because of its 46m length. The church has a weathervane, sundial, carved choir pews and a 14th-century effigy of Sir Robert de Stiverton, the lord of the manor. It also has wrought-iron lions on the gate, which are reputed to get down after dark to go to the canal for a drink, nothing at all to do with those who have been to the White Lion for a drink themselves. Tamer wildlife includes small frogs, pied wagtails, white dead nettles, ground ivy, butterbur, cow parsley and blackthorn. Over the river is a Grade I listed packhorse bridge of 1306, one of the best in Yorkshire, 42m long with a 200m causeway to the south, of ribbed arches of hewn stone. It is still in use although it has been widened on the downstream side. The canal has a narrow, skewed aqueduct over the approach road. The A629 arrives alongside to give traffic roar for a while. At first the traffic here was transhipment wagons as this was initially near the terminus.

Kildwick Hall is a 17th-century manor house. Lower down, life has been more cramped for the peasants. Riparian residents at Farnhill have even been known to move sofas in by narrowboat.

Farnhill Hall, on an eminence around which the canal twists, is a 12th-century dwelling with four battlemented towers.

Approach by vehicle is safe from the A629 from the north but any other move is better avoided because of the oblique entry onto a fast road at a blind corner.

From Farnhill, the canal cuts along the bottom of a steep beechwood, at the top of which is Victoria's Jubilee Tower. A memorial by the canal recalls the 1943 crash of a Wellington bomber with the loss of its seven crew.

The canal winds away from the A629 to visit Low Bradley with a swing bridge by a picnic area.

Moorings lead towards Low Snaygill. The A629 is replaced by the quieter A6131. The millstone grit is replaced by Craven lowlands Carboniferous limestone, up to 1.8km thick.

The Hannover International Motel & Club and Henri's Bar & Bistro add a slightly incongruous touch.

Swans usually nest on the offside of the canal but a

Looking north-west to the moors from Lower Holden.

The canal circles around Farnhill Hall on its rise.

Looking west over Cononley towards Elslack Moor.

Skipton's stone canyon.

Belmont Wharf and the Springs Branch.

pair have nested next to the towpath, making life very difficult for dog walkers and other pedestrians. Moorhens, mallards and magpies prove to be more agreeable feathered friends.

Skipton is a stone town, approached through a stone-building canyon. The canal is crossed by a freight railway line, the A6131 by Bodrum Kebabs & Pizzas and the Gallow Footbridge and then the A6069 by the Lock, Stock & Barrel and Bizzie Lizzie's Family Restaurant. Skipton – Scip-tun, Sceptone or Sheeptown – is the gateway to the Yorkshire Moors. It is one of Yorkshire's oldest market towns, with a market most days in the Georgian high street. The Black Horse Hotel hosted the first canal company meeting in 1770. The town hall is Victorian Palladian. There is a Craven Museum of local history and George Leatt's corn watermill from 1200 is an industrial museum. Holy Trinity Church is Perpendicular, partly 12th-century, enlarged in the 14th Century, with a fine oak roof and carved Jacobean font cover.

The town's pride is Skipton Castle, dating from 1090, one of the best preserved and most complete English medieval castles, still roofed. It has a Conduit Court, a 15m long banqueting hall decorated with seashells and a dungeon. It was built by the Normans to defend against the Scots, was overcome in the 12th Century, was rebuilt in 1311 by the 1st Lord Clifford and withstood a three-year Civil War siege. Butcher Clifford led the Lancastrians in the Wars of the Roses and his son, the Shepherd Lord, led the men of Craven at Flodden in 1513. Other features include a large kitchen, six 14th-century round towers, a 17th-century yew and a moat. Behind it is a battery. The castle and a limestone quarry quay can be reached along the Springs Branch from Belmont Wharf, which has the remains of a crane plus shops in old canal warehouses.

It seems that some residents would prefer to take sides with the nesting swans as dog mess has been a local issue; one young lady got herself sponsored in 1993 to clean it up to raise funds for a Yorkshire Schools Expedition trip to China.

The canal moves away from the centre of town, past Heriot's Hotel, Aireville Park, with its swings, and the station where Jimmy Savile's dad was stationmaster.

Dog rose, woody cranesbill, meadowsweet, red clover, angelica, hop trefoil and dog daisies are found by the canal and there are bird feeders before the A629. This makes a sweep over the canal on a long concrete viaduct, a contrast with the mood of the nearby ridge and furrow system of former times. The A6069 runs alongside for a while and there is a low swing bridge at the entrance to a farm. The A59 crosses on a bridge that is used by house martins for nesting. There are swallows and curlews in the area.

The A65 comes alongside for a while, the border of the Yorkshire Dales National Park, and the canal's scenery

is at its best for the next 16km. Rolling moorland has occasional stands of trees, a rookery here, Canada geese and oystercatchers there, but generally just wide open countryside. Johnson & Johnson's medical products factory standing alone on the outskirts of Gargrave therefore looks awkwardly out of place.

The A65 crosses over, the proposed Gargrave bypass never having been built, and the canal climbs through Holme Bridge Lock. This is the first of the well spaced Gargrave Locks. It crosses over Eshton Beck. A watermill on the left still has its wheel. Ray Bridge, which follows, is the canal's most northerly point, as the route turns from north-west to south-west. The canal has minnows, perch, roach, bream and pike. Meadow pipits make themselves busy above.

Gargrave is a grey limestone town with the 16th-century St Andrew's church and fragments of Saxon crosses, a Roman villa and a prehistoric camp. The Pennine Way crosses before Gargrave House in an attractively groomed area of town and countryside. The A65 crosses back below the lock at the Anchor Inn and leaves.

The top lock of the flight is Stegneck, where a mentally handicapped trip boat got caught on the cill in 1998 and then dropped free, drowning four of the party. This was the first inland accident investigated by the marine accidents people and resulted in the Chief Inspector of Marine Accidents, Rear Admiral John Lang, having to do some hasty reading on canals.

Immediately above this, the Leeds to Settle and Carlisle railway crosses and leaves up the valley of the River Aire. The river is crossed on Priest Holme Aqueduct, next to which is a roving bridge.

The six Bank Newton Locks are some of the most picturesque in the country, set in Pennine scenery. They have iron hooks for towrope guides, the top lock with mason's marks visible. Kingfishers and lapwings appreciate the environment.

Ferns on a bridge at Stegneck.

The roving bridge near Priest Holme Aqueduct.

The unique double-arched bridge at East Marton.

The Leeds & Liverpool Canal is a contour canal and around Green Bank it is second only to part of the Oxford Canal for its contortions as it winds around drumlins. A TV repeater mast twists from one direction to another. Vertical wooden rollers are installed for horse towropes.

Once the canal straightens out, it enters a wooded cutting and is followed by the Pennine Way for a kilometre, during which it is crossed by a unique double-arched bridge at East Marton. The lower arch carried a packhorse bridge and a higher arch has been built on top to take the A59. The Cross Keys Inn refers to St Peter's church, which has a battlemented Norman tower, a fragment of an intricately carved Saxon cross and a Norman font.

As another aerial is passed on the left, the canal moves from North Yorkshire into Lancashire in an area frequented by goosanders. A Greenberfield Locks Branch was planned to Settle, from where a further extension to Lancaster would have given a 50km pound, which would have been an extremely scenic route. This was anticipated by the Liverpool House restaurant in Settle but it never happened. Perhaps this is one for the new canal age, especially with the Ribble Link open. The three locks of 1820 replaced a staircase that was wasteful of water in the 10km summit pound, which begins above these locks. The original road bridge has been converted to a stable and the Lock Shop feeds humans. The locks lie on the line of the Roman road from York to Preston. The Pennines are a north–south anticline and it was the view from here that inspired the Easter hymn *There is a Green Hill Far Away*.

The first Lancashire industrial town of **Barnoldswick** is entered past an array of nest boxes and pirates. Somehow, this small Pennine town would not be the expected place to find Rolls Royce Aerospace but they have two factories here, the first in an old cotton mill by the B6252 crossing. Indeed, in the RB engine series the 'B' refers to Barnoldswick. Silent Night have a bed factory

in an old textile mill. Barnoldswick Mill Engine is marked by a square chimney with a cylindrical upper half.

A branch on the east side of the canal was the Rain Hall Rock Canal, which served limestone quarries. Both ends of it have now been filled in, leaving only 200m. It has a tunnel with a footpath, the only such tunnel on the Leeds & Liverpool Canal.

Also gone is a railway branch line that crossed before the B6383 bridge.

At Salterforth, the Anchor Inn predates the canal, the change in ground level resulting in the bedrooms becoming the bar and the bar becoming the cellars. The latter now have an impressive display of stalagmites and stalactites. This was the former county boundary, where wool traditionally gave way to cotton because Lancashire was wetter than Yorkshire and cotton requires more moisture than wool.

The return to civilisation comes with a speed camera warning notice, hopefully just a poor joke.

At **Foulridge** Wharf, the former Colne & Skipton Railway crossed on a trestle bridge. Limekilns have been restored and there is a restaurant in old stables. This is at the end of the Foulridge Tunnel. Despite being constructed by cut and cover methods, it does not have a straight bore. As a hat-making community, perhaps it was considered appropriate to cover the top of the canal.

The 1.5km tunnel is traffic-light controlled with traffic allowed south-west for the first ten minutes of the hour and the other way half an hour later. There is no towpath through. Steam tugs operated through the tunnel from 1880 to 1937. In 1912, Buttercup, a cow, swam through and was revived with brandy. Although it does not say so, unpowered craft are not allowed through the tunnel.

The walking route is complex and not clearly marked at all turns. It involves negotiating the B6251 and a series of housing estate roads, rather less safe than going through the tunnel for many people. Indeed, what signs there are avoid following the most direct route, along the B6251, instead leading down to one of British

Foulridge Wharf with its wharf building.

Foulridge Tunnel south-west portal showing the settlement.

British Waterways' workboats moored in Nelson.

Waterways' canal water supply reservoirs, which are located around the tunnel. The reservoir is used for sailing and few will want to portage beside it when 800m on the water across to the clubhouse is more direct.

Two sights that would be missed by using the safer tunnel route are a field of llamas and the hillside topped by the 1890 folly of Blacko Tower. On returning to the canal, it is important not to pick the former railway line instead of the footpath as the two are separated by a canal feeder stream with a high concrete wall that prevents it from being crossed, the signposting being absent again here.

The seven **Barrowford** Locks drop the canal from its summit level to a 39km pound. Again, the walking route is not user-friendly. Animal traps frequently have a gap in a fence which is narrow at the bottom to retain animals but with the sides sloping outwards so that people can get through. The ones here slope the wrong way so that animals are not stopped but anyone portaging a boat has difficulty getting past each one. This design of obstruction will be met again.

As the lock flight bends round, across the field by the A682 is the Pendle Heritage Centre. The centre is 6km from Pendle Hill but on Pendle Water, which the canal is to follow. The centre features a 15th-century cruck barn, a 17–18th-century farm labourer's cottage and information about the Pendle witches, ten mostly poverty-ridden females from children to octogenarians who admitted witchcraft and were hung in Lancaster in 1612 after a witch hunt.

The canal is crossed by the B6247 and the M65, which follows the canal for most of its length. The mood changes dramatically and almost instantaneously to industrial mill towns built on coal measures. Large amounts of debris floats, especially plastic bottles.

Large mill at Brierfield.

The canal crosses Colne Water on the three-arched Swinden Aqueduct, opposite Park Hill. Colne, from the British Celtic *calna*, means noisy river. Pussy willow quietens the industrial backdrop. There is a conspicuous Asian population.

A bank of chicken runs is passed as the A6068 and A682 pass over on their respective routes to the M65's Junction 13. The canal winds its way past the terraced houses and weaving sheds of the textile mill town of **Nelson**, which takes its name from a Lord Nelson public house in place of its former names of Greater and Little Marsden. It passes under the feeder for Junction 12 of the M65.

At **Brierfield**, the B6248 has single-line, traffic-light controlled passage over the bridge. A convenient grocery shop opens long hours.

Brierfield is left past a mill with canopies over the water and a church with an unusual clock. Surprisingly quickly, the canal is free of the built-up area and meandering along the contours with extensive views to the west across the Pendle valley. Moorhens and mallards paddle about below willows and silver birches. The Colne to Blackburn railway follows the canal, crosses on the approach to **Burnley** and continues to follow the canal for much of the way to Blackburn. Terraced houses and textile mills return where the coal from Reedley Colliery used to be loaded.

Mill with loading canopy at Brierfield.

The canal makes a marked 'S', passing first under the A682 and then up to a notice that instructs boats to hoot to warn those canoe training. In this area of coal-mining subsidence, the canal curves round Thompson Park with its boating lake and blue tits. It crosses a two-arched aqueduct over the River Brun by the Queen Victoria, at the same time crossing the line of the former railway to Bank Hall Colliery.

Ahead is another of the wonders of the waterways, the Burnley Embankment, 1.1km long and 18m high. It runs across the valley, wide and straight, perhaps less impressive from above than from below. During the

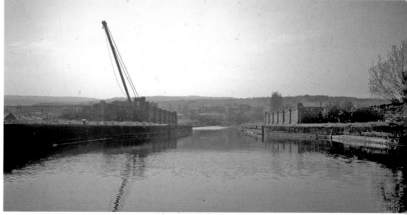

Looking onto Burnley Embankment with crane for stoplogs.

Some of the old Weaver's Triangle buildings.

The Victorians would recognise much of this scene in Burnley.

Gannow Tunnel. The portage leads up the slope on the right.

Second World War it had to be drained at times for fear of flooding the town if it was bombed. This is Britain's most outstanding canal embankment. It gives views over the slate roofs of the town, the green copper dome of the town hall, the Comfort Inn and Burnley Football Club. It also looks away to moors and Coal Clough Wind Farm to the south-east and to 557m Pendle Hill to the north-west, the tautological name of which means hill in Ancient British, Anglo-Saxon and Modern English respectively.

Domestic and Canada geese mill around a loading crane and Sainsbury's join limekilns at the foot of the embankment that crosses the A671 and the River Calder. At the Superbowl end it passes under the cast-iron Finsley Gate Bridge and turns in front of a derelict wharf.

A start on restoration of the wharf has been made on the other side of the A682, where the Inn on the Wharf is in an old warehouse as part of the Weaver's Triangle which conserves the industrial district and engineering works. The Toll House Museum covers the cotton industry, the canal, a weaver's dwelling and a working model fairground. The Burnley Mechanics Art & Entertainment Centre is a base for Mid Pennine Arts with an Easter National Blues Festival, all in well restored buildings.

The Paulinus Cross was erected in 800 but Burnley developed because of the coalfield. By the 18th Century, weaving was the main industry. Between 1800 and 1870 the village of Burnley grew tenfold in population with the Industrial Revolution. By 1914 there were 100,000 looms in the town, making it the world's largest cloth producer. The town has many mosques. Planned is a Travelodge, to be built over the canal on columns.

The A671 and railway cross again and the M65 dives underneath. Birches, alders and conifers have been planted as Lancashire Tree Aid, to fund a forestry project in Nepal. Magpies are among the home beneficiaries and there are ash trees which have planted themselves.

Gannow Tunnel is much shorter than Foulridge, just 511m long, straight but with no towpath. Steam tugs were used at the turn of the last century and there are mason's marks and symbols on the tunnelmouth. Again, unpowered craft are banned although there is nothing to

The canal turns in front of grand old warehouses at Church.

say so. The walking route is obvious at first, straight up the hill, but seems to have been obliterated by the building of the A671/M65 Junction 10 roundabout. It is necessary to reach the centre of the roundabout, either down steps and round a tight corner in front of shops or by dodging between the busy traffic on the roundabout. The centre is worth visiting if only to see the single tree labelled 'Forest of Pendle', which helps enhance the status of the National Forest further south. In the greater landscape the name remains around the valley of the Sabden Brook to the north, even if the trees don't. From here the unmarked route back to the water is through another subway tunnel, walking on an almost continuous carpet of broken glass fragments. Once again, the tunnel appears to be the much safer option for many people.

Beyond the tunnel is the Gannow Wharf public house and Rose Grove Wharf, where British Rail had one of their last three steam depots, the engines being filled with water from the canal.

The A646 crosses. The M65 and railway each cross and cross back before the canal winds round Junction 9 and then into a reach where the canal has been moved sideways to allow room for the motorway. A field of caravans precedes the Bridge House at Hapton, with its terraced houses, believed to have been the first street in England to have had electric street lighting, although it is not alone in making the claim. There are the remains of Hapton Castle on the other side of the motorway.

Time after time the canal turns north and then south again to find its way blocked by the motorway, like a fly trying to escape through a glass window. At the 17th-century Shuttleworth Hall, with its mullioned windows, it loops round Junction 8. Dog's mercury, gorse, hawthorns, alders and willows provide some vegetation. Skylarks enjoy the open grassland. The view to the north is generally open across the Calder valley while 399m Hameldon Hill stands to the south. A chimney on the far side of the motorway is the remains of Huncoat power station. Altham Barn Bridge was a packet boat stopping point.

After the A678 was Moorfield Colliery, where an accident killed 67 in 1883. It continued to operate. Eric Morecambe worked here in the Second World War as a Bevan Boy. The A680 crosses by the Albion Ale House in **Clayton-le-Moors**, the clay in the name being used for brick making. The village was the home of Dr Lovelace's soap factory which made floating soap. It now has rucksack firm Karrimor. Hyndburn Sea Cadet Corps also have their base for getting afloat.

The town is left past rough ground that is the legacy of clay removal for brick making. Dunkenhalgh was an Elizabethan mansion, much altered in the 19th Century. It is now a hotel in 6ha of gardens and woods, complete with lapwings and kingfishers. The towpath was set on the east side here to stop boatmen poaching their deer but, even without some of the present bridges, it should not have caused the average boatman too much difficulty getting across the canal. These days pike, trout and perch are more likely to be poached.

The Peel Arm was to have been extended to Accrington but it never happened.

The towpath is brought back across the canal on Church Kirk changeline bridge, which has had one of its horse ramps replaced with steps. St James' church is reckoned to be the midpoint on the canal. With a 15th-century tower and font, Burne-Jones Art Nouveau window, ceiling and balcony on slim pillars and ornate gravestones, it provides a substantial centrepoint in its setting of terraced houses. The first church was a temporary one set up by Oswald, King of Northumbria, on his way to battle. The current massive tower was used as a watchtower. The village formerly made red dye and printed calico. Sir Robert Peel's family had a cloth works here.

The railway crosses twice from an embankment that began life as a timber-trestle viaduct. On the canal bank are the Aspen Coke Ovens or Fairy Caves, ancient monuments that are the remains of beehive ovens used for making coke, all that remains of the coal industry here. The adjacent community of **Oswaldtwistle** was probably named after Oswald's *twistla*, King Oswald's boundary in the 7th Century. It has been an industrial area for over two centuries and its contributions belie its size. Stanhill resident James Hargreaves invented the spinning jenny in 1764. There have been many other textile inventions, including Terylene in 1941. Stockley's Sweets invite visitors to watch confectionery being made.

Beyond a golf course, the canal passes back over the motorway and is crossed by the A678 in **Rishton**, Saxon for rush village, the first place to weave calico in 1766. It had cotton mills, brickworks and coalmines in the 19th Century. The B6535 leads to the town's 1877 Gothic-style church. Reed buntings, curlews and bullfinches are present around Rishton Reservoir, which was built in 1828 to solve canal water shortages. It has been enlarged subsequently several times. In places, the canal has ramps to recover horses.

The railway crosses for a last time. There are the remains of Hodson's canal boat repair yard, closed in 1961. A retail park has been built on the site of the former Whitekirk power station, itself sited to benefit from Burnley coal supplies and cooling water from the canal. The M65 now moves away as the canal enters Blackburn.

Blackburn had Bronze Age inhabitants, the Romans built a fort here and it has been a cotton weaving centre since the 14th Century, much rebuilt since the Second World War. It was a newspaper report on the state of Blackburn's roads which inspired some of John Lennon's most discussed lines.

Imperial Mill is one of the few 20th-century cotton mills in east Lancashire. It is a listed building. By Sour Milk Hall Bridge are a bowling alley and horse chestnuts. Tesco have taken over an old chemical works as a store so it must be their trolleys which accompany such canal debris as bike frames and armchairs. Another 20th-century

The listed, 20th-century Daisybank Mill.

239

Eanam Wharf was once one of the busiest on the canal.

Sculpture on the Blackburn lock flight.

listed mill is the Daisybank Mill, built as a flourmill but now a business centre and studios for Granada. The Wharf public house accompanies the A677 crossing at Eanam Wharf, one of the most important on the canal and now restored. A major new road crossing is planned here. Almost opposite is Tommy Ball's Shoe Emporium, possibly the biggest shoe shop in the country.

The Blackburn skyline once had 200 chimneys, one of 95m probably being the highest in the UK. In addition, there are a dozen church spires. The cathedral was the 1826 parish church of St Mary, with medieval and contemporary glass, a magnificent lantern tower restored in 1998, a medieval pax, a John Hayward sculpture of Christ the Worker and a tower topped by a gently tapering cone with a cross on top. Despite the cathedral, Blackburn remains a town. The Museum & Art Gallery in a Grade II listed Arts & Crafts building has medieval manuscripts, local history, militaria and early textile machinery. The cotton mills of the Lewis Textile Museum have models of the spinning jenny, spinning mule and flying shuttle. Residents of Blackburn have included George Ellis – the father of the modern brass band – Kathleen Ferrier, Russell Harty and writers Josephine Cox and Alfred Wainwright. Barbara Castle, one of the Kinder Scout open land access trespassers, was MP for 34 years, bringing in legislation on national speed limits, compulsory car seat belts and equal pay for men and women.

By the station there are go-karts in a former goods shed. Blackburn ice arena and Waves Water Fun Centre with its Alien Encounter, children's shipwreck slide, wave machine, fountains, waterfall, jet sprays and whirlpool offer alternative entertainment. Between the A679 and A677 crossings are a mosque and the Asda store where work on a car park caused the canal embankment to subside in 1991.

The six Blackburn locks bring the 39km Burnley pound to an end and seem to be successful at sifting most of the floating debris from the water. Interest ranges from a toothed segment on a lock gate to a modern statue of a cyclist at Nova Scotia Wharf. There were formerly canal stables. Opposite is an infirmary, the first hospital in east Lancashire, supported by subscription, including that of the canal company. The Atlantic, Royal Oak and Moorings can provide refreshment in a town that hosts Thwaites brewery.

The Blackburn to Bolton railway crosses before the canal heads out onto Ewood Aqueduct. This crosses the B6447, beneath which the River Darwen is culverted. The large Ewood Park football stadium of Blackburn Rovers is prominent. Further points of refreshment are the Navigation and the Mill Bar.

To the north is the 1.9km² Witton Country Park with the British Small Mammal Collection in a 19th-century stable block and also 246m Billinge Hill. There are pigeons nesting in King's Bridge at Mill Hill, great and blue tits, stoats, coltsfoot, azaleas, rhododendrons

and willows as the canal runs into Cherry Tree, clearly the higher class end of Blackburn.

At Feniscowles there are stone quarries and a boatyard with a slip for sideways launching. There is a house balanced on the trunk of a tree and one house has phone and letter boxes next to the canal as garden features.

The A6062 crossing marks a change as the canal returns to a much more rural character, heading past wooded banks with celandines, butterbur, coltsfoot, spotted woodpeckers, kingfishers, swallows, sandpipers and grey wagtails. A turn is made before the newly returned M65 as the canal passes an old paperworks.

The A674 passes over Stanworth Bridge to reach Junction 3 of the M65.

To the north of Riley Green, beyond two masts, is Hoghton Tower, a fortified mansion of 1562–5, owned by the de Hoghton family since the time of William the Conqueror and still owned by them, the second oldest baronetcy in the country. Shakespeare is said to have worked here for a time as a servant. Better documented is the visit of James I in 1617, where he enjoyed a loin of beef and probably a quantity of wine so much that he drew his sword and knighted the meal. The name has stuck. Sirloin steak is on the menu of the Boatyard Inn.

With the exception of Wigan and several motorway crossings, this is a very rural and quiet section of the canal, hidden in the landscape rather than exposed on a broad cloth, as it is over the Pennines.

From Riley Green the M65 soon makes a final pass over the canal and leaves the cut in silence, apart from the larks, mallards and Canada geese. Gorse, hawthorns and oaks break up the undulating grass grazing and there are bream, carp, tench, perch and gudgeon in the canal.

Withnell Fold was an industrial hamlet where the Parke family opened a paper mill in the 1840s to produce paper for banknotes. There is a hall. The village has a square around a sundial to house the papermill workers. There are medieval stocks. Old filter beds and sludge lagoons have become a nature reserve. A high bridge of 1894 carries water from Thirlmere to supply Manchester.

Higher Wheelton and Wheelton are not even noticed as the canal eases through fields and woods.

The Top Lock public house is at the start of the attractive Johnson's Hillock Locks, seven locks taking the canal down to join the Walton Summit Branch. Faced with the problem of crossing Preston, the Lancaster Canal stopped on the far side of town and built an 8km tramway to Walton Summit as an interim measure, then continued the canal from Walton Summit to Westhaughton to carry Wigan coal north and bring limestone back. The tramway was abandoned in 1862 and the separate canal section was taken over by the Leeds & Liverpool Canal. Finally, most of the 5km of what had become the Walton Summit Branch was used as the route of the M61 in 1970, leaving just a 300m spur in water. During all this, spa water was found at

On the Johnson's Hillock Flight.

The Walton Summit Branch, left, was planned as the start of the Lancaster Canal.

Whittle Springs, always good for producing travelling customers.

A current watering hole is the Malt House Farm, by the B6229 crossing.

Botany Bay is said to take its name from the fact that the living conditions were as bad as they were for the convicts in Australia. These days, the Victorian Botany Bay Village five-storey mill and the Puddletown Pirates, the largest indoor adventure play area in north-west England, draw the public in and are surrounded by a helicopter, jet, anti-aircraft gun and fire engine spread around the car parks. The most attractive feature has to be a fairground organ belting out honkytonk music across the canal, a surreal contrast with the solitude of the previous hour.

The A674 crosses to Junction 8 of the M61, beyond the Railway the B6228 crosses and then the M61 itself crosses and heads away, leaving the canal to form the eastern boundary of Chorley.

Chorley has had its Flat Iron market since 1498, named after the weights used to hold down its cloth. The town itself was *ceorl leah*, Old English peasant glade. It had its market chartered in the 1250s and has become known for Chorley cakes, flat pastry buns filled with dried fruit, a recipe possibly dating back to the Crusades. The most prominent son was Sir Henry Tate, who moved to Liverpool in 1832 to set up his sugar company, also becoming a patron of the arts.

The canal turns over a stream and continues to Cowling, through blackthorn and bluebell woods with moorhens and sandpipers. The Preston to Manchester railway crosses, the canal crosses the River Yarrow and passes a moat and the A6 crosses. A nearby bridge has been used as a kestrel nest site.

Heath Charnock boasts Frederick's Ice Cream Parlour & Coffee Shop of 1896, with over 70 flavours of ice cream. Rhododendrons line the bank.

The former coal and textile town of **Adlington** is the home of Adlington carnival in August. Beyond the White Bear Marina, the largest on the canal, and the Bridge, the Red House Aqueduct takes the canal from Lancashire to Greater Manchester as it crosses the River Douglas. This has already picked up water from several reservoirs including Anglezarke and Rivington. Plant life varies from daffodils to wood sorrel and birches while birds range from blackcaps and jays to curlews.

Wigan Golf Club is based at Arley Hall with another moat into which to hit balls. A railway used to pass under the canal here and the B5239 went over them both at Red Rock. There are lily pads on the water, roach and bream in the water and woodpeckers in the Lombardy poplars.

The Crawford Arms is a rather strange canalside public house with no pumps in the largest bar but carpet in the toilets. Geese hang around because they know they will be fed by customers. There are also swans.

The canal retains its level as the River Douglas drops away steadily, resulting in ever more extensive views to the south-west of the canal. The prospect has not gone unnoticed and Haigh Country Park, edged with willows, is set around a pre-Tudor mansion rebuilt in 1830–49. The park has a golf course, mini zoo, model village, 381mm-gauge steam railway, tropical house, formal gardens, nature and geological trails and 1km² of woodlands,

largely planted to give work to unemployed cotton workers in the 1860s during the American Civil War. The site of a Victorian iron foundry, it had a summerhouse built out of hard Cannel coal.

Beyond the B5238 is a bell-towered house and a slagheap planted with pines.

The top of the Wigan Locks is well hidden on the right. This was originally the plan as the spur ahead was to be the main line, joining the Bridgewater Canal near Westhoughton. It only runs for about 200m, offering no more than a view to the Fiddler's Ferry cooling towers, 20km away.

The 23 locks of the Wigan flight drop the canal 65m in the better part of 3km, over 5 hours to work a boat through the locks. The Kirklees Hall and Commercial Inn come quickly but the rest of the flight is dry in that sense. The major problem with the flight's towpath is the series of fences across with kissing gates that have to be negotiated. There are bike-shaped slots in the fences although cyclists find it easier to lift their bikes over but those portaging do not have a choice. If the intention is to stop people racing down on wheeled transport there should be a way of doing it that makes life less difficult for everyone.

Were it not for the fences, the walk would be quite pleasant. Looking down the hill, a spire rises prominently above the haze of Wigan. This is the 14th-century All Saints church with a Roman altar, fine stained glass, many monuments and effigies and, as becomes clear from further down the flight, a leaning spire, all set in rose gardens. The towpath is of good quality. Hills of colliery spoil rise above the canal in an area of mining subsidence

Haigh Country Park.

Paved towpath on the Wigan flight.

Lock mechanism where there is no room for a balance beam.

The drydock in Wigan.

but there is also a copse with elder, alder, white poplar, birch, cherry, guelder rose, figwort and scentless mayweed. A former railway crossing has gone but there is a lock with segment gearing as there is not room for balance beams. There *is* room for Twiss' Bakery & Café across the A577.

The A573 crosses near the bottom of the flight, followed by the Manchester to Wigan railway and then the West Coast Main Line. **Ince-in-Makerfield** is named after the Welsh *ynys*, island, British Celtic *macer*, masonry ruins, and Old English *feld*, countryside. A gleaming Girobank building was formerly Wigan power station, the last business to transport coal on the canal. Beyond it is the Leigh Branch. The section from here to Newburgh was originally built as a branch to the Douglas Navigation from the southern route to Liverpool. The River Douglas crosses here and is followed by the canal.

To the Romans, the town was Coccium although **Wigan** is from the Saxon for rowan trees near a church. It was incorporated by Henry I in 1100 and received its charter in 1246 from Henry III. King Arthur was reputed to have carried out exploits in Wigan and it features in Walter Scott's *The Betrothed*. Cromwell pursued the Royalists through the streets in the Civil War. In 1651, the Earl of Derby was defeated in the Battle of Wigan Lane and killed, the mayor still being preceded by a sword of 1660, given by Charles II for the town's loyalty.

No. 1 Wigan Pier warehouse.

The replica Wigan Pier.

The Orwell has a commanding position in the basin but is not Wigan Pier, despite what it says.

At one time there were ten blast furnaces and 700 coke ovens. More recent fame came as the original source of Uncle Joe's Mint Balls. Beecham's pills were first made here.

After a playground on the right comes a lock with a dry dock on the left, still in use. Relaunching of small craft is difficult because of the high sides. There is a lower section on the left but it is within a British Waterways area that is kept locked. The next best option is the far corner of a wharf on the right next to Trencherfield Mill. The current mill was built in 1907 to spin cotton imported to Liverpool and brought by canal. It has the world's largest working steam engine at 1.9MW. This drives an 8.1m, 70t flywheel, powered 84,000 spindles in the mill and was used until 1968. The mill now displays cotton spinning machines, colliery fans and a rope walk, as well as the engine. It has a concert hall that hosts an annual international jazz festival.

Opie's Museum of Memories in the mill features 40,000 images of 20th-century social history. Regardless of its history, it was George Orwell's *Road to Wigan Pier*, written in the 1930s, that made the town's name. Displayed between the A49 bridges, it develops a joke by George Formby in a comparison with the piers of Blackpool and Southport. George was born here, as were Roy Kinnear, Ted Ray, Frank Rendle, Sir Ian McKellan, Angus Fraser, Sir James Anderton and Joe Gormley.

British Waterways' montage of crane, tug and lock gate.

The former windmill beside the canal in Parbold.

The Terminal Warehouse dates from 1777. Wigan Warehouses of 1890 house the Way We Were exhibition, covering 1890–1914, a Victorian classroom, a public house, a cottage and a 19th-century coal mine. There are also the Orwell and the Pier Nightspot. Just about the least significant thing is the pier itself. The original was dismantled in 1929 but the current replica was made in 1984 by students, not too onerous as it is, effectively, a couple of pieces of bent rail on the towpath. The original was a tippler that loaded coal in railway wagons from Meyrick Bankes' collieries into narrowboats. Even British Waterways' display, opposite, of a crane, a Bantam tug and a lock gate, is more impressive. There is also a 7.6m-diameter ventilating fan from Sutton Manor Colliery in St Helens.

There is another roving bridge with wooden rollers and a horse pull-out. After the Seven Stars Hotel, the Liverpool to Wigan railway crosses and the canal enters an area with a wasteland feel to it. Pallets are stacked beside the canal and there are industrial units. Pagefield Lock is faced by the large JJB football and rugby league stadium with its car parks.

Over the next pound, subsidence areas have resulted in pools; some are lined by anglers while others are filled with reedmace and arrowhead, lined by alders and used by coots, herons and wrens. An industrial estate on the south side of the canal includes a large Heinz factory, home of the baked bean.

The Wigan to Southport railway crosses and follows the canal to New Lane.

The next lock, Hell Meadow, has the most difficult take out on the canal for small craft. The live chamber on the left has high walls on each side. Beside is an empty chamber, then an overflow channel with a notice forbidding landing on the lower ground to the right of this. The only safe way is to take out before the walls begin to rise and face a long portage.

At Crooke, the John Pit was served by 2.1m boats legged from the coalface. Starting beyond the Crooke Hall Inn are oak woods that follow the canal for a considerable distance and provide an environment suited to the kingfisher. Another works at Gathurst is served by a railway operating across the canal before the Navigation.

The M6 passes high overhead, the railway crossing obliquely below it in the intervening space. The former lock connection to the river no longer exists but Dean Lock is a double lock. On the right side is a golf course. The other side gives extensive views westwards down the Douglas valley. Closer at hand, ivy-leaved toadflax grows in stone crevices, above which grey wagtails bob. Celandines and butterbur grow along the towpath edges.

At Shevington Vale, the canal returns from Greater Manchester to Lancashire and passes between the Water's Edge on one bank and the Railway on the other.

Appley Bridge was a linoleum manufacturing centre. Appley Lock has a deep lock on the left and two water-saving shallow ones on the right. Pheasants squawk across the valley. There were squawks of panic in 2003 when a breach of the embankment resulted in the canal being drained as far as stop planks placed at Parbold, also closing the adjacent railway temporarily.

Parbold Hill is 120m and topped by Parbold Hall and Ashurst's Beacon, built in Napoleonic times to be able to warn of French invasion. Reeds build up beside the reach past Priors Wood Hall, crossed near the end by the A5209 as it drops down from Dangerous Corner.

A tight corner on the canal in Parbold was intended to be a T-junction but the planned link on the right to the Ribble didn't get very far at all. A sculpture like a ribcage at the corner means it now gets even less attention. Formerly a flour milling centre, a windmill now stands armless beside the canal.

The towpath carries the Douglas Way footpath but

the River Douglas crosses under and leaves northwards at Newburgh. The soil is rich and black, like soil in the fens. Fields of sprouts and potatoes stretch out; lapwings, common gulls, longtailed tits, fieldfares and snipe provide a cosmopolitan mix of birds although vast expanses of polyethylene sheeting in the spring limit their foraging. Pillboxes defend the approaches to the River Tawd aqueduct from each direction. The West Lancashire Tree Nursery gives some variation in the environment for the wildlife. On the canal, angling competition position numbers are painted along the towpath.

At Hoscar, a recent canalside playground has been developed, a pirate ship of old timbers with a poetic message inscribed around it, a sculpted circle of two stone fishes and horseshoe footprints set in the towpath as part of a sculpture trail. There are a small parking area and another pillbox. Broken glass suggests modern pirates break into parked cars.

From the playground at Hoscar, a reach has hawthorns, blackberries and birches as well as birdlife from moorhens to black-headed gulls. It leads to the Ring o Bells in a former boatmen's community. Dorset Horn sheep are to be found.

After a low swing bridge, an imposing arched towpath bridge of 1816 crosses the start of the Rufford Branch.

The branch is followed by a cricket field before the Preston to Hunts Cross railway passes over. On the far side of the embankment is an old mill in the dark brick of the area with an ornate wrought-iron awning. **Burscough** was a packet-boat staging post, had stables and was where many of the boatmen lived. Another historical survival until recently was an Easter pace egging procession, a local version of a mummers' play.

The A59 crosses at Burscough Bridge by the Waterfront public house. Beyond is the storage area of a fairground company that has evidently not been on the road recently. Ashes are among the trees on this reach, which is home to mallards and Muscovy ducks, lapwings and starlings, red admirals and specked wood butterflies.

There is a slipway at the low swing bridge before the Farmer's Arms at New Lane Bridge. The farming of winter wheat gives way to market gardening, particularly celery and lettuces, while non-cultivated plants include red campion, pink convolvulus and rushes. Herons watch quietly.

After a stand of pines, the B5242 crosses Heaton's Bridge, beside which is the Heaton's Bridge Inn. A notice warns of shooting in a beechwood alongside, at the back of which is Scarisbrick Hall. In Victorian Gothic, the hall was designed by Pugin for Thomas Eccleston, who drained most of Martin Mere by steam power in 1787 to leave a lot of farmland and what has become an important waterbird reserve. The house is now a school.

There used to be a packet boat point by the Red Lion at Pinfold, used by carriages from Southport along what is now the A570, crossing at Scarisbrick Bridge. Once the mere was drained, Wheelwright's Wharf was employed to receive night soil from Liverpool to use as fertiliser, no doubt adding flavour to the leeks grown here. Scarisbrick Cross, made from a single slab of stone, is one of two lines of waymarkers and shrines laid through the marshes to Ormskirk and Burscough Priory.

A road crossing past the Saracen's Head on one side of the canal and a monkey puzzle tree on the other leads to Halsall, where the Grade I listed St Cuthbert's church of 1290 is the oldest and one of the best in Lancashire. It has a 15th-century octagonal tower with spire, original medieval door with an ornate top and a choir vestry that was the grammar school of 1592. Views from here are extensive across the flat farmland with its fine soil. Although the canal's inaugural meeting was in Skipton, it was here, in this much easier topography, that digging of the canal began.

Pirate ship, stone fish and artificial horse hoofprint in the towpath at Hoscar.

Decorative mill awning at Burscough.

The ancient Downholland Hall and a swing bridge.

A garden entrance beyond the public house is crowned by a raised castle. It is followed by a brick pillbox on the other bank. The canal enters a cutting with bluebells, oaks and sycamores and there are green woodpeckers and kingfishers. Tench and roach occupy the rest of the canal.

After the Ship Inn, the A5147 crosses at Haskayne, a village that dates from Norse times. It has cruck-framed

Entering the attractive village of Lydiate.

thatched cottages and the Old Post Office & Craft Centre plus football fields.

The canal's use of contours results in a wildly wandering course over the rest of its journey as it makes exaggerated loops.

Beyond an electricity substation, the ancient Downholland Hall stands near the canal. The A5147 crosses back by the Scarisbrick Arms at Downholland Cross. Gorse is seen increasingly on the banks, together with figwort and grain fields. Tanpit Farm takes its name from hide tanning.

Lancashire gives way to Merseyside but the 14th-century Scotch Pipers Inn may be the oldest in what was Lancashire before the county was broken up. The A5147 makes a final crossing over Lollies Bridge at the start of Lydiate, a surprisingly attractive village of small but well kept houses with long smart gardens running down to the canal. More low suspension bridges follow, including one near the Running Horses in **Maghull**. The Mersey Motorboat Club is said to be the oldest of its kind in the country.

In 1994 a brook culvert collapsed, breaching the canal, dropping the water level back as far as Liverpool and flooding 150 houses to a depth of 1.5m. Anglers tried to get TV coverage of the incident to say that the major concern was the loss of fish from the canal, as a consequence of which anglers are not particularly popular in this area.

Grey wagtails and geese are other canal residents.

The B5422 crosses Red Lion Bridge and then the A59 passes over again. St Andrew's church of the late 19th Century was built in 13th-century style to blend with Unsworth Chapel close by, the oldest church on Merseyside, dating from 1290. Under a beech in the churchyard is the ornate 1936 tomb of Frank Hornby, who invented Meccano in 1901 and whose company also produced Dinky Toys, clockwork Hornby Railways and electric Hornby Dublo trains. As such, he did more for the country's engineering skills base through the 20th Century than anyone else and his contribution to Britain's economy is inestimable.

The Preston to Liverpool railway crosses again, followed

by the M58. Small lilies float in the water, reedmace grows alongside and there are black-backed gulls and Canada geese, cormorants and coots.

Melling was a settlement on a sandstone hill in the marsh. Beyond another low bridge is Waddicar, where British Insulated Callenders Cables built their works on the site of a 19th-century jam stoneware pottery site.

The Aintree horse race course from the canal bridge.

The M57 crosses. At the end of this straight the canal doubles back at Aintree, after the River Alt aqueduct, as it is crossed by a very low swing bridge. Barbed wire ahead indicates the start of the Aintree horse race course, home of the Grand National, the world's most famous horse race, a race that has been organised annually since 1839. It also has a horse racing simulator, a motor racing circuit that hosted Grands Prix in the 1950s and, in the centre, the longest 9-hole golf course in the UK. Canal Turn, on the race course, is next to the canal and Beecher's Brook is a jump over a local watercourse. Blue Anchor Bridge takes a road across the canal into the course, where there is still plenty of barbed wire.

The name of the settlement is from the

245

Sheet piling with a story to tell at Netherton.

Well kept bridge in Bootle.

British Waterways excavating shopping trolleys and other debris in Bootle.

Attractively painted bridges lighten the mood of what could be a gloomy industrial landscape.

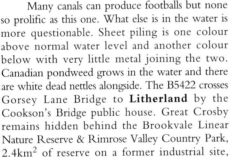

Clocktower facing the Mersey in a disused area of the docks.

Old Norse *eintre*, lone tree. A passenger service by canal to the centre of Liverpool began in 1814. After the A59 crosses Old Roan Bridge and the Preston to Liverpool railway crosses back, there is a stand of poplars. These and willows soften the defences. There are swans and kestrels. The A5036 crosses Netherton Bridge and there are elders, sedges and a thick carpet of azolla in the autumn.

Many canals can produce footballs but none so prolific as this one. What else is in the water is more questionable. Sheet piling is one colour above normal water level and another colour below with very little metal joining the two. Canadian pondweed grows in the water and there are white dead nettles alongside. The B5422 crosses Gorsey Lane Bridge to **Litherland** by the Cookson's Bridge public house. Great Crosby remains hidden behind the Brookvale Linear Nature Reserve & Rimrose Valley Country Park, 2.4km² of reserve on a former industrial site, reedy pools, scrub and grassland providing space for goldfinches, ten species of breeding warbler including reed and grasshopper, reed buntings, red rumped swallows, snipe, jack snipe, water rails, woodcocks, barn and short eared owls and hobbies. A smart new footbridge leads across from Ford to the reserve and picnic tables.

At another footbridge there are distant views to wind turbines and the Seaforth Container Dock Terminal of the 1970s, taking bulk carriers up to 75,000t and using modern handling methods, extremely unpopular with the unions at the time,

There is wild clematis about a well-landscaped towpath but it is now a city canal as it heads into **Bootle**, *bothl* being Old English for large dwelling house. The water is clear. Powered craft wishing to proceed beyond the Aintree area are required to give British Waterways 24 hours' notice, a relatively worrying requirement. Lilac and iris add splashes of colour in places but the setting is industrial. At the Litherland Moorings there is a traditional wharf crane. The site is heavily fortified against intruders.

The ornate wrought iron Litherland Road Bridge with its coat of arms is the first of several interesting road bridges in good decorative order. New housing has been built in the area and British Waterways have a boat with a lifting bucket on the front like a JCB to excavate

shopping trolleys and other debris from the canal. Between the Southport–Liverpool railway bridge and the A5058 bridge there is a coal wharf with the remains of a crane. Red clover adds colour in season. The canal is now running parallel with the docks, 11km of them having been built behind a massive wall along the front between 1824 and 1860, replacing a fashionable seaside resort. By the 1880s **Liverpool** ships were carrying 40 per cent of the world's trade but the docks still mostly use the old handling methods and have declined greatly, some now being used by breakers. At Sandhills a large wharf building with awning over the canal shows that inland water transport was also important. Maple, rowan, hawthorn, privet, bracken and roses soften the austere remains of industry.

The Preston–Liverpool railway crosses over and then there is a towrope-damaged post before a wrought-iron bridge, a following ornate bridge with holes in the wall from Second World War enemy fire and another bridge with a rope roller.

The 130m of canal beyond Eldonian Village was bought in the early 1990s by Merseyside Development Corporation, filled in and used as a housing estate site. A further section is dry and walled-in except at the northwest corner. The canal used to continue to Exchange station. It stopped by the A5053, the appropriately named Leeds Street.

The canal now turns right at Vauxhall. Stanley Dock Cut drops through four locks to the River Mersey although the connection was not made until 1846. Nearby are the former Tate & Lyle sugar factory and a listed parabolic warehouse.

The line passes under the Preston to Liverpool railway although there is no longer a towpath. On each side are derelict warehouses and ahead a fixed bascule bridge before abandoned docks. Turrets top several towers, including a clocktower in the centre of the docks, and the warehouses have much character, despite all the smashed windows. In due course the area of restoration and redevelopment must move north along the line of the canal link and these buildings be made into attractive features.

The docks continue south over the Kingsway Tunnel to the heart of Liverpool. The new 690m Liverpool Link to the new terminus passes under a tunnel and over the Queensway Tunnel to run along the Mersey waterfront past the Three Graces which consist of the Royal Liver

The new Liverpool Link past the Royal Liver Building, Cunard Building and Port of Liverpool Building.

Building with Britain's biggest clock faces, the Cunard Building and the Port of Liverpool Building. The new Museum of Liverpool is a lower building but is very obvious from the canal.

The new terminus is in the Albert Dock, location of the Merseyside Maritime Museum, the Beatles Museum, the International Slavery Museum, the Tate Northern and various other sports and cultural experiences. The Pumphouse is now a public house and there are plenty more in the Albert Dock buildings. This is the UK's largest group of Grade I listed buildings. Among the moored craft is the *Planet*, the former Mersey Bar and Channel light vessel.

Livpool that received its charter in 1207 from King John and became Britain's largest commercial seaport in Victorian times, handling sugar, spices, tobacco, cotton and slaves. It was the main 19th-century gateway for America. The name is from the Old English *lifer*, sludge, but the river is now being cleaned up, following on from the appointment in 1841 of the world's first public health engineer, Dr William Henry Duncan.

The city's 250 monuments and 2,500 listed buildings include the largest collection of Grade II buildings outside London and the city includes Europe's oldest African and Chinese communities.

The busy and fast-changing skyline in the centre of Liverpool, including the new Museum of Liverpool, the GWR port facilities, the Three Graces and assorted marine equipment.

In 2008 Liverpool was the European City of Culture. Liverpool has been used for filming *The Virgin of Liverpool, Between the Lines, 51st State, The Hunt for Red October, My Kingdom, In the Name of the Father, Letter to Brezhnev, Backbeat* and *Priest* and it is the most filmed city outside London. Its musicians have had 56 number one music hits, more than any other city. Local artists include Alan Bleasdale, Clive Barker, Beryl Bainbridge, Linda Grant, Adrian Henri, Roger McGough, Jimmy McGovern, Nicholas Monsarrat, Brian Patten and Willy Russell. Eleanor Rathbone was the first woman councillor, fighting for better pay and conditions for Liverpool workers, votes for women and, as an MP, the family allowance. Both Littlewood's and Vernon's football pools began here.

This is the heart of Liverpool, the fishing village of

Billy Fury and a wellwisher's flowers benath his foot on the waterfront.

Distance
208km from the Aire & Calder Navigation to the Albert Dock

Navigation Authority
British Waterways

Canal Society
Leeds & Liverpool Canal Society http://townsleyb. members.beeb.net/llcs

OS 1:50,000 Sheets
102 Preston & Blackpool
103 Blackburn & Burnley
104 Leeds & Bradford
108 Liverpool
109 Manchester

The Albert Dock, home of the Tate Northern and other attractions at the new terminus of the Leeds & Liverpool Canal.

47 Aire & Calder Navigation

England's premier commercial navigation

The River Aire received an improvement Act in 1699 and it was England's premier navigation by 1704. It is England's leading canal freight route to this day, wide with few locks, able to take barges up to 700t. From the 1860s until 1986 it used Tom Puddings, 40t compartment boats to carry coal, usually towed in trains of 19, to be lifted and tipped into ships. In 1913 there were over a thousand in use. More recently, they were replaced by 210t units, moved in threes by Cawoods Hargreaves, although the exhausting of local coal mines has resulted in the winding down of the trade. The locks were enlarged from 1884 and push tugs were introduced in the 1960s. Coal and grain were the main freight, up to 2,500,000t per year, plus aggregates, fuel oils, lubricants, steel and timber. Effluent was also carried for disposal until this was terminated in 1997 by the EU.

Victoria Bridge, of 1839, replaces an earlier one swept away by floods. Leeds Bridge of 1873 is at a crossing served by a ferry until the 14th Century. In 1888, Louis Aimé August le Prince produced the world's first moving pictures of a bridge here, presumably the traffic being the moving subject matter. Flyboat House recalls former fast boat traffic. British Waterways' regional office occupies what was the headquarters of the Aire & Calder Navigation Company. Asda have their headquarters and Tetley's Brewery Wharf has shire horses and a museum of local brewing and public house history. This is a stone's throw from where, in 1847, the Revd Jabez Tunnicliffe founded the teetotal Band of Hope by the canal. The elegant cable stayed Call's Footbridge of 1993 crosses to Call's Landing and other busy city watering holes. The ornate Crown Bridge of 1842 carries the A61.

The River Aire emerges from Leeds' Dark Arches carrying the railway.

Call's Landing is a popular venue in the centre of Leeds.

Fuel and aggregate transport still runs at about 10,000t per day, so other users should have full competence at boat handing and keep clear of larger craft. Commercial use by tankers is most likely from Lemonroyd Lock downwards on Mondays and Tuesdays.

The navigation begins in **Leeds**, at the foot of River Lock on the Leeds & Liverpool Canal. The River Aire has just emerged from the Dark Arches beneath the station, a series of long parallel tunnels where the water flows fast with the occasional small fall, the tunnels not being straight and some having complicated exit routes. A roadway and footway run across the insides of the tunnels at high level, parallel to the railway. There are no fish in the river.

Leeds is the UK's second financial city and Yorkshire's second largest city, having developed in the 14th Century on wool and textiles. It was where Michael Marks launched Marks & Spencer in 1884. The Henry Moore Institute is the largest purely sculpture gallery in Europe; Leeds City Art Gallery, Leeds City Museum, the Northern Ballet and Opera North are also resident. The city has been used for filming *Harry's Game*, *Jimmy's*, *The Good Old Days* and *A Touch of Frost*. The city has 8km² of parks.

The Hilton has been built on the former Co-op Coal Wharf. A wharf crane and section of track have been preserved. This is just one of the new buildings to go up. There has been extensive regeneration of Leeds Waterfront with old wharves being restored, retaining some of the industrial detailing. Bridgewater Place, also known as the Dalek, is the tallest office and residential block outside London, 110m high with 32 floors. Inside is a 17.5m sculpture, the tallest in Yorkshire.

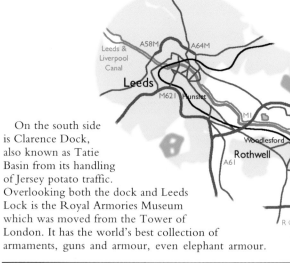

On the south side is Clarence Dock, also known as Tatie Basin from its handling of Jersey potato traffic. Overlooking both the dock and Leeds Lock is the Royal Armouries Museum which was moved from the Tower of London. It has the world's best collection of armaments, guns and armour, even elephant armour.

Call's Footbridge is a recent addition to Leeds' crossing points.

The Royal Armouries Museum overlooks Leeds Lock.

Thwaites Mill has become an industrial museum.

Demonstrations include jousting, falconry and wild west gunfights.

The lock bypasses a large weir on the river. On the other side, the Trans Pennine Trail follows the navigation as far as Mickletown. Shags suggest the city centre is being left. There is a sand and gravel wharf. The Hunslet Mills of 1838 processed flax and are listed.

The navigation makes its first major break from the river at Knowesthorpe, where Knostrop Flood Lock is usually left open. The Lord Merlyn-Rees Riverside Garden is a new amenity. The navigation is extensively piled, often with high sides. Large wharves front the navigation in places, including Leeds Oil Terminal, and British Waterways are considering building an inland port at Stourton.

Traffic lights control movements at locks, red flashing lights meaning the navigation is closed because of high water levels. At the foot of Knostrop Fall Lock there are massive supports for a former equally massive swing railway bridge of 1899 which was never opened. An industrial

Navigation Company to safeguard their water supplies. The current seed, flint and chalk crushing mill of 1823 was also used for preparing putty, pottery, bread and medicines at various times and had china clay delivered direct from Fowey. In 1976 it suffered flood damage.

Moorhens, coots and mallards paddle beneath the hawthorns, sycamores and willows.

A pair of concrete bowstring bridges formerly carried a railway. Two sets of powerlines serve Skelton Grange power station. There are various works and a container terminal at Stourton.

Gorse and magpies appear as the navigation moves progressively into open country. It is now followed by the Leeds to Castleford railway as far as Mickletown. A newer transport link is the M1 extension, which crosses at the east end of its Junction 44.

Rising land to the south hides Rothwell and the former mining village of John O'Gaunts although spoil tips can only be landscaped so far. To the north is Temple Newsome Country Park, one of the largest floral parks in Europe, 5km^2 having been laid out in the 1760s by Capability Brown around a Tudor/Jacobean mansion of *circa* 1500. Described as the Hampton Court of the north, it was the birthplace of Lord Darnley and once owned by the Knights Templar. Lakes and country park have been established on opencast workings. At Fishpond Lock there is an extensive rubbish tip but picnic tables are set out below the lock to gain some amenity.

Woodlesford Lock has bird hides. It also has numbered angling platforms. Bentley's brewery no longer exists but the A642 crosses Swillington Bridge towards Levensthorpe Hall and England's most northerly vineyard.

Fleet Cut, on the north side of the navigation, served Fleet Mill. There was formerly a lock to the River Aire. Alongside is Bayford's Thrust oil storage depot, served by boat. Lemonroyd Basin on the south side precedes the new Lemonroyd Lock.

To the north, Savile Colliery is closed and St Aidan's

estate to the north and an anti air-craft gun to the south stand along from the former railway line.

A steam derrick crane draws attention to Thwaites Mill Industrial Museum. There was a fulling mill here between the navigation and river. It had four wheels from 1641 and was bought by the Aire & Calder

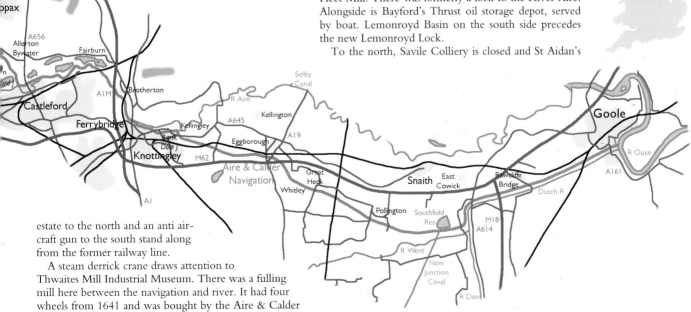

opencast site is being restored. Neither is visible from the water as the vegetation grows up. In 1988, the River Aire breached into the mine. Although the navigation was not affected, British Coal paid £20,500,000 to have 3km of the navigation moved sideways and the river diverted into the navigation. The work included a new Lemonroyd Lock and a dangerous-looking new weir on the river, abandoning the former Kippax Locks. The lower end of Lemonroyd Lock is surrounded by high concrete.

The river reach seems to be particularly popular with swans although there may also be terns, black-headed gulls, herons, great crested grebes and swallows.

Mickletown claims to be the second-largest village in England, a refreshing change from all those claiming to be the largest.

The old line of the River Aire is joined above Allerton Bywater, which had coal staithes to serve the last deep mine in the area, closed in 1992. The Victoria Hotel remains prominent.

Powerlines cross before the confluence with the River Calder, bringing in the Wakefield Section. The cut becomes the boundary between Leeds and Wakefield. The river loops away towards Castleford and Allinson's flourmills, grinding by water power at a weir. The navigation leaves as the Castleford Cut to complete a water crossroads, protected by Castleford Flood Lock, which is usually open.

On Ermine Street, from Lincoln to Tadcaster, the Roman fort of Lagentium guarded the Aire crossing point, the most difficult river crossing on the Great North Road. It was the site of a battle between warring factions in 947. There are fort and bath house remains

and a museum has archaeological finds. There was famous glassware and pottery in the 18th Century but **Castleford**'s claim to fame these days is as the birthplace of Henry Moore. The A656 crosses by Total Butler's oil distribution terminal. Beyond this is Hargreaves' depot with coal compartment boats moored several abreast.

Poppies and oilseed rape brighten up the bank to Bulholme Lock, where the cut rejoins the widened river. Beyond a bowstring railway bridge and powerlines are *ings*, Old Norse for meadows, although rather altered by mining spoil. Castleford Ings, on the south side, face Newton Ings and Fairburn Ings, well endowed with lakes. The latter is on what has been an industrial site since Roman times. It has suffered from coal subsidence to produce a wetlands reserve for the RSPB. Some 180 species of bird include reed and willow warblers, lesser whitethroats and garganey in summer, whooper swans and goldeneye in winter, water rails, common sandpipers, greenshanks and sand martins in passage and gadwall, shovelers, pochard, snipe, redshanks, gulls and lapwings all year. To the north are the remains of an abbey.

Canada geese and cormorants might be seen as the river wanders past giant hogweed.

Leeds gives way to North Yorkshire before the river is crossed by the Castleford to Sherburn in Elmet railway and the 2006 A1M bridge, cutting between power station ash lagoons and crossing banks where rabbits dig.

Ferrybridge is known for its power stations: Ferrybridge A, below the railway since 1927, and Ferrybridge B, above the railway from 1953, have both closed. Today there is just the 2GW Ferrybridge C from 1967, the largest in Yorkshire, burning 1,000t of coal per hour. Deliveries by water ceased in 2002. The tippler remains,

Hargreaves' boats await orders at Castleford.

Heavily reinforced brick railway arches with the new A1M bridge beyond at Fairburn.

The coal tippler at Ferrybridge. Barges have been replaced by coal trains arriving alongside.

able to empty a large container in 9 minutes but with no back-up when it frequently broke down. The coal now arrives by train, the merry-go-round behind the tippler. Sloping square tubes containing conveyors take the coal away to the power station. In the mid-1960s Ferrybridge was subject to a cooling tower collapse that led to a rethink on wind effects on groups of structures. For years, the river was marred by a thick layer of foam, which disappeared during the 1984 miners' strike. Inevitably, powerlines fan out in various directions. The Ferrybridge B site has been selected for a plasterboard factory.

After all the former mine sites and power stations, the darkened stone church tower near the river in Brotherton comes unexpectedly, as does Brotherton Bridge, carrying the York to Rotherham railway. Built in 1840 as an impressive tubular bridge on massive abutments, it was replaced in 1901–3 by towering Whipple Murphy trusses. At one time it formed part of the only line between Scotland and England.

A different notable bridge follows, John Carr's listed masonry bridge of 1804 to carry the Great North Road, now disused. Its arches on a tight bend do not help navigation, one arch having been blocked in recent years by a barge that broke loose in flood conditions. Beyond and apparently adding to the confusion is the replacement bridge that carried the A1 until 2006.

Ferrybridge was a Norman ford site and has one of the most interesting round barrows in Britain.

Located so as to be visible through a bridge arch is the traffic light for Ferrybridge Lock, a flood lock normally open beyond the Golden Lion as the navigation leaves the river for the last time and takes to the Knottingley & Goole Canal. This has more reliable water levels than the river. It has protective concrete panels in front of the entrance, erected after floods in the 1960s.

Knottingley, named after the Dane Cnotta, was chosen by Rockware Glass to make bottles and glass. John Harker's shipyard was also sited here. The canal has a gentrified urban feel to it at first, unique for this canal. Near Mill Bridge there is a branch to a weir on the River Aire. A new-looking loading chute has a safety barrier across the top, suggesting it has not been used in anger. The Steam Packet Inn recalls passenger transport on the canal. Cow Lane Bridge is the first of three crossings by the A645. St Botolph's church has Victorian stained glass and a low ceiling. A unique feature of the canal is 500m of new sheet piling on the north side of the canal, which retired

John Carr's Great North Road bridge with the recently superseded A1 bridge beyond.

Ferrybridge's coal loader.

Just part of the sheet-piled linear flowerbed.

Freda Turner has adopted, having planted the whole length next to the towpath with mixed flowers: one long flowerbed which is continuous colour in the summer. Sheet piling is a regular feature the rest of the way to Goole. It results in the drowning of deer which get into the water and then are unable to get out.

Bank Dole Junction is not marked as such. The narrower bridge to the left is the Selby Branch, leading to the River Aire and the Selby Canal. The Knottingley & Goole Canal of 1826, leading to the right under Skew Bridge, proved a shorter route to the Humber. Beyond chemical works, the character changes again to flat scenery and sweeping curves through agricultural land, mostly elevated. A nature reserve is sited between the canal and the river.

Powerlines loop south over the canal and round Kellingley Colliery. This had staithes to supply Ferrybridge power station. The Wakefield to Goole railway crosses before Stubbs Bridge where the canal finally leaves Wakefield and is entirely in North Yorkshire.

Another set of powerlines pass over. The north bank of the canal is edged in concrete with ramps at regular intervals to help wildlife to climb out. Pairs of owl boxes appear along the rest of the canal, seemingly of more interest to wood pigeons. The birds heard here are skylarks. A windmill stands towards Kellington.

A flat-topped hill on the far side of the M62 is a power station ash disposal area.

By the wharf at Low Eggborough, a mill used to grind barley for pig food. These days its purpose seems to be to carry the aerials of every mobile phone company imaginable.

The A19 crosses Whitley Bridge, heading for a prominent power station. It is followed by the M62 crossing. Whitley Lock, which has a bypass weir in its old chamber, has moorings below and was affected by mining subsidence in 2006, the head gates and lower of the two chambers having to be raised. Surprisingly, as the country becomes flatter, the number of recreational craft seems to increase.

The banks are lined with cow parsley in the summer. Black-headed gulls and swallows fly over the canal.

The road bridge at Great Heck is also settling significantly, giving tight clearance for empty barges. Heck Basin is used by South Yorkshire Boat Club. Hecking usually involved blocking rivers in connection with floating logs yet there is no river in the vicinity, the River Aire and the River Went each being some 4km away.

A dismantled railway had a rolling lift bridge from 1885 but it never had to be opened so it was not fitted with an opening mechanism. Still in place is the bridge carrying the East Coast Main Line, just south of where the driver of a Land Rover and trailer fell asleep on the M62 in 2001 and drove down onto the track, causing a crash that killed 10 and injured a further 82 people. This

One of the owl boxes near Eggborough.

Pollington Lock greets visitors.

resulted in protective barriers approaching railway bridges being reviewed throughout the country.

North Yorkshire gives way to the East Riding of Yorkshire.

Arrival at Pollington is between the imposing hall on the north side and a diminutive but conspicuous church spire to the south. In 1992 Pollington Lock, which has a large greeting sign, was the first British Waterways lock to have gates made of West African opepe instead of oak and greenheart. More moorings follow.

Beyond a set of powerlines, the canal crosses the line of a dismantled railway that formerly crossed on a swing bridge.

The canal has been converging on the River Went but never actually meets it. Instead, it is met by the New Junction Canal. Built in 1905, it was the last new canal in England, until recent years, and runs dead straight for 9km.

On the other side of the canal is Southfield Reservoir, opened in 1884 to meet increased water demand, especially

from Ocean Lock. It is used by Beevers Sailing Club. There are two very shallow entrances, the first with no superstructure across it, creating minor rapids as flows pass through the gaps. Concrete angling platforms face the canal at regular intervals.

Beyond Beevers Bridge, the canal swings to the north to run alongside the tidal River Don. Rich vegetation on the banks ranges from willows and hawthorns down through stinging nettles to buttercups, deadly nightshade and red clover. Birdlife includes swans, greylag geese, herons, harriers, magpies, lapwings, terns and pied wagtails.

The River Don was not one of Vermuyden's successes. In 1626 he drained the Thorne and Hatfield marshes into an exisiting artificial channel that ran north to connect with the River Aire but this resulted in flooding around Snaith, followed by lawsuits. Consequently, he diverted it 9km east to the River Ouse via a sluice that was washed away in 1688 and not replaced. Consideration must have been given to using this new Dutch River for navigation with a lock at the end and a bypass weir but flows are fast, especially in times of flood, and it was decided to run the canal alongside but unconnected.

Heading towards the windmill at East Cowick, the canal and river both turn sharply through 90° to pass under New Bridge, carrying the A614. A tall, dark, square, brick chimney is passed, standing alone by the canal, and then the M18 passes over, immediately before its northern end as it joins the M62.

Formerly an industrial eyesore, Sugar Mill Ponds have become a local nature reserve. Rawcliffe Bridge is served

Sobriety cruising towards Pollington.

Summer flowers beside the canal near New Bridge.

The whole of the New Junction Canal.

Southfield Reservoir has connections to the canal, forming small riffles as the flow passes over shallows.

by the Black Horse, near which there is a mooring area. The latter is across the canal from a mill that emits a pungent odour when the wind is from the north. The Wakefield to Goole railway arrives from behind the mill, to run beside the canal. A pheasant may fly low across from one bank to the other.

A large glass factory stands near the edge of a trading estate, which merges into **Goole**. The railway joins the Doncaster to Hull railway. This passes over the canal on a bridge with decorative railings beneath. Some designs have suffered the attentions of vandals.

As the canal eases into the docks, it is increasingly affected by tides, even though there are locks onto the

Ornamentation on the handrailing approaching Goole.

River Ouse. Those locks also restrict the times of movement of large vessels.

One of the first premises on the south side of the docks is the Sobriety Waterways Adventure Centre & Museum. It gives the canal's history and offers a café. The namesake boat is a 1910 Humber keel and there are also the *No. 58* grain barge, converted as a conference centre, the *Opportunity* floating photo workshop, the enclosed boat *City of Hull*, the *Audrey*, converted from a 1915 Humber lightship, the *Wheldale* push tug, a jebus false bow from Tom Pudding trains and two samples of compartment tubs from the latter.

Beyond the British Waterways depot are old timber ponds, used for moorings.

Goole Docks were created in 1828 for the canal, to export coal and textiles, and are Britain's premier inland port. They are the furthest inland port yet they trade with the Baltic and other parts of Europe. Nine docks offer 5km of quays, silos, a container terminal and a steel terminal with canopy. A Tom Pudding tippler, virtually unchanged since 1865, was in operation until 1986, one of four used on land and accompanied by another floating one. Hydraulic rams in pepperpots at the top tilted compartments through 125° to empty the coal. A liquid biomass fuel shed is being built at the Caldaire Terminal, to supply power stations along the canal by barge.

The A161 is taken over the South Dock swing bridge of 1899. Two notable water towers stand to the north. The smaller is Grade II listed, a 9.1m diameter cast-iron tank holding 140m³ of water on a tall, fluted, brick

Moorings above Goole.

The Sobriety centre with a jebus moored in front.

column of 1885. The other, in reinforced concrete, has a 27m-diameter tank 6.7m deep, holding 3,400m³ on a 37m-high lattice frame. It was the largest in Europe when built in 1927. Also on the skyline is the spire of St John's church, built in 1843–8 with canal help and containing memorials to Goole ships lost at sea. Near it is the Museum & Art Gallery, which includes maritime paintings, many by Goole artist Reuben Chappell.

Although there is barge access at any time, Ocean Lock only operates for large ships from two and a half hours before local high water to an hour and a half after local high water. A high intensity flashing white light indicates that a ship movement is due through the lock. Flows in the River Ouse are to 11km/h or more and high mud banks are a feature.

The Goole docks skyline with the two water towers in the centre.

Distance
55km from River Lock to the River Ouse

Navigation Authority
British Waterways

OS 1:50,000 Sheets
104 Leeds & Bradford
105 York & Selby
111 Sheffield & Doncaster
112 Scunthorpe & Gainsborough

48 Aire & Calder Navigation, Wakefield Section

The world's largest cast-iron aqueduct

The Wakefield Section of the Aire & Calder Navigation runs north-east across West Yorkshire from **Wakefield** to Castleford. It links the Calder & Hebble Navigation with the main line of the Aire & Calder Navigation, part of the trans-Pennine route to the Humber. William Hamond Bartholomew was the resident engineer on this broad canal. Sunday is a quiet day because the locks are not manned except by arrangement.

The Calder & Hebble Navigation ends at Fall Ing Lock at Belle Vue. The Wakefield Section uses the River Calder at first. On the opposite side is Wakefield Old Lock at the end of the short Old Wharf branch.

The water is dirty with unsightly debris along the banks. Swans do not seem to be deterred. Ropes hanging from the freight railway bridge downstream show that local children swim in the river.

A power station stood at the end of the disused Barnsley Canal that linked with the Dearne & Dove Canal, a significant portion still being present and restoration helping to improve its prospects.

High ground on the right, topped by the hall at Heath, pulls back. A railway bridge crosses. The navigation carries on as the river bends right. It is crossed by a bridge carrying pipes that have been running along the right bank.

Broadreach Flood Lock protects the navigation as the river begins several kilometres of meanders.

The navigation then runs straight for over 2km. Large dump trucks rumble past on the right bank as they extract gravel from the adjacent pits. Opposite, a colliery found its main market in Ferrybridge C power station.

The hamlet of Stanley Ferry is the major point of interest on this canal. Ramsdens Bridge is a swing bridge with a stone keeper's house that has been badly modified. A red brick-porch stuck on the front and a concrete extension to the chimney owe more to utility than aesthetics.

Looking down the River Calder from Fall Ing Lock.

The Calder aqueduct: the largest cast-iron aqueduct in the world and one of the most ornate.

The Stanley Ferry inn, positioned next to a spur which acts as a marina, is a popular venue.

The jewel in the crown, however, is the arched aqueduct over the River Calder. The original design of six arches was rejected because of spate flow volumes in the river. Instead, George Leather (with assistance from Thomas Telford) came up with this design, one of the most interesting aqueducts in Britain. Each side has a cast-iron arch in seven segments, from which the iron trough is hung by 35 wrought-iron rods. Built in 1836–9, it weighs 1,700t, holds 940t of water and is the largest cast-iron aqueduct in the world. Its full glory is not seen by passing over it, however. It is best viewed from the 2,300t concrete aqueduct built alongside in 1981, from where it is seen to have a dramatic Renaissance style with closely spaced fluted columns along each side and pediments in each of the four corners of the aqueduct. Its replacement has, of course, nothing to do with increasing barge loads as has been claimed, the aqueduct load remaining constant while carrying water regardless of boats passing over. The new aqueduct is wider, reducing the potential for collision damage. British Waterways have a large maintenance yard adjacent. To their credit, they floodlight the aqueduct each night, such is its splendour.

Also worthy of closer inspection is the church at Stanley. Its twin towers are ornately fashioned in stone although it is not possible to see any detail from the canal.

Large and boldly executed housing stands on the bank of the River Calder between Stanley and Bottom Boat, seen well across the flood plain from Birkwood Lock. This lock is the uppermost of the mechanised locks. Like those below, it has a brick control tower and a gantry with traffic lights over the lock chamber.

Altofts is an outlier of Normanton, a settlement based on the canal trade. Down the hill from the village is King's Road Lock. Further down on the same side are the remains of Foxholes Side Lock, which served Foxholes Basin, now filled in. The M62 Trans Pennine Motorway bridge follows.

Below the farm at Penbank, marked with a sign like that of a public house, the Fairies Hill Lock Cut still exists on the right, formerly taking in Altofts Lock, Altofts Basin and Fairies Hill Lock. These have been replaced by the 4.1m deep Woodnook Lock. This lets the navigation rejoin the River Calder above the railway viaduct rather than below it. The stonework bears black oily deposits. Below the lock is a convenient platform with two heights above river level. Floodwater deposits sand on the planks and this sweeps up onto the banks with cornices through which it is possible to step inadvertently. The river rejoins past a small sewage works and the navigation is now wider, not that the canal section was undersized.

The Fairies Hill Lock Cut and a colliery basin emerge together on the right, midway between the railway viaduct and another around the following bend. They are followed soon after by a further bridge for a dismantled railway, such was the proliferation of lines in this area.

On the other side of Pottery Bridge is a large sluice gate and a barrage of notices stating which anglers may enter and to beware of the bull.

The river takes a relatively straight line in the vicinity of Castleford Mere, cutting between various oxbow lakes in former meanders. A tank farm is passed on the right before arriving at a water crossroads with the River Aire at **Castleford** Junction. There is a direction board and batteries of traffic lights pointing in all directions. On the right side of the lock the stone wall has been damaged, presumably by a vessel failing to slow down at the junction.

Distance
12km from Fall Ing to Castleford Junction

Navigation Authority
British Waterways

OS 1:50,000 Sheets
104 Leeds & Bradford
105 York & Selby
110 Sheffield & Huddersfield

Castleford Junction with direction signs, traffic lights and Castleford Lock.

Birkwood Lock, the first to be mechanised, with control tower and traffic light gantry.

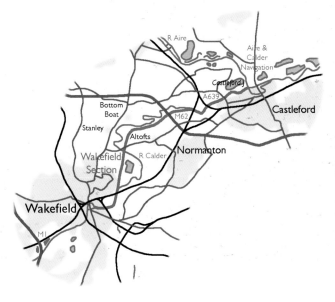

49 Leeds & Liverpool Canal, Rufford Branch

Long predating the main line

From 1740 until late in the 18th Century boats wishing to travel from Wigan to Preston did so via the River Douglas. The Rufford Branch of the Leeds & Liverpool Canal, also known as the Lower Douglas Navigation, provided an alternative route for much of the distance. It was opened in 1781, long before the main line of the Leeds & Liverpool Canal was completed in 1816.

Running north, it drops quickly at first from the main line at **Burscough** Bridge to cross Burscough, Mawdesley and Croston Mosses. This is flat but fertile farmland which forms the Lancashire coastal plain, a completely rural route.

Turning off the main line, the Rufford Branch passes under an imposing arched bridge of 1816 and immediately enters the most interesting area on the branch, the canal settlement around the basin at Lathom, a conservation area. On the left is a drydock with all its boat support saddles in place. The dock is filled by allowing water in from the basin and emptied by permitting it to drain out further down the lock flight, a very neat and simple system. A stone building between it and the bridge is shaped like half a hexagon with a pointed roof. The first two locks come in quick succession. Lathom Top Lock, forming the end of the basin, has footbridges across both ends, the upper one being a swing footbridge. The one at the far end of the lock is fixed but still interesting in its shape. Around the basin are a group of cottages, two of which have the bottom halves of corners cut away to allow towing horses to pass unhindered.

Beside Lathom Bottom Lock is the Ship Inn. It has a rather higher class of customer than in earlier days, when it went under the name of the Blood Tub.

The bywash at each lock is unusual in that it is exposed and runs past the lock in large section culvert at high level, discharging down to the lower level in the vicinity of the bottom gate. Paddle gear for filling locks is also unusual, taking the appearance of a couple of miniature balance beams above the top gates, having to be lifted to open them, a process that is complicated by their having to be padlocked down to prevent non authorised interference. Some of the lower locks, instead, have a screw arrangement, using a very coarse pitch of threaded rod.

Stone is used in the handling of water flows and it is worth looking at the ditch on the right just before the

The lock flight at Lathom. The end houses are cut away to assist towropes.

The dry dock and surrounding cottages at Lathom.

first road bridge. The road on the right side is carried over the ditch on two high but narrow stone slots. The bridge itself is a steep arch and a pipeline attached to one side of the bridge faithfully follows the line of the arch.

Runnel Brow Lock follows.

The Southport to Wigan railway crosses on a metal and blackened stone bridge before the canal arrives at Moss Lock. Large glasshouses at New Sutch House Farm are the first of many in the vicinity of the Douglas. Away to the right are a line of hills that edge the plain. A lone kestrel might be seen hovering over the fields or flights of wildfowl might pass from the Martin Mere wildfowl centre 4km away to the north-west, particularly pink-footed geese, for which this is home for a quarter of the world's population.

The canal is edged with reeds and greater reedmace, bracken, brambles, hawthorns, rosebay willowherb, arrowhead and water lilies, generally low vegetation with an absence of large trees. Canadian pondweed and freshwater mussels are plentiful in the clear water. German's Lock, Baldwin's Lock and Prescott Bridge follow.

The Liverpool to Preston railway crosses at the same point as a farm track. Strangely, this results in a level crossing over the canal on an unusual X-shaped bridge.

The canal follows the A59 fairly closely all the way but now pulls right alongside at a former wharf site. The adjacent Marsh Moss bridge is the first of the low swing bridges.

Depending on the wind direction, the smells on the canal may include pig manure and effluent from the sewage works at Croston. Much more pleasant aromas derive from the market gardening closer at hand and one particularly noticeable smell in the autumn is that of celery which is being cropped. So strong is it that it verges on being pungent but certainly gives an interesting variation from the usual canal odours.

After the penultimate lock at Rufford, the canal passes a small Italianate Victorian church with many monuments to the Hesketh family, who were local landowners. One sculpture, dating from the late 1450s, includes no less than 11 offspring.

The canal swings into a long straight lined with ancient trees of a wide variety of species.

On the left is Rufford Old Hall, a timbered house, one of the finest 15th-century buildings in Lancashire, with Jacobean extensions, ornate hammer-beam roof, intricately carved, immense, moveable, oak screen and fine collections of 16th-century arms and armour and

Old wharf building adjacent to the A59 at Bank Bridge

17th-century oak furniture as well as a folk museum, 6ha garden, shop and tea room. The house was once owned by the Hesketh family. There used to be a swing bridge opposite the house. Although it has been removed, the route is still used by the Old Grey Lady, a spectre crossing from the Old Hall at this point.

Rufford New Hall dates merely from 1760. It still retains its icehouse in the garden.

The A581 crosses on a bridge with red sandstone abutments, assorted textures having been hacked into the stonework. Apparently a recent addition, the reason for the patterning is less than obvious.

A moat is to be found close by, next to the A59. The canal has its banks edged with sloping concrete and interlocking concrete blocks which make this reach rather unattractive. This all changes at Sollom, with its ducks. The canal now occupies the original line of the River Douglas while the river has been diverted to a new cut below Rufford, an unusual juxtaposition of what normally happens on navigations. Even after two centuries, the river character is obvious with tall reeds, oak trees, yellow iris, foxgloves, convolvulus and an occasional heron.

The two channels flow side by side under the A59 at Bank Bridge. A large but conspicuously security-proofed old wharf building stands on the downstream side of the bridge between the two channels.

Town End Bridge at Tarleton is the final low swing bridge at the start of several factory units.

The boat activity on this canal emanates from the boatyard in the final reach. For seagoing craft, they build and fit out steel and wooden boats. The sky is filled with jostling masts and there is a tremendous assortment of craft from a drifter through a wartime patrol craft to narrowboats and cruisers.

The final lock has diamond gates at the bottom to act as tidal doors. These only open onto the tidal River Asland or Douglas at high water to give access to the River Ribble and Preston or the Irish Sea. It is significant that the banks are high, steep and muddy and that the tidal river flows very fast. Timing is important for ongoing trips and advice is given locally.

Distance
12km from the main line to the River Douglas

Navigation Authority
British Waterways

OS 1:50,000 Sheets
102 Preston & Blackpool
108 Liverpool

Looking north from Runnel Brow Lock as a container train heads from Wigan to Southport.

Beyond a marina the Italianate Rufford church stands with its Hesketh monuments.

50 Lancaster Canal

The Lancaster Canal ran southwards through Cumbria and Lancashire from Kendal to Preston, a broad beam contour canal designed by Rennie and constructed between 1797 and 1819. It had been intended that it should join the Leeds & Liverpool Canal at Wigan but the final link across the River Ribble was never built and the canal remain isolated, except for a branch in the centre to the sea. Promoted by Lancaster merchants, it carried coal, lime, slate, timber and food.

It also carried passengers. Despite the hilly terrain, Rennie managed to produce a 92km run with only a flight of eight locks at Tewitfield, allowing express packets to travel from Preston to Kendal with frequent changes of horse, averaging 16km/h including changing horses, in the astonishingly quick time of seven hours in 1833. The service ceased in 1846 because of rail competition. The last coal barges ceased their trade 101 years later and the canal was abandoned above Tewitfield in 1955. Culverting of several points above this, mostly by the M6, means that reopening of the canal will be more difficult than would previously have been the case. There is little midweek activity on the canal.

The absence of powered craft at the top end means that the water is clear but weedy. The canal has bream, chub, pike, roach and tench.

The upper reaches, owned by Cumbria County Council, are dry and had always had leakage problems through limestone fissures. Nevertheless, restoration from Kendal is planned. In 1980, the 48th Field Squadron RE repaired the portals of the 345m Hincaster Tunnel for the Lancaster Canal Society.

The water begins suddenly at Stainton where ownership passes to British Waterways. A lane passes over the end of the dry section on a weathered-stone bridge that is so typical of those to follow, a squat arch with vegetation attempting to smother it. Despite the flat arches beneath these bridges, they are usually steeply humped on top, the sounding of horns of cars approaching blind summits being a frequent disturbance of the peace down the canal.

In springtime there is a conspiracy of yellow flowers, cowslips, daffodils, gorse, primroses, celandines and marsh marigolds.

The first reach crosses Stainton Beck, a minor feeder, and then winds past a deer park and nature trail near Endmoor before turning close to the A590. A small sprawling scrap tip down the canal embankment does not improve the ambience. Agreeable leftovers are the stone mileposts which are found not infrequently down the canal.

Larches are the most common trees, accompanied by hawthorn hedges along the banks and duckweed forming on the surface of the canal.

The first boat on the canal is moored in front of the Crooklands Hotel. The B6385 crosses on a bridge, beneath which is a fence across the canal but not the towpath.

Crooklands Aqueduct crosses Peasey Beck, the main feeder from Killington Reservoir, which was built to serve the canal. Coke used to be made at Crooklands with a long wharf and coke ovens by 1819. There is still a coalyard here but former stables are derelict.

The canal is usually full-width but a narrow section runs alongside a highway depot with stacks of roadsigns beside the canal.

The first M6 crossing is at the northern end of Junction 36, here on high embankment. The canal disappears into low culverts and the walking route is under the adjacent A65 bridge.

The current head of the canal at Stainton.

Approaching Farleton Fell, the highest peak passed.

The canal passes under the A65 in culvert twice in quick succession. On the second occasion the road is on a significant embankment and a box culvert has been inserted to take the towpath through.

Curlews fly audibly overhead, their sounds to be replaced by the tumble of Lupton Beck dropping under another aqueduct. Despite being on the edge of the Lake District, the scenery is not as hilly as might be expected except at Farleton Fell, an extensively quarried limestone crag rearing up some 400m above the canal with an aerial on top and peaking only 800m from the canal.

Farleton was an important canal village, its stable and packet house now disused. The former coaching inn of 1630 by Duke's Bridge has become a private house.

The M6 crosses back again. This time the culvert is slightly higher. The motorway embankment is much lower than before, however, so the problem of installing a full-sized bridge will be greater as the line of the motorway will need to be raised over some distance, humpback bridges not being approved for motorways. Restoration will also involve provision of considerable funding, disruption of motorway traffic and admission that an oversight was made in the first place by the planners, the latter probably being the major problem. Those not able to get under the culvert face an 800m walk over the road bridge to the north. There is a footpath beside the motorway on the west side but on the east side the only legal option is to go back to the previous bridge across the canal. In the circumstances, the least the authorities could do is lay a level path along the motorway verge and fence it off so that pedestrians have a route round the obstruction.

The next culvert under a minor road at Holme is very much simpler.

Sloping concrete canal edges with chain draped along one side to help swimmers climb out provide an unusual canal lining in this rural setting.

Holme Mills was a factory colony with houses and a millpond. A flourmill had been established by 1790 and it was one of the earliest mills in the area to use steam power, already converted by the time the canal's construction was completed. It had to be rebuilt in 1860 after a fire.

Gates were formerly installed under the next bridge, intended to slam shut in the event of a breach.

Two high crossings over roads are made unnoticed on New Mill Aqueduct and Burton Aqueduct.

Like the M6, the earlier West Coast Main Line followed the line of the canal and now comes close alongside as the canal leaves Cumbria for Lancashire, passing behind Burton services on the M6. The M6 crosses again at Cinderbarrow, this time requiring a 400m walk over a bridge that will be in the way when the motorway has to be lifted to clear the canal. A recent communication aerial stands on the adjacent hillside.

After only 400m comes one of the best known

abandoned lock flights in the country because the M6 runs close alongside although increasingly being screened off by trees. Disused since 1952, the locks lack gates but the iron bollards and other metalwork are painted. A pair of lock gates have information panels attached by the towpath above the top lock. The canal has been all at one level so far. Now the eight stone-chambered locks drop the canal 23m to its other long level pound which reaches all the way to Preston, the longest level pound in Britain at 66km. Was Rennie just lucky or was it skilful engineering that enabled him to group all the locks in 1km? Perhaps a solution to the level problem at the

The canal is truncated by the M6 again at Cinderbarrow.

All the locks come together in one flight at Tewitfield.

motorway crossing is to rebuild the top lock to the west of the motorway and deepen the cut for several hundred metres. Despite the proximity of the traffic, the flight is popular with both walkers and orange-tip butterflies.

The final culvert comes immediately below the bottom lock. The A6070 crosses on a high embankment next to the Longlands Hotel. This time the towpath is diverted under the A6070 bridge across the M6, to emerge on the far side in Tewitfield's marina. Suddenly there are narrowboats moored right up to the final obstruction, picnic tables, a children's playground and full British Waterways facilities. This is the most northerly point in England that can be reached by large canal craft. Gone are the weeds but also the clean water. There are a gaggle of aggressive geese that launch waterborne attacks.

The canal moves away under Tewitfield Turnpike Bridge towards Borwick. Borwick Hall was an Elizabethan manor house built around a 15th-century peel tower, the gateway including a stone dating it at 1650. Charles II stayed here in 1651. These days it is a Lancashire Education Authority youth club training centre. The house next to the canal has an interesting end wall with a large number of bricked-up vertical slits in addition to the windows.

Across the valley to the right are Pine Lake and other sand and gravel pits. Warton Crag towers above with Warton at its base, including the 15th-century tower of St Oswald's church which has stars and stripes on the coat of arms of the family ancestors of George Washington, preceding their use in the American flag.

Quarry Branch, also known as Lovers' Creek, served Webber Quarry, which produced limestone with boat-loading chutes and the remains of loading cranes. An ivy-coated brick wharf building on the main line bears the remains of another crane and a more recent television aerial.

The Leeds to Lancaster railway crosses at Capernwray and then follows the canal for a while. The canal crosses the Keer Aqueduct, a mill with a rusty wheel having been used to pump water up from the River Keer. Lanes lead to a gamekeeper's tower, a packhorse bridge and a disused quarry. Lapwings and oystercatchers are black and white birds of contrasting habits and characters in the locality.

Over Kellet limestone quarries are hidden behind Junction 35, around which the canal has been realigned, passing under the M6 and A601M in turn, now with boat-sized bridges.

Signposts on the canal point not only along the water channels but also to town centres and features of interest, as does the one by the playground in **Carnforth**. Carnforth is very much a railway centre, the point where the Furness and Leeds lines depart from the West Coast Main Line. Steamtown was one of the last depots servicing steam engines during the change to diesels in 1968 and is a railway museum with over 30 steam locomotives including one of the most famous, the *Flying Scotsman*. In 1945 the station was used as the location for filming *Brief Encounter*, adapted from Noel Coward's *Still Life*. The town had a steelworks until 1931. It still has a marina, situated at a widening in the canal.

Cinder ovens were formerly located at Crag Bank, near where the sounds of a rifle range may be heard. The A6 and then the railway crowd in on the canal. There are fine views past a caravan site and out over the wide expanse of Morecambe Bay to the southern fells of the Lake District. Inevitably, this section of canal is exposed to westerly winds.

The canal winds past wooded hillsides and parkland at **Bolton-le-Sands**, the Royal Hotel and the Packet Boat Hotel by a former wharf.

At Bolton Town End, a school complete with bell stands beside the canal. The A6 crosses and goes on its

A derelict wharf crane at Capernwray holds a boat TV aerial.

Fields sweep to the canal at Mount Pleasant, Bolton-le-Sands.

Looking over roofs at Hest Bank and across Morecambe Bay to the southern hills of the Lake District.

way. Hatlex swing bridge is usually left blocking the canal.

The high-water mark at **Hest Bank** is only 200m from the canal. It is hidden from the cut by houses at its closest point, a nature reserve. A nesting swan does not seem to have need of any such protection. The bay is one of the top five wildfowl feeding areas in Europe with up to 50,000 waders visible, some of which find their way onto the canal. A stone pier, previously used for transhipping freight between barges and coasters, became disused after 1831 when Glasson Dock was established. A footpath across the Kent estuary to Kents Bank, probably much older, still runs out from here. During the Second World War, water from the canal was used to supply water troughs on the adjacent railway. The Hest Bank Hotel is an old coaching inn standing near the canal.

A bridge at Slyne led to former stables. Below the canal are a golf course and Bare, an extension of Morecambe. The two great nuclear power station halls at Heysham, in the distance, feed powerlines that cross by Foley Farm, formerly Folley Farm.

Racing-pigeon lofts follow in gardens as the canal wanders past Beaumont Hospital and into the outskirts of Lancaster.

The A6 recrosses before the canal turns sharply past an overflow outlet to the River Lune and crosses one of the finest aqueducts in Britain. Built in sandstone from 1794 to 1797, it has excellent proportions and was Rennie's first aqueduct incorporating hidden inverted arches to spread the stresses. The foundations were made with pozzolana imported from Italy. The 183m Lune Aqueduct stands 19m above the river on five 21m semi-circular arches that rise 18m from the riverbed. Curved wing walls and Gothic ends on rustic pillars complete the canal's major engineering work, a structure that Turner

Rennie's masterpiece, the magnificent Lune Aqueduct carries the canal over the River Lune at Lancaster.

chose to paint. Bulk Road Aqueduct, by way of contrast, crosses the A683 and was only built in 1961. Rising above the golf course at its end is the Ashton Memorial of 1907–9, built in neoclassical style by Lord Ashton in the 15ha Williamson Park of 1896 in memory of his wife. The Edwardian era is shown on screen, the Old Palm House has tropical butterflies and Dukes Playhouse has films and theatre. The Promenades are among the most spectacular open air theatre in Britain.

Uphill from the large tidal weir on the Lune, the canal enters the fully built-up part of the city.

Lancaster is the red rose city, centre for the Lancastrians in the Wars of the Roses. It takes its name from *lune*, Celtic for healthy, and *ceaster*, Old English for fort. The market square is where Charles II was proclaimed king in 1651.

Terraced houses back onto the canal, somebody having added a dash of humour with a sculpture of giant maize plants in green and yellow in a canalside garden.

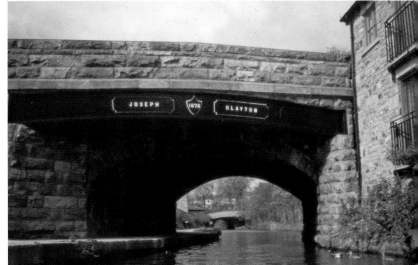

Engineering details on a Lancaster bridge carefully highlighted.

The canal winds past high, brick mill buildings. The local sport seems to be throwing bottles across the canal from the towpath under one bridge to smash on the opposite abutment. A layer of broken glass lines the ledge but most of it, clearly, lies on the bed of the canal. An 800m section of canal in Lancaster is the only length of the entire canal to have the towpath on the east side.

The Romans built a bath house in Lancaster. The 15th-century Perpendicular Benedictine Priory church of St Mary includes a Saxon wall and doorway on the site of a Roman fort and contains 13th-century carved choirstalls, which are some of the earliest and finest in England, in addition to fine needlework and Abyssinian Coptic crosses. A Saxon church of about 600 was replaced

Tall, brick mill buildings crowd the canal in central Lancaster.

261

with a 1380–1430 model, the memorial chapel of which has the colours of the King's Own Royal (Lancashire) Regiment.

The Norman castle is now a prison, past residents including the ten Demdyke witches of 1612, its implements including a clamp last used for branding criminals in 1811. The Turret or John of Gaunt's Chair has a view to the Isle of Man and was used to signal the approach of the Armada. The massive gatehouse is 15th-century but most of the castle was restored in the 18th and 19th Centuries. Its two-storey keep was heightened from its original form and there remains part of a bailey curtain with a round and two square towers.

Less dramatic buildings include the Baroque Music Room of the 1730s with ornate plasterwork, the Judges' Lodgings town house with Gillow furniture and a museum of childhood. The former town hall of 1783 contains a regimental museum. The city museum and the Cottage Museum are in an artisan's house of about 1820. St Peter's Roman Catholic church of 1859, in Geometrical style, was one of Paley's finest, with ten bells, notable architecture and stained glass and one of the finest organs of its type. It became a cathedral in 1924. The Lancaster Maritime Museum has a set of 37 sheets of plans of 1880, covering the canal at 1:1,584.

Refreshments can be obtained in various cafés including the Water Witch, which is in former canal stables. The footbridge to the Water Witch was erected more easily than expected in January 1987 as workmen were able to walk across the canal on the ice. On the other hand, maintenance in 1988, when the canal was drained down, was made more difficult by the discovery of a wartime bomb under Penny Street Bridge.

Extensive redevelopment is going on around the former docks near the Water Witch.

By the infirmary is the British Waterways depot with its hand-operated crane and the roofless packet boathouse, built with a strange, skewed entrance. Boats are floated inside and lifted to the first floor for repair.

The West Coast Main Line crosses for the last time but is to follow the canal for much of the rest of its journey to Preston. Also crossing is Aldcliffe Road footbridge, made in 1954 from a shortened ship's gangplank, with a postbox in the road side of the bridge.

The canal moves south from Lancaster past a neat fence dividing it from an adjacent road. It passes an old castellated brick boathouse, smothered in ivy, as it enters the 3km long Burrow Heights cutting. This is lined with trees and wild flowers, especially bluebells, wood anemonies and primroses in the spring. In the centre of the cutting is Broken Back Bridge. It takes its name from the unusual lines of its stone courses. Hidden is Burrow Beck, which passes beneath in a syphon. During construction of the cutting, a Roman find of two lions, four heads and a stone statue of Ceres was unearthed, now in the Lancaster City Museum.

The canal emerges from the cutting to run past coppices from where terns fish, before arriving in Galgate.

From the bridge in Galgate, the canal crosses the Conder Aqueduct and winds round a cricket pitch to meet the A6 and the railway. The latter frames the far

Old canalside buildings refurbished at the White Cross.

Broken Back Bridge in the cutting through Burrow Heights.

The old packet boathouse with its strange skewed entrance.

The Glasson Dock Branch leaves at Lodge Hill, leading straight into a flight of locks.

side of the village with high embankment, bridges and a viaduct. Beyond the viaduct is the tower of the church of St John, by which is the oldest silk-spinning mill in England, built in 1792. The stone village comes to an end with Galgate Basin Wharf picnic table area, moorings and basin excavated in 1972–3. Canalside moorings are extensive with boats moored at an angle on the east side as well as parallel along the west bank.

At Lodge Hill, the Glasson Dock Branch of 1825 leaves to run down the Conder valley to Glasson and the Lune estuary, until recently the canal's one contact with the outside world. At the junction are a narrow stone bridge and a lock keeper's cottage, frequent locks on the branch contrasting with the absence of them on the main line.

Opposite Junction 33 of the M6 is the fine Italianate villa of Ellel Grange, built 1857–9 on top of a ridge and surrounded by parkland set with beeches and oaks. The adjacent church of St Mary is also charming. Local bridges include an ornamental bridge with balustraded parapets across the canal and a double bridge with a wall down the middle to separate farm and estate roads. Plovers and herons add their contrasting flight styles to the canal. This seems to be something of a watershed for farm livestock. North of here sheep predominate but these now give way to Fresian dairy cattle.

An aqueduct carries the canal over the River Cocker, which flows down through Bay Horse where a Euston to Glasgow train ran into a local train in 1848 when two different companies were running their trains over the same lines.

From Potters Brook can be seen the distinctive mushroom-shaped tower of Forton services on the M6.

The canal is crossed by the Lancashire Cycle Way at Forton. Forton's stirring contribution to the war effort during the Second World War was the removal of the parapets of Stony Lane bridge so that enemy troops might be seen crossing it. Views are frequently extensive from the canal. From here, the expanse of the Fylde, the coastal plain that is Lancashire's market garden, can be seen. Inland, the fells are steadily rising again around the Forest of Bowland. Transport routes follow the fault line between the Triassic sandstone and the hard Carboniferous rocks of the Forest of Bowland.

Water lilies are seen in increasing numbers in the canal and meadowsweet graces the banks.

There is a caravan site at Cabus Nook and another caravan site fronts the canal at some length at Cabus. A few bends later, the canal arrives at Snape Wood Farm, marked by a less than beautiful concrete water tower. Away to its right, at Winmarleigh, lies an agricultural college in a red-brick hall, originally built in 1871 but reconstructed in 1927 after a fire that destroyed most of the building.

Tank traps beside the canal at Ford Green show that the canal's defensive role was taken very seriously.

Brick bridge abutments by the canal just before Garstang are those of the former Garstang & Knott End Railway, alternatively known as the Pilling Pig.

The canal passes under the A6 again at Cathouse Bridge, **Garstang** formerly having been a staging post on the A6. Among other bridges, it passes under a pipe bridge of 1927, which carries water in a sweeping arch on its way from Barnacre Reservoir to Blackpool, the former located by a trio of aerials. Thursday is Garstang's busy market day, the market having received its charter from Edward II in 1310. The town hall with its diminutive belltower, opposite the cobbled market place, dates from 1680. The church of St Thomas is an 18th-century structure. The Old Tithe Barn, in stone and timber beside the canal, was restored and reopened in 1973 as a canal and agricultural museum and restaurant. It was in Garstang that an ice-breaking barge sank in 1945.

The canal leaves Garstang over the Wyre Aqueduct, a

An ornate bridge across the canal at Ellel Grange.

The remains of the 500-year-old Greenhalgh Castle, badly damaged in the Civil War.

single 16m arch passing 10m above the River Wyre. It passes round Bonds to turn just short of Greenhalgh Castle. This was built in 1490 by the Earl of Derby and was a Royalist stronghold until destroyed by the Roundheads in the Civil War. Now it is little more than a ruin.

Powerlines pass backwards and forwards over the canal for some distance. More traditional power is seen at a coalyard where the railway and M6 come back alongside the canal for a final 3km.

Catterall Basin was formerly the site of a papermill. These days a wooden benchseat suspended on ropes as a garden swing beside the canal suggests a more leisurely pace of life. The River Calder is passed in a syphon under the Calder Aqueduct.

Beyond an aerial in Catterall is the tower of St Helen's church in Churchtown, a magnificent parish church dating in parts from about 1300 and known as the Cathedral of the Fylde. Claughton Hall was originally a church neighbour. All except one wing was dismantled and moved to its present hilltop site.

A copse on the right of the canal at Stubbins is a sea of bluebells in the spring, a marvellous sight. Dredging from the canal shows plenty of mussels among the spoil.

The A6 comes alongside and then crosses for the last time as the canal prepares to move westward, leaving the other more recent transport routes and going round three sides of a square to enter Preston in an easterly direction. Terns and black-headed gulls fishing in the canal are a reminder that tidal water still lies only a few kilometres to the west.

First comes the Brock Aqueduct, where a levelling error resulted in a tunnel having to be built under the canal for the River Brock. Just down the river valley stands the Lancashire College of Agriculture on the outskirts of Bilsborrow.

The canal breaks free of the modern transport routes by Guy's Lodgings Hotel.

The Lancashire Cycle Way crosses just before Hollowforth Aqueduct, which clears the **Barton** Brook. It is followed by a swing bridge that is closed by means of a chain lying on the bed of the canal.

Increasing numbers of rushes make their appearance around Moor Side. Woodplumpton Aqueduct crosses the Woodplumpton Brook.

The notable feature at Swillbrook is the Swill Brook Basin hire fleet, tea room and shop in converted canal stables and cottages.

The cutting leading to Salwick.

Increasing signs of modern activity begin at Blackleach. From here, the aerials of HMS Inskip can be seen grouped around a disused airfield.

The canal turns sharply south, is crossed by the M55 and eases into a long cutting frequented by orange-tip butterflies. This is crossed again by the Lancashire Cycle Way near the Hand & Dagger public house and not far from a windmill. The cutting ends as the canal turns eastward past Salwick Wharf and Salwick Hall with its moat. In 1858, 60m of embankment collapsed at Salwick Marsh. It had originally been planned to build an extension to Kirkham and the Fylde but this never happened.

Prominent beyond the Blackpool to Preston railway line these days are British Nuclear Fuels' Springfield works at Lea Town, a golf course and numerous lines of pylons converging on Howick Cross electricity substation on the far side of the River Ribble. The canal crosses the approach line of the main runway at Warton Aerodrome, where the Eurofighter was unveiled.

Easing its way into Preston, the canal passes Haslam Park and crosses the Savick Aqueduct. Canal enthusiasts have canalised the Savick Brook to provide a canal link with the rest of the canal system via the River Ribble, River Douglas and the Rufford Branch of the Leeds & Liverpool Canal. This 4km Ribble Link is tidal at its lower end, the eighth and final lock with radial gates drowning out at the top of the tide and the footpath giving up the struggle. It retains many of its bends and passes under an existing brick arch carrying the Blackpool to Preston railway. The non-tidal section has a towpath and a large wooden sculpture of a contemplative man, entitled Water, at the top basin. Another, entitled Crow, is above the A583 crossing in a section like a heavily overgrown jungle.

The city largely ignores its canal and does not show its best face. A slender church spire near Moor Park is striking.

Preston was the second oldest borough in England with a Member of Parliament since the 13th Century and a Merchants Guild that has met every 20 years since 1542. It was a market town that developed in the Industrial Revolution, not least because Arkwright was born here in 1732 and his spinning frame made the town a cotton spinning centre for 150 years. Dismal, darkened brick mills are prominent from the canal, notably Tulketh Works and Shelley Road Mills. Between these two is an ugly lattice bridge in reinforced concrete that looks as if it has been built with a canal widening scheme in mind. Instead, the canal comes to an abrupt halt near the A583/B6241 junction. The last 3km of the canal have been filled in to the former terminus at South Basin Wharf. The Lancaster Canal Company built 16km of what is now the Leeds & Liverpool Canal as their South End but never built the final 8km to join the two parts together. Instead, a tramroad was constructed to connect with the Leeds & Liverpool at Johnson's Hillock. The tramroad was closed in 1879 and the canal sold to the London & North Western Railway in 1885, all except the South End, which was leased to the Leeds & Liverpool. Building work in the intervening years means that construction of the direct link is now impractical.

The current end of the canal has been landscaped but the surroundings are a significantly industrial area that has seen better days.

Distance
81km from Stainton to Preston

Navigation Authority
British Waterways

Canal Societies
Lancaster Canal Trust www.lancastercanal trust.org.uk, Lancaster Canal Restoration Partnership www.thenorthern reaches.co.uk, Ribble Link Trust Ltd www.clegg.fsnet.co. uk/rlthome.html

OS 1:50,000 Sheets
97 Kendal & Morecambe 102 Preston & Blackpool

The canal at Ingol on the outskirts of Preston.

The first lock on the Savick Brook, now the Ribble Link. Water, the oak statue, is known locally as the 'Preston Pisser'.

The tidal section of the Savick Brook as it crosses Lea Marsh to the River Ribble.

51 Pocklington Canal

The Pocklington Canal was built in 1818 by George Leather Jr, linking the market town of Pocklington on the edge of the Wolds with the River Derwent and thence to the industrial towns of West Yorkshire. For 30 years it prospered, carrying heavy loads despite being an agricultural waterway; coal, lime, fertiliser and industrial goods were taken one way and agricultural produce the other. In 1847 it was sold to the York & North Midland Railway Company, after which trade declined. The last commercial barge use was in 1932. The last pleasure craft were a fleet of ten motor cruisers at Melbourne two years later. The swing bridges were fixed in 1962. A proposal to use the canal as a tip for chalk sludge raised local anger and the Pocklington Canal Amenity Society was formed in 1969 with the intention of restoring the canal. With the assistance of British Waterways and Humberside County Council, they have completed much of the task and all the canal has some water.

Strangely, the canal stopped 1km short of Pocklington at Street Bridge on the Hull to York turnpike. This was formerly the Brough to York and Newcastle Roman road, just north-west of the Roman fort at Hayton, the road now being the A1079. The canal is fed by Pockling-

Lockhouse Lock restored; not so the pound to get to it.

fan-leaved water crowfoot and narrow-leaved water plantain in the water and have cut down trees on the bank so that walkers can see them and so that they will get more light, even though existing light conditions obviously suited them. It is claimed to be one of the most important British canals for wildlife and has been declared a Site of Special Scientic Interest down to Church Bridge, just above Melbourne.

Winding gear on the ground paddles.

The basin at Canal Head with the converted wharf building and the Wellington Oak beyond.

ton Beck. It flows past the airfield at Barmby Moor, used for gliding.

Canal Head is a pleasant spot with parking, an information centre, a grassed picnic area, a local resident selling ice creams, daffodils in the spring and ducks. Mellow brick warehouses stand close by and the Wellington Oak public house, half-timbered with herringbone brickwork, looks down the line of the terminal basin.

The nine locks, with substantial iron wheels on the paddle gear, are mostly restored, eight of them Grade II listed. Lockhouse or Top Lock controls the basin and has been rebuilt. Below it, the banks have been excavated but the water is thick with reedmace. The next lock, Silburn, is in a less happy state. The reaches between the three upper locks have had their restoration stopped by English Nature, who have found rare plants such as

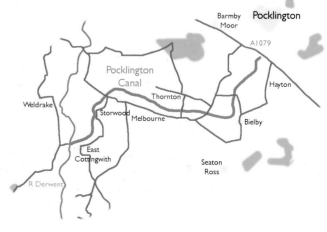

265

Some significant restoration still to be done.

Some lock gates have unique paddle-control wheels.

Bielby Creek is disused and identified as an aquatic reserve.

Below this, all locks have been restored and the rest of the canal is fully in water.

The line is completely agricultural. Hawthorns, horse chestnuts, oaks, brambles, nettles, oxeye daisies and betony grow on the banks while the water may have flowering rush and water forget-me-not, duckweed and watercress.

Sandhill Lock has a blaze of gorse alongside and the yellow flower theme continues along the right bank in the spring with celandines, cowslips and daffodils.

Coats Lock, with its benches, has a water level gauge. Beyond are the first road bridge and then a set of power-lines.

Bielby Creek is now an aquatic reserve with a disused watermill, formerly serving Bielby, and has a Wesleyan chapel of 1837.

Thornton Lock has balance beams that seem too thin.

Pretty as a horticultural exhibit; less useful as the navigation for which it was cut.

Several attractive bridges over the canal show true craftsmanship in brick.

The Pocklington Canal Amenity Society's trip boat New Horizons at Melbourne.

The main line of the canal turns through a large bend here and moves from a southerly to a westerly direction. The Wolds, hidden at the head of the canal, are more easily seen from here. Water lilies make their appearance in the clear water and clumps of marsh marigolds nestle among the stalks of the bulrushes. Alders grow on the banks and, for a while, form a tunnel towards Walbut Lock. Fish include bream, carp, perch, pike, roach and tench. Otters and water voles are present. The loam in the adjacent fields provides such a soft seedbed that tractors need to have double wheels to spread the load sufficiently.

From Walbut Bridge to Thornton Lock is a reach crossed by powerlines.

The bridge between Thornton and Melbourne is one of four brick bridges on the canal of a particularly attractive design. They have flowing curves and a rounded pillar at each corner, perhaps least appealing to the motorist faced with a narrow width and a sharp blind summit. All are scheduled ancient monuments.

Melbourne has a single street of handsome Georgian houses with a conspicuous white school at the far end. The basin has been renovated and now provides moorings, not least for the Pocklington Canal Amenity Society's trip boat, *New Horizons*. There are seven swing bridges downstream of here, that have been renovated.

Boats are moored at intervals or drawn up onto the banks.

Gardham or Bramleys Lock is a complex structure with a swing bridge across the centre. Like some of the other locks, it has thin rail balance beams which look inadequate when compared with the massive wooden beams used on most canals.

The land here is flat and low, just 10m above sea level. To the south is Ross Moor. To the north is Thornton Ings, an SSSI wetland with wild flowers, breeding birds and wintering wildfowl, farmed as hay meadows. Barn owls and bats are present from dusk. A notice proclaims that there is only private shooting. Pheasants keep their heads down and run but plovers and skylarks fill the air with their songs. The canal follows the Beck. Thornton Ings are crossed by Sails and Blackfoss Becks, all joining together before Hagg Bridge. Just beyond the bridge is a large rectangular pond in a field. A notice bans horses

Swing bridge at Melbourne as the canal leaves through ings.

from the towpath but it seems that some riders are prepared to disregard it and remind us what the path was built to carry.

Storwood has a Viking ring to its name although great woods are noticeably absent these days. Instead there is a low swing bridge. It is followed by a gaggle of geese and a flock of sheep by another low swing bridge, heading across to a former moat on the left. On the right are Weldrake Ings, another SSSI in an area crossed by several streams including the old course of the River Derwent. These flood each winter, attracting Bewick's swans, Canada geese, curlews, tufted ducks, coots, moorhens, wigeon and hundreds of other wildfowl, even the bittern. Smaller birds include reed buntings, grey wagtails, sedge warblers, whitethroats, turtle doves and kingfishers. Red-eyed damselflies are among a wide variety of insects.

Cottingwith Lock used to be the tidal limit before a barrage was built at Barmby on the Marsh. Immediately below it, the canal joins the Beck. An arm goes off on the other side to the brick agricultural village of East Cottingwith's inn and red-brick Georgian church, surrounded by rich arable and pasture land.

The canal becomes the East Riding of Yorkshire/North Yorkshire boundary as it moves out onto the River Derwent with Drax power station visible in the distance.

Distance
15km from Pocklington to the River Derwent

Navigation Authority
British Waterways

Canal Society
Pocklington Canal Amenity Society
www.pocklington.gov.uk/pcas

OS 1:50,000 Sheet
106 Market Weighton

52 Driffield Canal

The Driffield Canal was opened in 1770 and extended in 1805 to connect Great Driffield with sufficiently deep water on the tidal section of the River Hull to allow boats up from Kingston upon Hull. It fell into disuse after the Second World War. Restoration has been commenced by the Driffield Navigation Amenities Association.

flights of external steps and other interesting features. A warehouse across the road advertises seeds and peas, the Riverhead Mill advertises its own presence on the east side of the basin and a couple of hand-operated cranes stand next to the canal. Water feeds from a crystal clear chalk stream that has flowed through the town and emerges under the road in a low arch between the

Former warehouses and a mill around the canal terminal basin in Great Driffield.

The canal basin is in **Great Driffield**, the capital of the Wolds. As befits a market town, it has brown-brick warehouses around the terminal basin, mostly restored to form an excellent set of private dwellings, retaining their original mood, with cobbled pavements outside and intriguing sets of semi-circular windows, grills,

The canal is lined with hawthorn bushes below Town Lock.

Blue Bell and the highway depot. Undisturbed by larger craft, the first reach of the canal has quite clear water. The line ahead leads away past public lawns and picnic tables down to Town Lock.

The landscape of Holderness is generally flat boulder clay over chalk but there is enough height change to require six locks over the course of the canal, all but the last being derelict. The main point of interest at Town Lock is the bungalow on the right bank. The walking route involves removing a section of fence which unbolts and then crossing the lawn, the right of way having been the cause of legal battles between the owner and ramblers. The River Hull comes close and there is also a sewage works, the effluent from which ensures that the water is turbid downstream from here.

The banks are at their highest here and are edged with hawthorn bushes and, after Whinhill Farm, the B1249, which follows the left bank. Whin Hill and Wansford Locks are separated by an almost straight reach of over 1km with the bushes absorbing most of

The lock at Wansford awaiting restoration.

Frodingham Beck section of the canal at Emmotland.

the noise from passing traffic. After Great Driffield, Wansford is the only significant settlement along the line of the canal. This village has a post office and store by the canal, a very low concrete bridge below the lock, a former swing bridge and the Trout Inn, the name of which is significant.

Reeds narrow the canal down and provide spring nesting sites for swans.

The road turns away just before Snakeholm Locks, leaving only the occasional lane leading down to the canal over the rest of its line. It is a flat landscape with limited numbers of vehicles, buildings or landmarks. The locks are a two-chambered staircase with most of the water being diverted through a large trout farm alongside the locks.

A dike flows in from the north-east under a gracefully slim brick arch. The area is riddled with dikes and drains and the water appeals to both mallards and cormorants. This 6km pound is the longest on the canal. Where there is a slight rise, such as the one bearing a farm at Brigham, the impact is dramatic.

Dairy farming is the local speciality and herds of Fresian and Charolais cattle gaze at each other from opposite banks of the canal.

Moorings beyond the bridge at Brigham have an assortment of craft. There are two sailing clubs, Brigham Scow Club and Brigham Sailing Club, the latter with a race control tower like a miniature lighthouse, but hedges and high banks steal much of the wind. Sailing boats have to be rigged high. Powered craft are restricted to 8km/h.

Somewhere to the south, gunfire can be heard on a still day but only a moat in a field gives an indication of turbulent times nearer at hand.

The canal takes to the course of Frodingham Beck, which is also part of the canal up to the B1249 bridge. The first reach flows directly away from the spindly church tower at Church End.

A small rise of land as Brigham is approached.

After a few more bends, the confluence with the River Hull is reached. The canal uses the river's course for the rest of its journey.

A number of exotic ducks roam the river. They have a large duck house located among the moorings clustered about the swing bridge that serves Struncheon Hill Farm.

Scurf Dike emerges under another of those graceful brick arches as the river runs through a straight cut for 1km to the Struncheon Hill Lock staircase, the old river line curving round to the west. The staircase has been replaced by a single lock with steel gates that prove rather difficult to use. It is the only operational lock on the canal, leading down onto tidal water.

The lock is separated by a large waterworks from the head of the Beverley & Barmston Drain. The drain follows the river closely but passes under all its tributaries and does not become tidal for another 20km. Not only does it flow to the Humber but, as the name suggests, there is also a continuous route via Frodingham Beck, Old Howe and Barmston Main Drain to the North Sea at Barmston.

The waterworks continue for over 2km with two large reservoirs built between the River Hull and the Beverley & Barmston Drain.

Occasional outcrops of white rock show on the river bank but concrete-filled sacks are more frequent and the banks are now almost continuously levéed. River banks have teasels, marsh marigolds and stinging nettles in season. Any copse of trees is seen from afar and farms are obvious landmarks. One at Baswick Steer has a building from the 19th Century with a new wooden tower on the roof and a roller door facing the river.

Watton Beck emerges through sluices. By the end of the Leven Canal, at Aike, the River Hull is sufficiently large for no further canalisation to be necessary.

Here the river is 2km from the end of one of the runways at Leconfield airfield and the skyline includes the minster at Beverley and the tops of the towers of the Humber Bridge, the longest suspension bridge in the world when it was built. It is possible to see considerable distances over the flat land of Watton, Leven and Arram Carrs, where some of the land is below sea level.

Distance
18km from Great
Driffield to Aike

Navigation Authority
Driffield Navigation
Ltd

Canal Society
Driffield Navigation
Amenities Association
www.driffield
navigation.co.uk

OS 1:50,000 Sheet
107 Kingston
upon Hull

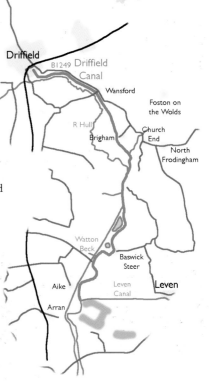

Scurf Dike enters under the remains of a brick arch bridge.

53 Louth Canal

The Louth Canal is now disused and likely to remain so, having been closed in 1924 and not being linked up to the cruising network. It runs northwards from the market town of **Louth** at the foot of the Lincolnshire Wolds to the mouth of the Humber, entering opposite that prime spit, Spurn Head. On the way it crosses the fens, flat land reclaimed from the sea. Designed by John Grundy, it was opened in 1770.

The locks, now without gates, all come in the first 6km.

The canal is fed by streams running down from the Wolds and emerging together from a large culvert by the Woolpack on the north-east side of Louth. Large old mills in brick retain an air of the bygone agricultural industry of the area. Landscaping has made features of these but also included practical items such as parking space.

Some industrial buildings are still operational. The canal passes between Gardners' Plant Hire centre and a pet food warehouse which claims that it is going to the dogs. The industry, such as it is, is left behind at the first sluice, a 2m drop. Locks have no gates on this canal. This first sluice is followed by a 500mm gabion weir.

Housing, with gardens dropping down to the canal and sheltering trees continue to the first proper lock at Keddington. Another lock fronts Abbey House, with the remains of the abbey behind it. A third is at Keddington Corner. Each chamber had sides consisting of four brick arches capped with stone and tied back at the ends with timber that must have given them great character when they were new. Now they are in varying stages of disrepair, surrounded by broken glass and nettles and looking rather sad. In 1985, three of them were given provisional Grade II listings and are to be repaired, despite the fact that the canal is to remain closed to powered craft. There is something of a mystery why six of these locks should have been built to this almost unique design while the sea lock and top lock were conventional. It is thought that the replacement of Grundy by James

The bank of the canal has been landscaped within the town.

Hogard part-way through the construction in 1767 could have resulted in fresh thinking in the design of the locks for the final climb up to Louth.

The canal has the occasional oil drum or fertiliser sack in it but lack of craft means that it is generally a clean waterway. Freshwater mussel shells lie on the teasel clad banks from time to time.

Beyond the sewage works is another lock, the left wall of which has collapsed. The lock at Alvingham is the only one fronted by private buildings.

A bend past a church with a large square tower fails to reveal from the canal a striking feature, another church with tower right beside the first. Behind them both is a water-powered mill. On certain limited occasions the mill is open to the public. Demonstrations are given of corn grinding using this traditional power source.

Like the roads in the fens, the canal seems to be a series of interconnecting straights. One runs from Alvingham directly to the prominent and distinctive High Bridge House, turning sharp left under the bridge itself at the last minute, bringing the canal from a north-easterly to a northwesterly direction.

Warehouses at the head of the Louth Canal as it emerges from a culvert.

Houses near the canal are screened on the outskirts of Louth.

Weir at Tetney Lock, almost drowned out by the tide.

Tunnel mouth by the bridge at Austen Fen.

The lock near America Farm is the last, its left side battered back after the lock wall has collapsed.

Unlike other canals, the sides are high levées which must have caused some problems with horses high on the towpath on the east side.

Drainage outfalls become common, often in matching pairs on opposite sides of the canal. Black Dike flows in from the direction of Yarburgh and another drains from Covenham St Mary to arrive through a long brick tunnel at Austen Fen. These are both much more obvious than the channel from the large Covenham Reservoir, near Fire Beacon.

Windmills at Marshchapel and Fulstow cannot be seen from the water.

The illusion of pastoral tranquillity can be rudely shattered, however, by jets. This is only a few seconds' flying time from the RAF's Donna Nook bombing range.

At Thoresby Bridge, a single large tree by the canal is almost dwarfed by a monstrous shooting tower.

From New Delights, the canal moves back to a north-easterly course. It joins the Waithe Beck at Tetney Lock, protected by a pair of flood doors under the road and by

recent sluices. These latter may be completely drowned out at high tide but at lower stages of the tide they can have dangerous weirpools below them. Anglian Water's control building has nearby amenities including a telephone box and the Crown & Anchor.

Side channels are now terminated with flood doors. Two sets of flood doors stand across the canal near Stonebridge Farm, guarded by miniature pillboxes.

The canal becomes flanked by saltings, designated as a nature reserve. Obviously out of place is a large pipe crossing above the water, connecting the Tetney oil terminal with the Tetney Monobuoy in the centre of the Humber estuary, so that tankers can discharge their cargoes directly into the pipeline without having to dock.

A channel leads off under a small bridge into Tetney Haven, an almost landlocked area of saltings, lagoons and islands.

The canal continues northwards between two training walls which cover at high tide. Their extremity is marked by a beacon that also locates a sewage outfall. Needless to say, this is a favourite gathering point for the teeming estuarial birdlife of the area.

Prominent features beyond are the Haile and Bull Sand Forts. Less obvious are the numerous wrecks that dot the outer edge of Tetney High Sands, uncovered at most states of the tide. Up the estuary lies Cleethorpes, dominated by a water tower on the waterfront. To the east lie dozens of large ships heading to and from the North Sea, tankers, freighters, fishing boats and more using this busy waterway, now all sailing past the once-commercial Louth Canal.

Distance
23km from Louth to
the River Humber

Navigation Authority
Louth Navigation Trust

Canal Society
Louth Navigation Trust www.
louthcanal.org,uk

OS 1:50,000 Sheet
113 Grimsby

Low Farm and a boat not destined for the canal.

Haile Sand Fort lies beyond the final marker at the canal end.

271

54 Stainforth & Keadby Canal

The Stainforth & Keadby Canal was built between 1793 and 1802 to extend the River Dun Navigation from near **Stainforth** to the River Trent. Both form parts of the Sheffield & South Yorkshire Navigation. Effectively, it was a ship canal for 200t craft and was to prove a financial success. Being cut across fenland, it had mostly long straight reaches and only needed two locks, the second of which was to deal with the tides on the Trent. Construction of the New Junction Canal has now removed most of its traffic.

It is a waterway heavily used by anglers, having numbered pegs over much of its length. Most bridges open and are low.

At first the canal follows the tidal River Don, on the other side of which is Fishlake with an armless windmill and the fine medieval church of St Cuthbert. The church is noted for its intricately carved late Norman doorway with animals and foliage and for a belltower with a notice banning the ringers from wearing hats or spurs.

Views to the north are not helped by levées each side of the River Don or by the trees that line the canal nearly to the M18, giving the canal its most attractive reach. Towards the M18, the trees give way to oilseed rape fields and skylarks. The River Don turns away to the north, to become the Dutch River.

Beyond the M18, the Doncaster to Hull railway crosses as the canal enters **Thorne**, the only town along its line. It quickly arrives at Thorne Lock and swing bridge, the first of three boatyards in the town and recent canalside housing. The Canal Tavern is tucked in next to the A614 as it crosses and the Doncaster to Grimsby railway line crosses and follows the canal, most of the way keeping close to the bank of the canal. Massed swans patrol the water. By the railway bridge is a water tower the shape of a flat champagne glass. Fenland is opening up, flat but fertile, and plants include bog rosemary. The rare large heath butterfly is found.

At Wike Well End there is a low drawbridge. Another quickly follows and Maud's Bridge is a third, with resident cobwebs. Deer are often drowned around Thorne because of the sheet-piled sides which stop them leaving the water if they get into it. Ramps have been added in places to help them escape and deer fencing has been erected beside the railway

To the north and south respectively are Thorne Waste and Hatfield Chase, former marshes, the largest and most diverse lowland peat area in northern Europe, now mostly cut commercially but still with rare plants and birds. Birds around the canal range from mallard, moorhen and great crested grebe to swallow, plover and black-headed gull. Bulrushes and lilies are commoner canal plants. High masts take powerlines over the canal. Drainage channels run parallel with the canal on each side for the rest of its journey, unseen. What is conspicuous is the railway. In some of the flattest landscape in Britain it seems perverse to have Trans Pennine Express units running past. The swing bridge at Medge Hall is very low. Guarding the track to the hall is a signal box. In its remote position and with so little road traffic the signalman's job must be one of the least demanding in existence.

The Old River Don is crossed, now no more than a drainage ditch, for which it would be hard to guess the original line. The diversion to the current line was undertaken by Vermuyden in 1625, hence the name Dutch River.

The canal avoids Crowle with its Georgian houses. Its church has a large chancel, a clerestory and a 2.1m Saxon, carved stone, believed to represent King Oswald and his son, Oswry. The Regal Motor Museum is housed in a former cinema. It has cars and motorcycles of 1902–30 and a Flying Flea aircraft of 1936. The A161 is busy enough to justify a high level crossing at Crowle Wharf on the edge of Ealand, an area edged by lakes. Four kilometres south is the Isle of Axholme, on which is built Belton, a true island when the fens were all marshes.

A heron fishes as the canal makes a straight line for the Trent across the fens. Approaching Keadby, there is a minor swing bridge and then the railway makes an oblique crossing on a sliding bridge of 1926, the last remaining in the world. The heavy girder structure rolls sideways, taking the track with it. As it closes, the track moves back to its position. Accompanied by assorted rumblings and clicks, the track is pulled down some 100mm into place and rods slide into position to lock everything together. This is a marvellous piece of machinery to watch in operation, somthing that does not happen frequently.

The final reach is dominated by a recent gas-fired power station which replaces a coal-fired model. Between the Auld South Yorkshire and the boarded-up Friendship public houses are the B1392 and Keadby

The lone Thorne Lock attracts spectators.

Coots' nest in the reeds near Crowle.

The magnificent sliding railway bridge at Keadby, one of only three in Europe.

Powerlines surround the gas-fired power station at Keadby.

Keadby Lock has a complex arrangement of gates.

Distance
20km from the River
Dun Navigation to
the River Trent

**Navigation
Authority**
British Waterways

OS 1:50,000 Sheets
111 Sheffield
& Doncaster
112 Scunthorpe
& Gainsborough

Lock. The lock is a two-way structure because the River Trent, onto which it opens, is tidal. The river has streams that may exceed 11km/h and has an eagre on equinoctial spring tides. Although the lock length is limited, longer boats can pass through when the Trent and the canal are at the same level so that all gates can be opened for clear passage through. At low tide, the river exposes high muddy banks.

In the distance, Scunthorpe breaks the skyline. Immediately south of the lock are the ends of the Three Rivers, three channels that run dead straight beside each other for 3km from a complex junction fed by the Boating Dike, Hatfield Waste Drain, River Torne, South Engine Drain and Folly Drain.

Also seen from the lock is the King George V Bridge over the Trent above Althorpe.

Built in 1916, it carries the railway and the A18. It was the Great Central Railway's largest bridge project. With five spans, it includes an unusual 50m rolling lift bascule span, the heaviest bascule span in Europe, fixed down in 1960.

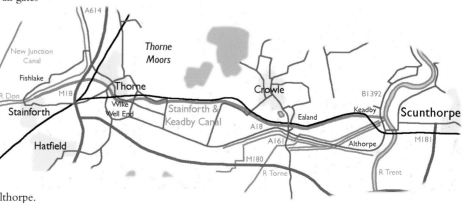

273

55 Chesterfield Canal

Despite being one of the early canals, the Chesterfield included some notable engineering. In particular, it needed 49 locks in the top 32km in order to get over a ridge of high land from the Rother valley. It also needed the Norwood Tunnel, still the sixth longest to have been built in Britain. The canal was designed and constructed by Brindley although he died before it was opened in 1777. It was two years late, following problems with the tunnel, which was 50 per cent over budget because of fraud. Traffic peaked at 200,000t in 1848, particularly stone, grain, lime, timber and ale, but decline followed quickly after purchase by the Manchester & Lincoln Union Railway. The section above Worksop was unusable by 1896 and the tunnel suffered a collapse in 1908 as a result of mining subsidence. All commercial traffic ceased in 1962 although the earlier closure of Walkeringham brickworks had been a major blow to commercial use.

The isolated canal runs eastwards from **Chesterfield** to the River Trent. It is a narrow winding canal, mostly quiet and rural and often passing through attractive wooded scenery. Much has been restored. Reconstruction of the tunnel will be a major task.

Initially the line uses the River Rother. A new terminal basin complex is being constructed downstream of the former terminus, infilled in the 1960s. The Nottingham to Sheffield railway runs between the river and a golf course. Voles scrabble up the steep banks as the river winds past a plywood works. There are alder trees at the foot of Castle Hill.

The canal leaves the river but follows its right bank for several kilometres. Tapton Mill Floodgate is at a small brick bridge. The view back to Chesterfield (named from the Old English for a Roman fort in open country) is crowned by the twisted and bent spire of St Mary's church, which was hidden at the start. On the right bank stands a Tesco store.

A recent tunnel carries a road across just before Ford Lane Lock. The lock itself, together with its lock cottage, now a canal visitor centre, has the feeling of being in the middle of a busy roundabout although traffic does not go all the way round. Artwork is displayed under the first bridge. A young sapling planted at the lock between the canal and the adjacent river is dedicated to prisoners of the war with Japan. The Trans Pennine Trail and the Cuckoo Way long distance footpath follow the canal. A milepost gives the walking distances to Chesterfield and Istanbul.

The railway divides and crosses the canal on a pair of bridges. They are accompanied by a Mercedes Benz building and a Sainsbury's store.

The canal passes some small industrial premises and the Mill public house on the B6050.

Mosaic panels are set into the towpath by Wheeldon Mill Lock. Between the canal and the river there was a railway at one time. Blue Bank Lock faces onto a reach that is rather open. Between New Whittington and **Brimington** was the Dixon Mine. Dixon Lock has been restored with a tiny humped footbridge, typical of this canal.

The next lock is Staveley Works or Hollingwood, Stanton & Staveley being major producers of iron pipes. Powerlines weave from side to side over the canal.

Section past a wooded bank near New Whittington.

Teasels climb up to the brick footbridge at the Dixon Mine.

A huge water-filled pit lay across the line of the canal into the mid-1990s but this has now disappeared without trace and the landscaping has nothing more intrusive than regular angling platforms. The water runs on to **Staveley**, where it finishes below a bridge.

Restoration of the 13-lock climb out of the Rother valley from Belk Lane Lock is likely to result in considerable workload. Irritation has been shown at the building of a couple of dozen houses on the line of the canal at Killamarsh.

Water reappears at Norwood, where there are extensive views down over the Rother Valley Country Park, from which the

Dead boat near the reconstructed Blue Bank Lock.

Hollingwood Lock has outlived the railway that formerly crossed beyond it.

Heading towards Staveley with no remaining evidence of the former mining lake here.

The Rother Valley Country Park as it appears from the canal at Norwood.

The east end of Norwood Tunnel at Kiveton Park.

honking of Canada geese emanates. The canal winds past uninspiring premises belonging to industrial users. Here water seem rather an overstatement, it being mostly a reedy marsh fed by effluent trickling from a pipeline.

A wharf on the north side has some interesting stone projections over the water.

Beyond the A618, which crosses the canal by the Angel Inn, the canal comes back into water properly for several hundred metres.

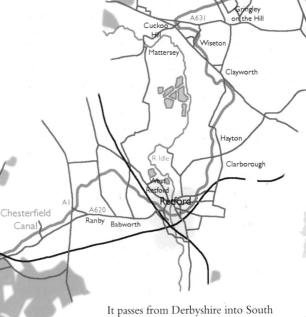

It passes from Derbyshire into South Yorkshire but comes to a halt at a winding hole below dried-up locks. These lead to the western end of the 2.8km long Norwood Tunnel. However the problems of mining subsidence collapse are handled, an additional problem to those of 1908 is that the M1 now passes over the top and may not be disturbed by restoration work.

Restoration of the eastern end of the canal has been completed up as far as the eastern portal at Kiveton Park, now bricked-up. The canal is followed closely by the Sheffield to Cleethorpes railway. Local industry includes Redland bricks.

The wharf at Norwood overlooks only reedmace.

Overhung by trees, the canal leads past Kiveton Park station.

Hawks Wood and Old Spring Wood offer rural canal scenery at its best. The former hides Thorpe Salvin with its hall. Steetley chapel has a reasonable claim to being the most elaborate and perfect Norman building in Europe with a triple rounded porch, slit windows and exquisite apse.

A winding hole appears alongside Old Spring Wood before the descent begins in earnest: 22 locks in 1.6km. The woods are extremely pleasant. Turnerwood is an attractive canalside hamlet with a towpath well tended by householders, only light aircraft landing at Netherthorpe

airfield breaking the solitude. Violets and celandines carpet the ground between the trees in the spring.

The canal passes from South Yorkshire into Nottinghamshire as it crosses the Shireoaks aqueduct over the River Ryton. At Shireoaks the Station public house faces the Steetley sportsground. Two features dominate the line as far as the A57, the spoil from the Shireoaks colliery, some of it subject to landscaping, and seven more locks.

A freight line crosses at Rhodesia, preceding the A57 viaduct and the Woodhouse Inn.

The canal is flanked on both sides by roads. Just before the A60 is another lock. Continuing the tour of pipe

Summer in the Hawks Wood reach.

Full restoration below the A57 at Rhodesia.

The hamlet of Turnerwood.

Cuckoo Wharf in Worksop, formerly a Pickford warehouse.

Turnerwood, one of the best wooded lock flights in the country.

Traditional canalside buildings in Worksop.

The pumping station by the Ryton aqueduct.

manufacturers, Hepworths make clay pipes by the canal.

After Morse Lock there is another sports ground, a winding hole and a caravan site. The canal eases into **Worksop**, which developed from the Old English Weorc's valley to serve the North Nottinghamshire Coalfield. Surrounding woodland is the remains of Sherwood Forest.

Just downstream of Worksop Lock is the restored Cuckoo Wharf with a three-storey Pickford warehouse built over the canal, now featuring a canalside bar with tables round a traditional canal crane.

There are some interesting old buildings on the right, behind the Canal Tavern. These include a museum and the priory, which has a 12th-century restored church and a 14th-century gatehouse with a skull in which an arrowhead is embedded.

A brick building with a distinctive chimney near the canal has all the hallmarks of a pumping station in the grandest Victorian tradition. The canal crosses back over the River Ryton on another aqueduct between the two Kilton Locks.

The mine beyond Manton has been thoroughly landscaped. Crossing road, canal and river is a brick railway viaduct that has been heavily and conspicuously reinforced with sections of railway line.

The views from the canal now become extensive over parkland. Owned by the Duke of Newcastle, who was one of the contributors to the name of the Dukeries for the area, Clumber Park's 10km² of country park reach nearly to the canal. Close on the north bank is Osberton Hall and its associated buildings with some graceful architecture and farmland in immaculate order. The church was built in 1830 and has been restored. Scofton is a private estate village for Osberton Hall, built about 1806 by James Wyatt. To the north lies a disused airfield.

The occasional pine copse or deciduous woodland follows, with a carpet of bluebells in the spring.

The next main transport link over the canal is the A1 before the canal turns into Ranby, where the A620 is now a bypass.

From the Chequers Inn the canal moves back to be followed by the A1 for a kilometre, perhaps not as noisy as might be expected because of the screen of trees. Indeed, the skylarks manage to make themselves heard quite effectively above the traffic roar.

The canal leaves the valley of the River Ryton and the A1 at the Barracks and makes a large semi-circle round a hill topped by a prison. This loop has some of the most attractive scenery on the canal. It includes the last remnants of Sherwood Forest, oak and sycamore woods with colour added by clumps of yellow irises and, at Forest Top Lock, a garden of red hot pokers.

Geese and swans follow as the canal winds round to Top Middle Lock, below which the stone canal wall has been built-up to look like sheet piling, presumably to break up washes for craft moored alongside. The remaining two Forest Locks are below Forest Farm.

After Babworth Park, surrounding Babworth Hall, the main East Coast railway line crosses the canal which then eases its way into West **Retford**. The screen of trees reduces the impact of the houses and hides the cemetery and hospital. Gamston Aerodrome, 5km to the south, does not intrude either.

The A620 crosses on a bridge that has been conspicuously widened, now being almost a tunnel of various diameters. It is immediately followed by Wharf Lock, most of which is covered by a huge chestnut tree.

An old wharf building has been converted to a house that stands prominently beside the canal.

A series of three aqueducts includes one over the River

The conspicuously repaired railway viaduct at Manton.

The canal turns away from the A1 near Ranby.

British Waterways barge dredging below one of the Forest locks.

Idle, the valley of which is followed over the rest of the canal's course. A long forgotten wooden drain plug was accidentally pulled out of a washout pipe in 1978, allowing an extensive section of canal to drain quickly. Old wharf buildings are accompanied by modern buildings in complementary design on the East Retford side of the river by Retford Lock. The Wharf restaurant follows just before the canal passes under the old Great North Road bridge, now reduced to the status of a minor town street. The canal passes King's Park. Notable town buildings include the flamboyant town hall of 1868 and a cruciform church with a peal of ten bells and protection in the form of an 11kg cannon brought back from Turkey. On one occasion it was removed for maintenance, leaving the intriguing sight of an empty plinth labelled 'Taken at Sebastopol'.

point for the packet boat to Retford, operational until replaced by a bus in the 1920s. It now houses Retford & Worksop Boat Club, who have done much to reopen the canal.

The most prominent building in Clayworth is the Hall, a large white mansion that looks down on the canal. The bridge to the north-west of the village is on the line of the Roman road from Lincoln to Tadcaster.

Overhanging hazel trees shelter swarms of insects. The greenery opposite hides Wiseton Hall. A bridge, with wrought-iron railings, serving the hall has been the most attractive on the canal before until becoming covered in vegetation. Also hidden beyond Wiseton Park is Mattersey Priory, founded by St Gilbert of Sempringham in 1185 but destroyed by fire in 1279. It housed the only wholly English monastic order of the Middle Ages.

At Hayton with a heron perched on an oak-tree branch.

Grove Mill in East Retford, a maltings-turned-public house and snooker club.

The Grove Mill is converted from a large old maltings beside the canal to a public house and snooker club. The canal winds out of East Retford past modern housing until it is crossed by the A620 at the Hop Pole, in the centre of a hop-growing area until the trade was ruined by hops being imported from Kent along the canal.

Whitsunday Pie Lock is the first of the wide locks. Retford helped finance the canal to Retford on condition that it was built wide enough to bring Trent keels up to the town. The lock is said to take its name from a huge pie baked for the navvies by the wife of a local farmer to celebrate completion of the lock, a fact recorded by the serving of a pie to boat club members each year. However, there are signs that the name is rather older, relating to the time of the tenancy change of a pightle of land.

Bonemill Bridge takes its name from a mill which used to produce fertiliser by crushing bones brought by canal.

The Cleethorpes to Sheffield railway line leaves as the canal curves northwards to run round the foot of a ridge of high land, the benefit of the longer route now being a 14km lock-free pound.

The Gate has a car park on a former weighbridge wharf at Clarborough.

A house at Hayton has a wall incorporating a row of cartwheels. The church dates from 1120 and has box pews. The Boat Inn also has a children's playground and there is a canalside picnic site.

The canal winds on through fields to Hayton Castle Farm. At intervals there are overflow weirs with narrow bridges for walkers and wider fords for towing horses.

A former inn at Clayworth Wharf was the departure

The canal turns sharply in front of the White Swan Inn with its small basin and wharf. Ahead is the end of the ridge of Cuckoo Hill, carrying the A631, a sandstone ridge which the canal has taken a large detour to avoid as far as possible. The last part of the ridge is passed in the 141m Drakeholes Tunnel, which has no towpath. It has a variety of linings, corrugated sheeting, concrete and brick, although in places the red sandstone is unlined. Travelling north, there is a clear view of any traffic approaching.

The canal turns through another right angle to follow the foot of the ridge, skirting Gringley Carr.

Gringley Top Wharf and Top Lock lie below Gringley on the Hill, with its windmill and church. The Beacon Hill viewpoint has views as far as Lincoln cathedral. The brickworks just before the lock were the last commercial

Bushes near Wiseton shelter swarms of insects.

Drakeholes Tunnel passes under the end of Cuckoo Hill.

users of the canal, clay being taken from the adjacent area, which is now a lake. This lock is followed by Gringley Low Lock and Low Lock Bridge.

A square iron-bound chimney near **Walkeringham** Bridge marks a wharf that was used by munitions traffic until the late 1940s.

Fountain Hill suggests that artesian conditions have been present as Gringley Carr becomes Misterton Carr, an area drained from marshes by Vermuyden in the 17th Century. Misterton has a military depot and a church with a stubby spire. The spire on All Saints church dates from the 1840s although the church is 13th-century and built on the site of an earlier one. The list of vicars dates from 1254. The stained-glass window in the north wall is over 500 years old. There are two locks: Misterton Top Lock and Misterton Low Lock. A bridge carries the Doncaster to Gainsborough railway over and then the canal runs out onto a straight embankment over a low lying area of ground.

The cut passes Stockwith Bridge and the Waterfront Inn to enter West Stockwith Basin, where seagoing ships used to offload onto narrowboats. There was also boat-building in the basin. Interesting buildings include the 1797 warehouse. Several Dutch style houses in red brick are very modern.

Trent Lock only opens near high tide and leads out to the River Trent. Hand winches were used to assist sailing keels into the canal when the river was in flood. Just downstream, the River Idle also enters the Trent.

The spire on the church almost opposite, at East Stockwith, is worthy of note.

Old warehouse building with the lock beyond at West Stockwith Basin.

Distance
81km from Chesterfield to the River Trent

Navigation Authority
British Waterways

Canal Trust
Chesterfield Canal Trust www.chesterfield-canal-trust.org.uk

OS 1:50,000 Sheets
*112 Scunthorpe & Gainsborough
119 Buxton & Matlock
120 Mansfield & Worksop*

56 Fossdyke Navigation

**The oldest
canal still
in use**

When Britain's canals vie with each other with claims of longevity, they use as their reference point the Bridgewater Canal, opened in 1761. There is, however, one canal with which none of the others can begin to compare. The Fossdyke Navigation was dug by the Romans in about 120 and is by far the oldest canal that is still navigable by larger craft.

The Fossdyke linked the Trisantona Fluvis, the tidal River Trent, with the River Witham, providing an inland route between the Humber and the Wash. While part of its function was for drainage, the fact that, with the Car Dyke and Cnut's Dyke, it linked right through to Waterbeach in a continuous channel, could only imply its use for transport.

The navigation leaves the River Witham at Brayford Pool, overlooked by **Lincoln** on its limestone ridge, a striking island in the flat Lincolnshire countryside. The city was the Celtic Lindon, the hill fort by the pool. Later it became Lindum Colonia, a Roman walled garrison housing the 9th and then the 2nd Legion in the 1st Century, before becoming a settlement for retired Roman soldiers.

The Danes used the navigation when invading England. When Lincoln became a Norman stronghold, the navigation was used for importing the stone for building the cathedral of St Mary, begun in 1072. The cathedral is Lincoln's crown, topping the ridge and being visible from Brayford Pool and from a considerable area of the surrounding countryside. It was subject to a fire and an

earth tremor in the 12th Century and has had two restorations. Between 1192 and 1250 it was rebuilt in Early English style, using Purbeck marble. The magnificent central tower was completed in 1311 and has profuse monuments and wood carvings. The Seamen's Chapel has a window depicting Matthew Flinders and the library has a collection of rare books and maps relating to exploration. It was used in filming *The Da Vinci Code*.

The Norman castle has an original copy of the *Magna Carta*.

Usage taxed the canal and in about 1121 it had to be scoured out. By the 17th Century it was almost impassable again and Acts were established in 1753 and 1762 to improve the navigation. The upgrading in the 18th and 19th Centuries of the fenland drainage gave the navigation its present form and it was used commercially until grain barges called for the last time in 1960, since when it has only been used by pleasure craft. There are masses of mute swans.

A slipway is located at the east end of the pool in front of the General Accident insurance company buildings. There is a disused lifting railway bridge at the south-east corner of the pool. Its neighbouring railway bridge is still very active and the railway line follows the navigation bank to **Saxilby**, with the A57 also following to Drinsey Nook.

A variety of buildings surround Brayford Pool, ranging from modern offices to old warehouses. The area

Lincoln cathedral overlooks Brayford Pool and the River Witham exit.

Assorted craft in Brayford Pool.

Saxilby faces the navigation with mown lawns.

Torksey Lock with its unusual capstan-operated gates. Tidal water lies beyond.

Cottam power station cooling towers seen in the distance.

around the Royal William IV has been made into a pleasant waterside pedestrian walkway, continuing along to the public toilets by the harbourmaster's office.

The Fossdyke heads in a westerly direction towards the Trent.

At first there is a line of houses on the right but these are soon left behind and there are very few other buildings alongside the rest of the navigation. Those that are, include some traditional brown-brick houses with rib-tiled roofs. On the other hand, the sheetpiling and the raised banks are to be almost continuous features. The horse race course and a golf course, both on the right, are among the first features they hide. Opposite is a marina.

A large watercourse emerges under a bridge from alongside the golf course and crosses the navigation at water level, being lifted out of the other side by a pumping station. Eventually, it finds its way through other channels back into the River Witham and the navigation.

The Pyewipe Inn, next to the navigation but below water level, is easier to find by boat than by road, despite being next to the A46 bypass bridge. Other bridges carry pipes across but the navigation is now in open country for some distance. Occasionally there are willow trees along the banks but more often there is nothing taller than reeds or grass. In addition to passing trains and boats, there are training jets climbing away from the airfield at Scampton.

The River Till joins under the A57 bridge at Odder. The major landscape feature here is the large fertiliser factory beside the canal. Notices state that lorries must be fully sheeted before they leave and a red windsock gives an indication of where and how fast the fertiliser would go without the covers. The factory has a wharf with mooring points along the navigation.

Saxilby is the only significant village along the Fossdyke and presents an attractive face with mown grass banks. Amenities include the Sun Inn, Ship and a chip shop. It is a village of bridges, being crossed by the A57, a pipe bridge, the railway turning off for Gainsborough and a neat footbridge. The arched pipe bridge has handrails leading up one side to the air release valve at the top, giving an undeserved appearance of dilapidation. Almost below it are the remains of a swing bridge, protected by heavy piling at each side of the navigation.

The windmill on the south side of the village is not seen from the water, just the arms of JCBs in a depot next to the navigation.

Pike House stands on the left of the waterway. Soon after, a signpost to Doddington Hall tops the embankment

as a tunnel appears under it. The sign is for the benefit of users of the B1190 as the Wigsley Drain diverges from it after 600m and moves off in a more southerly direction.

The navigation turns through a right angle at Drinsey Nook, into a 4km length that is almost straight, more what the Romans might have been expected to build. As it makes the turn, the cooling towers of the power station at High Marnham are in sight, a scene that is to be repeated. The largest building on this length is a battery chicken farm. Buildings of any sort are well spaced.

The banks have masses of cow parsley, together with deadly nightshade, blackberries, toadflax, yellow iris and mustard from time to time. The navigation is free of vegetation except for floating weed at the edges.

The navigation's only lock, Torksey Lock, is 400m from the end and acts as the tidal limit. Its gates are operated by unusual, elegant iron capstans but its major difficulty is that it makes no allowance for anyone not intending to lock through. It is possible to carry a small boat right, cross the narrow footbridge at the bottom end and climb over the wall onto the A156, getting in again on the far side of the road. An easier but longer portage is to go left by the Wheelhouse Restaurant if the gate is open, relaunching as before on a set of pontoons where a kingfisher might be found perching.

Torksey itself lies 800m to the north, its stone church tower overshadowed by the brick and stone remains of its Tudor castle. Overwhelming all, however, are the cooling towers of the power station at Cottam. Another set of power station towers rise downstream at Sturton le Steeple. These modern

Distance
18km from the River Witham to the River Trent

Navigation Authority
British Waterways

OS 1:50,000 Sheet
121 Lincoln & Newark-on-Trent

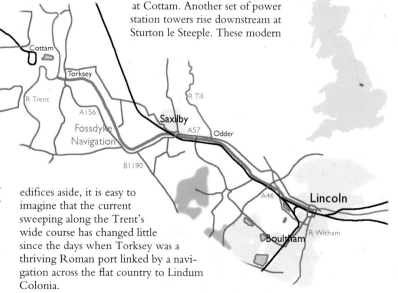

edifices aside, it is easy to imagine that the current sweeping along the Trent's wide course has changed little since the days when Torksey was a thriving Roman port linked by a navigation across the flat country to Lindum Colonia.

57 Union Canal

The Forth & Clyde Canal was constructed to bring coal to **Edinburgh** from coalfields to the west and to allow craft to operate from Glasgow to the Forth and thence to Edinburgh. Its problem was that boats using the Firth of Forth had to be seaworthy and narrowboats were not. Thus, goods had to be transhipped from narrowboats to sailing boats. To get round the problem, the Edinburgh & Glasgow Union Canal was constructed in 1818–22, linking the Forth & Clyde Canal directly with Edinburgh so that narrowboats could be used for the whole journey. With assistance from Telford, Baird engineered the whole canal east of Falkirk to be on one level. Swift boats were introduced in 1836 to fight back against railway competition, cutting the time to Glasgow to seven hours. The locks were filled in 1933, severing the route. The canal was closed to through navigation in 1965.

Today the Union Canal is complete once again. It runs as a top pound at a single 73m level for 50km from Edinburgh to Falkirk and locks take it down to the Falkirk Wheel and the descent to the Forth & Clyde Canal. As a contour canal, it was nicknamed the Mathematical River.

Its start is at Edinburgh Quay, previously called the Lochrin Basin, just 800m from Edinburgh Castle. Until the 1920s the canal continued to Port Hamilton and Port Hopetoun, since filled in. New office buildings surround the basin which, for decades, has been home to Forth Canoe Club, Scotland's leading canoe club, with slalom practice gates slung across the basin as there was no passing traffic. Leamington Lift Bridge of about 1906, a hydraulic vertical-lift bridge, was moved from Fountainbridge in 1922, when the Port Hamilton and Port Hopetoun basins at the end of the canal were filled in. It now serves the McEwans brewery on the right.

Initially the tenement houses are old and tall, their stone walls close to the canal. Bridges are of plate girders rather than of the stone that appears later.

A railway passes under the canal here and unseen trains roar beneath.

George Watson's College Rowing Club boathouse appears on the left and rowing boats can be quite devastating here, their blades reaching nearly to the banks as students appear backwards at speed. Slower rowing boats

Leamington Lift Bridge is the first canal crossing.

It was from Gray's Mill that Prince Charlie successfully ordered Edinburgh to surrender in 1745. Prince Charlie Aqueduct crosses the A70. After the Water of Leith Centre, it is followed by the eight-arched Slateford Aqueduct just round the corner, 183m long and 20m high over the Water of Leith. It uses 15m arches, which are the same size as those on the other river crossings, and uses the same iron troughs, that are wider at the invert than at the top. Telford's hollow piers are used although he did not agree with Baird's need for masonry supports to the iron troughs.

The canal is still in the built-up part of the city but plenty of vegetation is found on the banks. There is also a large Tesco store. The Waverley to Glasgow Central railway passes under.

The canal cuts through between the high rise blocks of Wester Hailes with the Pentland Hills as a backdrop, giving little indication of the difficulties of restoring this section of canal, which had a housing estate built across it. It passes under the A71 and immediately turns over the A720 on the Scott Russell Aqueduct.

In 1834, John Scott Russell, the engineer, mathematician and natural philosopher, was watching a barge being drawn by two horses in the vicinity of a bridge at Long Hermiston. The horses and barge stopped but a 300–400mm high wave broke free from the front of the barge and carried on at some 14km/h. Russell followed it on horseback for 2–3km before losing it round a series of bends. As a result of this incident he published his Theory of the Solitary Wave or the soliton, a principle that is now known to have widespread application in the natural sciences.

The new residents of Edinburgh Quay.

More decorative than most anti traffic fences.

Slateford Aqueduct crosses the Water of Leith.

The Bridge Inn at Ratho played a central role in the restoration.

Restored section through the housing blocks of Wester Hailes.

There are some surprisingly fragile items of artwork displayed in public areas around the canal at Ratho.

A log frog seat at Wester Hailes.

This rural hillside at Ratho hides the M8.

At Ratho, the Bridge Inn has its own restaurant barge, the *Pride of the Union*. For years this inn was managed by Ronnie Rusack, a prime mover in restoration of the canal and a promoter of the Seagull Trust, with two charity boats here.

Bonington Aqueduct takes the canal over the B7030. A 5km lade, the canal's main feeder, complete with tunnels, arrives next to the 5-arched 128m long Almond Aqueduct, which carries the canal 23m above the River Almond. From here the views are excellent, the Forth road and rail bridges being visible down the valley. Nearer is the 3km long Almond Valley railway viaduct in two sections on the Waverley to Queen Street line, far enough away to show its sweeping grandeur without showing the unsympathetic repairs resulting from ground settlement caused by oil shale workings.

A quarry at Craigpark has resulted in spoil being tipped along the canal bank with some of it being washed into the canal.

The National Rock Climbing Centre is located in Ratho Quarry on the north side of the canal.

Wilkie's Basin used to load stone. The short

Muirend is 4km from the end of the runway at Turnhouse and in direct line. The noise of jets taking off is then replaced with that of vehicles as the canal passes under the Waverley to Glasgow Central railway and the M8.

The canal turns north from Powflats and the British Waterways depot, passing by Port Buchan Basin at **Broxburn**, before heading north-east.

Beyond Broxburn there is a boat ramp on the right, used by a paddle-wheeled weed remover and others.

From Broxburn, red shale bings are frequent, scheduled monuments resulting from oil shale extraction for paraffin manufacture in Broxburn. At Niddry, the canal turns back to the north-west and the ruined block of Niddry Castle squats at the foot of another heap of waste, the castle having sheltered Mary, Queen of Scots before her final capture. The Waverley to Queen Street railway follows the rest of the canal, mostly closely. The M9 is usually about a kilometre to the north as far as Polmont.

Red shale bings loom above the canal north of Broxburn.

Niddry Castle and more oil shale waste tips.

A wooded reach near Craigton.

Decorative handrailings on a bridge at Clifton Hall.

A date on a stone block of the Almond Aqueduct goes back to its construction.

Winchburgh is the next village to front the canal. The canal then cuts between two large stacks of waste, some of it perched precariously above the canal.

The aqueduct over a minor road at Philpstoun is above a bridge over the Haugh Burn. After an estate, the canal returns to rural surroundings which contain such birds as the yellowhammer.

The Park Bistro enjoys a rural setting. The canal crosses the B9080 on a single arch

The **Linlithgow** canal basin is a delightful spot in a setting of low canal buildings with former stables, multi-

level bridges and a museum with canal and Roman remains. It is also the home of the *Victoria*, a diesel-driven replica of a steam packet, owned by the Linlithgow Union Canal Society. At the back of the public garden opposite is a dovecote, like an oast house-sized beehive with an entrance at the back big enough to crawl through but little more. It has holes in the top and a boxed finish to the walls on the inside.

Beyond is the red sandstone block of Linlithgow Palace, birthplace of Mary, Queen of Scots. The 1960 aluminium crown on St Michael's church to the right looks strangely out of place, a modern metal sculpture on a Gothic 13th-century sandstone building.

A golf course on the left bank surrounds an old quarry which could well have provided stone for the canal bridges and for its stone lining.

The A706 follows and crosses at Woodcockdale Stables, where stone was loaded onto canal boats. They are now used by the 1st West Lothian Sea Scouts. Then the Avon Aqueduct uses twelve stone arches 26m above the river and the cast-iron-lined waterway has a cobbled towpath on each side. At 247m, it is Britain's second longest of its kind. A wooded reach and another bridge follow. This wooded reach is at the northern end of the 69ha Muiravonside Country Park, with its gardens and parkland.

Near it is Causewayend, the site of the Almond Ironworks, bringing passengers from Glasgow and coal from the Slamannan coalfield to be taken to Edinburgh.

Looking across the Almond Aqueduct.

The Victoria moored by the old stables in Linlithgow basin.

The stables at Woodcockdale, now used by the Sea Scouts.

The dovecote near the basin in Linlithgow.

Linlithgow Palace, as seen from the canal.

Slamannan Dock is 46m square and had railway lines all round with turning points at the corners to rotate coal trucks.

The bridge on the stretch between Whitecross and Brightons is of Armco culvert, with rubbing strips inside for cruisers and narrowboats. Adjacent to this bridge is an open area of bank laid out with picnic tables for the public to enjoy the canal and its traffic. As the hilly scenery declines, there are views down to the oil refinery complex at Grangemouth.

Winding holes are surprisingly frequent, perhaps one every kilometre, and there are numerous remnants of loading platforms still present along the canal.

At Redding there is a swing bridge serving a chemical works on a small industrial estate.

Bridges are of sandstone, flat arches meeting vertical abutments almost at right angles. Each bridge has a

The Avon Aqueduct, the second longest of its kind.

projecting stringline just above the arch. It may be an optical illusion caused by the arch but each of these projecting courses appears to dip, making the whole bridge appear to sag.

A large young offenders' institution beside the canal has recently been extended. The line of an old drove road is crossed.

The Glen High Laughin' & Greetin' Bridge is named after the faces on the keystones, a happy contractor facing

A bank of buttercups at Brightons.

Limestone structures forming inside the tunnel.

Part of the young offenders institution at Polmont.

A swing bridge at Redding which has not swung recently.

Just inside the west portal a powerful water jet falls from the roof.

The Falkirk Wheel in the process of rotating.

the easy run to Edinburgh and the miserable face of the one needing to construct the 620m long Falkirk Tunnel under Glen Village. This was necessary as William Forbes of Callendar House would not let the canal pass through his grounds. The only rock tunnel in Scotland, it is wide and takes the towpath through. It is largely unlined so that water drips in places, pouring through in a jet just inside the west portal. Passage through is easy because the far end is clearly visible from the start. Lighting is placed through the tunnel, allowing the limestone features to be seen. Far from taking thousands of years to form there are curtains, 120mm straws and even longer stalactites that have formed since the tunnel was cut. Sloping rock beds on each side of the exit increase its grandeur.

Lady Kilmarnock had been the unwilling hostess of the Government army's General Hawley in 1746 but ensured that he had plenty of wine ahead of the Battle of Falkirk. Around the next corner on the left bank is the Battle of Falkirk monument. Here Bonnie Prince Charlie beat the English, who were facing driving rain and not expecting the Highlanders to make the first move.

On the right, across **Falkirk**, is the best view over the

The Falkirk Wheel is approached over a high aqueduct.

Firth of Forth to the hills on the Fife side. A 500m-extension added in 1823 shortened the distance passengers had to walk between the two canals.

The eleven-lock flight down to the Forth & Clyde Canal was lost under a road. After the Falkirk Wheel Top Locks the canal turns a corner and passes in the new 168m Roughcastle Tunnel under the Falkirk High to Queen Street railway and the Antonine Wall, known locally as Graham's Dyke. Rough Castle was a Roman fort. In place of the old lock flight is the iconic 35m Falkirk Wheel of 2002, the world's first rotating boat lift. With 35m arms, it is approached over an aqueduct and lowers boats in its gondolas to New Port Downie. The total load of 1,800t takes four minutes for the journey. A visitor centre describes this new landmark structure and offers facilities including an exhibition centre, café and bar at what is now Scotland's second-most visited attraction.

Golden Jubilee Lock takes the canal down to join the Forth & Clyde Canal.

Rhododendrons add a splash of colour at Bantaskin.

Distance
56km from Edinburgh to the Forth & Clyde Canal

Navigation Authority
British Waterways

Canal Society
Bridge 19-40 Canal Society www.bridge19-40.org.uk

OS 1:50,000 Sheets
65 Falkirk & Linlithgow
66 Edinburgh

58 Forth & Clyde Canal

Plans for a canal (originally known as the Great Canal by Smeaton, who wished to upstage the Languedoc Canal) to link the Forth and Clyde estuaries were considered as early as the reign of Charles II. It was eventually to be built from Grangemouth to Kirkintilloch in 1768–73 and extended to Bowling and Port Dundas in 1786–90. It was the world's first sea-to-sea route, allowing ships to avoid the north coast. The first major transport project in Scotland, it was finished with money confiscated from Jacobite estates after the failed 1745 rebellion. The line followed a possible former drainage path from the Holy Loch to the River Forth. It was designed for high-mast ships, no fixed bridges being used. The navvies were local men to ease unemployment. Many Roman artefacts were dug up.

The canal carried the *Vulcan*, Scotland's first iron boat, which was used for passenger transport from 1818. It was the birthplace of the Clyde Puffers and also had the first vehicle ferries, carrying loaded carts and railway wagons. A connection to Stirling was proposed in 1835

but never built. It carried coal and iron, passengers in express passenger boats and a series of Queen pleasure boats, the last of which ceased operations in 1939. Passenger traffic declined after the Glasgow–Edinburgh railway opened in 1842. The canal was bought by the Caledonian Railway in 1868. The canal carried freight until 1914 and the last vessel passed through in 1962. Closure came in 1963, possibly the most short-sighted of all our canal closures.

Interest was kept alive with regular trans-Scotland powered inflatable races and the Glasgow to Edinburgh canoe marathon. The whole line was restored by 2001 as the Millennium Link with Millennium Commission funding. Seventy per cent of the Scottish population live within an hour's drive of the link.

A canal running through the northern end of Glasgow's urban sprawl has the potential for being a derelict eyesore. The Forth & Clyde Canal isn't. It forms a linear park that is a credit to the city and is well used.

The old lock direct from the Clyde is now sealed off but Sealock gives access from the tidal harbour at Bowling, where boats are often laid up at the west end. The upper basin acts as moorings for a variety of small craft, including former fishing boats. The old custom house was built to receive the foreign ships that used to arrive when this was the major transport artery across Scotland. There is a disused, lattice, Caledonian Railway, swing railway bridge across the canal, a listed monument from which the local youth jump. Although the canal follows the bank of the River Clyde for 5km, it is not seen after the railway bridge. At this point, the river is still quite narrow, despite being used by commercial shipping. The banks are sandy. On the far side is a golf course next to an ex-servicemen's hospital.

This end of the canal includes several bridges of what are probably a unique design. They are double-bascule bridges, each half lifting separately. The operator needs to fit a windlass which operates a gear train leading to a toothed sector ring and a ratchet wheel, beautiful mechanisms from the Industrial Revolution. Working boat crews must have cursed the effort needed to raise them. The second lock at Bowling is flanked by railway arches, three of which are occupied by a café, a wood turner and a cycle shop.

The steep Kilpatrick Hills crowd in on the river and the canal is in close proximity to the A82, the Fort William to Glasgow railway and a cycle route along the former railway line.

Among the greenery is the first Roman fort site at the western end of the Antonine Wall. The Antonine Wall was built 61km from Bowling to Kinneil, circa 142, in the reign of Antoninus Pius. Constructed 20 years after Hadrian's Wall, it was less durable, had a 3.5m-high turf rampart on a 4.3m-wide stone base with a 12m-wide x 4.2m-deep ditch, a Military Way road and forts every 3.2km. It was intended to keep out northern tribes at the northerly

Tidal basin at Bowling, adjoining the estuary of the Clyde.

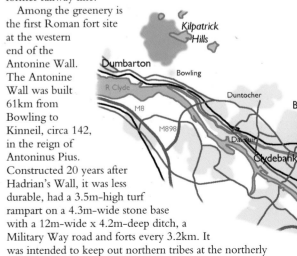

Bowling Lock and one of the bascule bridge mechanisms.

Interesting toilet at Bowling Lock.

The A814 Dumbarton Road leaves at a newsagent's shop. The canal continues on its way, largely oblivious of its surroundings, including the Duntocher Burn, which it crosses.

The line is mostly built-up as far as Bishopbriggs. There are fields alongside in places and plenty of greenery. The towpath is heavily used, particularly by cyclists, and it is surprisingly attractive, the water mostly in good condition. Swans' nests are frequent.

The A814 has to be crossed at the north-west end of the **Clydebank** Industrial Estate by the recent Dalmuir Drop Lock, Britain's only such lock, lowering boats 2.8m under the road and raising them again on the far side.

Ancient and modern bridges at Kilpatrick. Even the soaring Erskine Bridge was not safe from being bashed.

extremity of the Roman Empire. It was attacked twice so the Romans abandoned it and withdrew to Hadrian's Wall.

The Saltings, which lead down to the river, were a dump in the 1960s but have been developed as an ecology park.

The feature that cannot be ignored is the Erskine Bridge of 1967–71, one of Scotland's major bridges, carrying the A898 300m over the canal and the river with 55m of clearance to the canal. Despite the great height, the bridge was damaged on the underside by an oil rig being moved downriver from UIE's Clydebank yard to the Captain field in 1996. This soaring structure is, quite literally, on a different plane from the old canal buildings, which stand around its ankles. It is an unusually light structure at 690kg/m2. Ferry Road swing bridge gave approach to the older route across the Clyde.

The railway slips below the canal unnoticed at Dalmuir, the canal running past an industrial estate on the former Singer sewing machine factory site.

Above the buildings by the A8014 can be seen a Titan hammerhead crane, 46m long and 49m high, able to lift 150t. It is on the former site of UIE Shipbuilding, now occupied by Clydebank College. Live willow sculptures form tunnels on the south bank. The most remarkable reach on the canal follows. The architects have clearly given free reign to their ideas and the setting is completely surreal. Carefully landscaped gardens lead to the pedestrian Clydebank Shopping Centre with two footbridges across the canal. McMonagles fish and chip ship (no, that isn't a typographical error) is the world's first sail-through takeaway, an enormous boat-shaped restaurant, totally out of proportion to the canal, built on the bed of the cut.

Three kilometres to the south-west is Glasgow Airport's main runway so aircraft frequently approach over this

The fish and chip ship and two bridges it never passed.

section of canal. Unfortunately, that is not all that is in the air. From the direction of old factories on the right bank come some overpowering stenches and one can only feel sympathy for the local residents. On the opposite bank is an industrial estate with, among others, a Wimpy outlet and a Lidl supermarket.

Housing opens out beyond a bridge. Three locks at Whitecrook lead up past a newsagent's shop.

The dual carriageway A82 Great Western Road crosses. What was a major severance has been planted with trees which are now maturing, around which is mown lawn. On one side is a recreation area, then a small food shop. The other side has a children's playground, then tower blocks.

After the next flight of three locks at Cloberhill is a floating jetty, then the A739 **Bearsden** Road passes over.

A railway passes under before the Temple Locks, another runs along the left bank and a third formerly passed under the canal. In the centre of this railway triangle is the Lock 27 where the public house's benches are set out on both sides of the towpath so that it is necessary for those on land to pass between the tables of drinkers. Two large gas holders stand next to the canal at Kelvindale.

At the foot of the flight of five Maryhill Locks, ending the 48m climb from the River Clyde to the start of the 26km top level pound, is the Kelvin Aqueduct. Robert Whitworth's 120m-long stone structure, with its scalloped faces acting as lateral arches, passes 21m high over the River Kelvin and was the largest structure of its kind in Europe when built in 1790. It has an unusual number of masons' marks on the stonework. Contractor William Gibb was believed to have made a massive loss meeting his contract price. A tower block dominates one side of the flight. The Kelvin Dock liquor store overlooks the drydock. The dock built ships from 1789 to 1949, including the first Puffer and D-Day landing craft. This group of canal features is one of the finest in any urban setting in Britain.

Between two of the locks is a carefully fenced winding hole, around which the towpath is taken. The turning point serves a drydock. The canal then crosses over the A81 Maryhill Road.

The Glasgow Branch runs south-east from Stockingfield Junction and formerly connected with the Monkland Canal. A floating bridge used to carry the towpath across the junction but this is now routed under the aqueduct. Also gone is the huge Maryhill ironworks, its place taken by small businesses. From here there are views across to the centre of Glasgow with the city skyline. It has been a long time since the British Celtic *glas cau*, green hollow, applied.

A railway burrows under by Ruchill golf club and then the A879 Balmore Road passes over. Between these transport routes at Lambhill, a valiant effort has been made to smarten up some of the drab Glasgow housing.

The navigation runs past the marshland nature reserve at Possil Loch. Reedmace edges the canal. Electricity lines converge on a substation at **Bishopbriggs**.

Three golf courses are sited alongside this section of canal, the first opposite a colliery spoil tip. Caddermill, before the next bridge, is now a private house thoroughly smothered in ivy. The canal passes the church of 1820 at Cadder, built on the site of a 13th-century place of worship, complete with watch house and mortsafe, an iron coffin to deter body snatchers.

The canal cuts through the full-height Antonine Wall before another fort site and the A807 Torrance Road crossing.

At Glasgow Bridge, a slipway is almost opposite the Stables public house and moorings for several boats, including the canal society's craft. Adjacent is the high-level concrete A803 Glasgow Road Bridge on the line of what was once a steel swing bridge. The canal crosses the line of the Antonine Wall and passes a Roman fortlet, camp and fort.

Stockingfield Junction looking towards the Glasgow Branch and Glasgow skyline.

The bottom of the Maryhill flight of locks leads onto the splendid Kelvin Aqueduct with its scalloped faces.

The Antonine Wall, seen from outside the Roman Empire.

A new boathouse and a new bridge replacing an infilling embankment at Kirkintilloch.

The Forth & Clyde Canal Society's fleet moored at the Stables, Glasgow Road Bridge.

The Kilsyth Hills rise in the distance beyond Kilsyth.

Bar Hill at Twechar.

From its position to the south of the Campsie and Kilsyth Fells, **Kirkintilloch** was an agricultural and weaving town until the arrival of the Forth & Clyde Canal, when it became industrial with iron foundries as the pig-iron could be exported by canal. The Barony Chambers museum covers the social and industrial history of the area. The Auld Kirk of 1644, now museum premises, includes canal displays.

A new boathouse is built on a slipway used when Kirkintilloch was a shipbuilding centre from the 1860s to the Second World War, another source of Puffers. The 40-berth Southbank Marina opened in the former shipyard in 2008. The dominant building is the red sandstone church of St Mary of 1912–14, looking more impressive because of its position high above the canal. The A8006 viaduct follows.

The aqueduct of 1774 over Luggie Water was an engineering milestone, the first major canal aqueduct in Scotland and the prototype for the Kelvin Aqueduct. Unlike the latter, the navigation channel width is not reduced over the aqueduct. The 38m long x 27m wide structure is 15m high and has horizontal side arches for support. The builders of the Campsie Branch Railway, now a footpath, found there was sufficient room for them to get a line through under the same arch by building a twin-arched culvert for the river and running the railway across this.

The iron bridge dates from 1938 on the site of an earlier opening wooden bridge.

Another Roman fort built beside the Antonine Wall attempted to control the unruly at Kirkintilloch who were named from the Celtic Caerpentulach, fort at the end of the ridge.

A bing accompanies St Flanan's colliery on the right bank. Shirva Aqueduct carries the canal over the Shirva Burn. Near this is a ruined stable block.

Twechar was a mining village. The pivot for a swing railway bridge stands on the right bank of the canal. The railway company were allowed to build their bridge across the canal if they agreed to transport a large proportion of the coal by canal. Hence, this colliery was still moving coal by water when most other collieries had gone over to using trains. There is another Roman fort site beyond at Bar Hill.

A meandering section of canal at **Kilsyth** is followed by a swing bridge. Near Kilsyth a soldier on horseback was found preserved in a bog.

Croy Hill has a disused quarry and Roman fort site. Red cliffs surround a picnic site on the opposite side of the canal. Between them is a spur that is a marina, served by a hotel. This section of canal was being considered for an oblique crossing if this line had been chosen for an extension of the M80.

There are a couple of significant widenings at the start of a reach that runs dead straight and very broad for 4km with extensive views. A feeder discharges water from the River Kelvin and powerlines cross. Tufted ducks and herons frequent this reach. To the north is a motte at the end of a small loch that was the site of a battle in 1645, Kilsyth being associated with the violent struggles of the Covenanters.

A short spur leads off in the direction of Cumbernauld and its airport.

At Banknock, the canal leaves the River Kelvin and follows the right bank of Bonny Water.

The new Boathouse, built by BW's Waterside Pub Partnership at Auchinstarry marina in 2008, claims to be the country's first sustainable public house and includes geothermal heating.

To the south at Castlecary is a square tower, burned down by the Highlanders in 1715, and a Roman fort site.

Wyndford Lock begins the descent to the River Forth. The A80 is a very busy trunk road and not the obvious place for a picnic site. Severance of the canal here in 1963 terminated through traffic on the canal. Two Castlecary Locks follow.

The trees overlooking the second lock contain a rookery. Frogs and fish frequent the canal. Underwood Lockhouse is now a steakhouse. A Roman fortlet site is then passed.

The canal is crossed by a bridge near the M&M Rehearsal Studios. **Bonnybridge** Aqueduct carries it over a stream with a cobbled invert so that it could also be used by farm carts. The canal leaves Bonny Water. Gorse and beech trees overlook the canal as it departs from Bonnybridge.

Several powerlines cross from a substation as the Carlisle to Perth railway runs alongside before passing under the canal.

The Roman fort at Rough Castle is one of the best preserved. Charred wood suggests that the semi-circular mound was a signalling platform.

The unique giant **Falkirk** Wheel lifts boats 34m to an extension of the Union Canal's line in an area dominated by opencast mining spoil.

This reach is a broad waterway flanked by reedmace and willow trees.

Barr's works, next to the Union Inn, discharges into the canal a liquid which is definitely not Irn Bru.

The Union Canal formerly connected where some low white railings are located at Port Downie in Camelon. This wharf, above the top lock of the Falkirk Flight, was one of the most important points on the canal, not only as the junction but also because the line is only lightly locked between here and Glasgow. Both canals were served by the Georgian Union Inn. The world's first steam boat was launched onto the canal here in 1789 but had problems with its paddle wheels. The steam boat *Charlotte Dundas* conducted further trials from the Glasgow end in 1802, towing two 70t barges 31km. Experiments were abandoned for fear of the wash damaging the banks. From 1831 to the late 1840s swift boats operated a passenger service that took three and a half hours for the 40km. The boats travelled at 16km/h, pulled by horses that were changed at frequent intervals. Bringing the communications up-to-date, the towpath between Edinburgh and Glasgow contains fibre optics cables as the first such British canal communications route.

Falkirk takes its name from the Middle English *faw kirk*, mottled church. Locks are now frequent. The Canal Inn is located below the top lock of the flight. Landscaping, mown lawns and maturing trees make it a

Rosebank Distillery in the centre of Falkirk and one of the current breed of canal bridges.

pleasant area with a children's playground and bowling greens adding to the welcoming and open feel. A Beefeater restaurant, with canal artefacts, is in a former bonded warehouse. The former Rosebank Distillery, retaining its copper stills, is in the middle of Falkirk on the opposite corner of the A803, a quite busy road. In the spring there is a spread of purple, gold and white crocuses across a wide area of public lawn. The locks here have iron hooks instead of mooring bollards and the paddle gear is operated by hand spikes.

Schools and the Fratelli restaurant by the towpath precede the B902 at Bainsford Bridge in an area that formerly undertook iron casting. In 1298 Sir Brian de Jay was killed here by Wallace's Highlanders after the Battle of Falkirk.

The canal, its banks neatly mown, runs above the road past the Bankside Industrial Estate. Beyond the road are a scrapyard, an electricity substation and a timber yard.

The last 2km of the canal's original line have been filled in and all that remains is a kink in the bank of the River Carron and the main road through Glensburgh, which runs along the former line of the canal. The canal's line was severed by the M9 and A905.

The village, established where the Forth & Clyde Canal joined the River Carron, was called Sealock, subsequently changed to Grangemouth after the Grange Burn, which runs close by, now a town with a major oil refinery and chemical complex.

From Ladymill Weir, the Carron Cut runs north as a short branch to join the tidal River Carron beyond a new Sea Lock just above the M9. One benefit of this route is that Skinflats is the best place to see winter wildfowl and waders on the Forth.

This area between Falkirk and Grangemouth is to become the Helix, with 750,000 trees, paths, cycleways and public art. Displacing the water from the lock will be two 35m high, rocking, silver, kelpie heads, the world's largest equine statues. What giant shining horses' heads rising up beside the motorway will do for the accident rate remains to be seen. The canal will then be extended downstream on the east side of the River Carron for 2km.

Models of the kelpie heads to be placed next to the M9.

Distance
55km from the River Clyde to the River Carron

Navigation Authority
British Waterways

Canal Society
Forth & Clyde Canal Society www.forth andclyde.org.uk

OS 1:50,000 Sheets
64 Glasgow
65 Falkirk & Linlithgow

The low fence marks the former end of the Union Canal.

59 Crinan Canal

The Crinan Canal is isolated from the canal network. It is unlike virtually anything else in Britain. It was built to avoid craft having to face heavy seas off the Mull of Kintyre and cuts 137km off the journey north from the Clyde. In so doing, it provided a market for people in the Western Isles and, in turn, was used to supply them with salt and coal. Even now, it is surprising just how many work vessels on the west coast are still built to fit the canal's lock dimensions.

It was built by a private company launched by the Duke of Argyll. Work began in 1794 under John Rennie but there was inadequate supervision and skilled labour was in short supply in this remote part of Argyll & Bute. Ground conditions were difficult, with alternating sections of hard whinstone and peat moss. Thomas Telford and James Watt were later involved in the engineering. It was opened in 1801 with the help of Government loans and completed in 1809. Despite carrying heavy traffic, the anticipated profits were never achieved. The year's tally for 1854 was 33,000 passengers, 27,000 sheep and 2,000 cattle. By 1906 it was carrying mainly goods but the situation has switched back with the bulk of the traffic these days being yachts, cruisers and fishing boats. There are no inland craft. In July, Glasgow holidaymakers traditionally thronged the canal on steamers.

The Maggie, the 1953 Ealing comedy about an old Puffer, was partly filmed on the canal. Clyde Puffers may still be found moored in the basins, Puffers being built to the canal lock dimensions, taking coal to the west coast and bringing back whisky and other produce. An early user was the *Comet*, the first steam craft to go to sea, wrecked just west of the canal in 1820. In 1963 it carried a hovercraft.

The canal has been called the most beautiful shortcut in Britain. Surely nowhere else on the British canal system can match the sheer breathtaking splendour of the canal basin at Crinan, a basin that has not been ruined by commercialism as might have happened in a more accessible spot. Just a coffee shop and a hotel famous for its seafood have been added to the usual canal basin facilities.

Crinan Harbour lies on the open Loch Crinan. The sea lock at Crinan, as at Ardrishaig, opens at all states of the tide but other locks are only operated from 8.30am to noon and 12.30pm to 4.30pm, Mondays to Saturdays.

Crinan only acquired its name when the canal was opened. Until then it had been Portree, king's port. There are twin sea locks, only one of which is in use, with moorings in the basin for a small number of craft. The basin has a wonderful mixture of canal and nautical influences. The bridge keeper's house was named Puddler's Cottage as it was his job to seal leaks in the canal with puddled clay. Lobster pots are stacked up in a garden and a small red-and-white lighthouse acts as a beacon. Beyond all this is a magnificent panorama of islands across the Sound of Jura, Jura, Scarba and Luing. It is a basin worth visiting for the view alone.

Above lock 14, a low concrete platform projects into the canal. From here to Crinan Bridge the canal is at its narrowest, cut out of the granite of the hillside. One-way traffic is controlled by telephone. Blind bends and heavy usage by large craft result in an audible warning system being advised. However, between the rock outcrops there are several inlets, lined across with buoys to keep craft from entering but forming useful boltholes in an emergency for small craft. Normally, they are the preserve of wildlife such as heron.

Soon after lock 14 there is the first overflow weir to Loch Crinan, alongside which the canal runs.

Crinan Bridge is the end of the one-way section and the start of the B841. This connects by passenger ferry to Crinan Ferry. The canal is closely followed by roads throughout its entire length. Over most of its length it is edged on the south side by the flanks of Knapdale Forest: rich oakwoods, birches, larches and other firs with a copious selection of mosses, lichens, ferns and bracken. Deer come down from the forest in the winter and wildcats have been seen. Young alders line the banks and there are sections of reeds and gorse from time to time.

Steadily the canal widens to the dimensions of a broad river. Soon after an old-style gypsy caravan, the canal reaches its largest inlet, the sheltered Bellanoch Bay with pontoon moorings for sleek yachts. Here and further on, the instability of the hillsides is illustrated by

The canal basin at Crinan.

The terminal basin at Crinan, possibly the most beautiful spot on the British canal system.

remedial work after slips have taken place. For a while the banks drop enough to give views over the adjacent River Add as it winds across the Mòine Mhòr, one of the few remaining raised mires in Scotland, with rare plants and animals. On the far side is the outcrop of Dunadd, for hundreds of years the capital of the Scottish kingdom until a move to Scone under King Kenneth MacAlpine.

Bellanoch Bridge crosses both canal and river at Island-add before the canal leaves the river. The cast-iron bridge is cranked open by hand. Wildlife may include swans and midges.

The canal has many feeders from lochans in the hills of Knapdale Forest. The one from Lochan Dùin feeds in at low level before the canal begins the climb to its 21m summit level through the five Dunardry Locks. There is a

Only a track separates the canal from Loch Crinan at first.

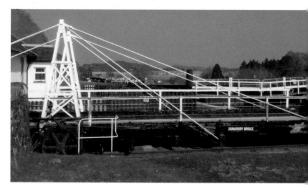

Dunardry Bridge is a cantilevered footbridge on rails.

Bellanoch Bay, the widest point on the canal, at the foot of Knapdale Forest.

294

The basin at Ardrishaig. The Clyde Puffer VIC 27 repainted for the BBC as Vital Spark, the star of The Maggie.

high landing platform before each lock. The steep banks grow primroses and daffodils in the spring and wild strawberries later in the year. The locks are separated by large ponds. Lock 11 is crossed by a rolling bridge that is wound open by hand; it was installed in 1900 to replace a swing bridge with unstable foundations.

The summit pound is connected to Loch a Bharain, which has two islets bearing pine trees. It also has the remains of a boathouse which housed the steamer *Linnet*. Almost opposite is the feeder from Daill Loch reservoir, which is flanked by a forest walk. At the far end of the pound there is a rather larger feeder, the sluice-controlled Carndubh Burn, which empties over a weir and under a stone bridge arch, bringing water from Loch na Bric, Loch an Add, Loch na Faoilinn, Cam Loch, Loch Clachaig, Dubh Loch and Gleann Loch reservoirs. Wildlife on this reach, from the mink to the wren, is noticeably tamer than in other parts of the country. Norwegian beavers are to be re-introduced in 2009, despite concerns that they might spread and damage the canal.

The first four locks down are at Cairnbaan; the fourth, just below the B841 swing bridge, is overlooked by the Cairnbaan Hotel. Near the hotel is Leacan nam Sluagh, the stone of hosting, with Bronze Age cup and ring markings. Carn Ban, the white cairn, is on the right after the final lock of the flight. It is a Bronze Age burial mound.

On the far side of the A816, near a fort at Achnabreck, there are more cup-and-ring-marked rocks plus petals, stars and spirals, the most extensive collection of prehistoric rock art in Scotland. Visible from the canal is Stane Alane, a 2.4m standing stone by the wall of a cemetery on the line of the older Lochgilphead coaching road. Just before the power lines at Badden there are signs of prehistoric workings. The present line of the A816 was only able to be used when a former loch was drained for construction of the canal.

A line of rhododendrons on the left bank leads to **Lochgilphead**. Built on the Dippin Burn, the town dates back to medieval times but owes its present size to the canal traffic, being mostly 19th-century and the only significant town on the canal.

The A83 is all that separates the canal from the west side of Loch Gilp. There was formerly a wharf at Oakfield, now used for boat repairs. Miller's Bridge of 1877 was named after a bridge keeper in the mid 19th Century, who also ran a coal business at the wharf. The manual swing mechanism is located on the west bank so that the owner of Oakfield House could decide when he wanted privacy.

A water waster of 1895 in a canalside building involves a pair of large buckets on a rocking beam, one of which plugs a drain hole. The device is ingenious but it is not clear how it is better than a simple spill weir.

Above a canalside distillery site was the Robber's Den, on the site of an Iron Age fortress. It was occupied by a MacVicar youth, who raised the alarm when the MacAlisters raided the cattle of the MacIvors, precipitating the Battle of Carse. The MacIvors sought vengeance. When they finally trapped him, they set fire to his hideout and then forced him to jump from the crag onto their waiting spears below.

The Royal Hotel was renamed after Queen Victoria's visit in 1847. Before this it was the Poltalloch Posting House.

Ardrishaig was largely created by the canal, near the ancient chapel of Kilduskland, which was reputedly haunted by its incumbents.

The four final locks come as a flight. The banks are steep.

Stances Inn took its name from the wooden stances on which fishing nets were dried. Today this sort of activity is gone. There is a small oil depot and the British Waterways workshops overlooking the cramped canal basin. More a seaport than a canal terminus, it has a lighthouse on the 1793 breakwater that established the port. The breakwater was extended in 1932, when a new sea lock was built, the old one being visible beside the opening bridge carrying the A83. There are views across Loch Gilp and Loch Fyne and on a clear day it is possible to see as far as Arran, a canal seascape outclassed only by that at Crinan.

Distance
14km from Loch
Crinan to Loch Gilp

Navigation Authority
British Waterways

OS 1:50,000 Sheet
55 Lochgilphead
& Loch Awe

Cairnbaan Lock and the Cairnbaan Hotel.

The Crinan Canal leads out onto Loch Gilp.

60 Caledonian Canal

Avoiding Napoleonic privateers

The Caledonian Canal, running north-east right across the Highland Region from the head of Loch Linnhe to the Moray Firth, is unusual for several reasons. Our most northerly operational canal, it is isolated from the rest of the canal system, has large dimensions, incorporates three major lochs and runs through dramatic scenery. It uses the line of Glen Albyn (Glen Mor or the Great Glen), a tear fault line dating from the Caledonian orogenesis, partially infilled with Old Red Sandstone, unlike the thrust faults found elsewhere.

Corpach Basin with Ben Nevis in the background.

The route was considered by Captain Burt in 1726 and it was surveyed by James Watt in 1773. Designed by William Jessop, it was expected to take seven years to build. The three lochs account for two thirds of its length. This was one of the greatest projects of Thomas Telford, who began the construction in 1803 and did not finish it until 1822, deepening from 4.6m to 5.2m then taking place before 1847. Funding was by the Treasury, Britain's first state-funded transport project. It had been planned to serve two political purposes. It was intended to get potential emigrants used to labouring instead of crofting at the time of the Highland Clearances but was not entirely successful because of absenteeism at harvest time or when the fishing was good. It was also to give a safe route for naval vessels away from the risk of attack by French privateers during the Napoleonic Wars, saving 560km on the sea route around the north of Scotland. The risk of attack had declined by the time it was finished.

It reached its most successful point in the 1880s, carrying fishing vessels, grain, salt and timber and running a regular passenger service. Gradually ships outgrew the locks and steamships became more able to round the north of Scotland so use declined. Much use was made of it during the First World War. The main users today are pleasure boats, yachts, fishing vessels and small coasters, with a high proportion of foreign craft. It loses money but is important to the Highland economy as a whole, especially to fishing boats. A 10km/h speed limit operates throughout.

Glen Albyn is followed by the prevailing wind and so the canal is very exposed. This is particularly so on the lochs, especially at the downwind ends, where surfing conditions may prevail, notably on Loch Ness. The alternative to the wind is the ubiquitous midge.

The canal is entered at Corpach from the head of Loch Linnhe, just east of Wiggins Teape's paper and pulp mill.

Corpach Sea Loch is the first of 29 locks on the canal, operated during normal working hours from Monday to Saturday, this one only being used within four hours of high water. When built, these were the world's largest locks, mostly 55 x 12 x 6.1m deep. Entry is made difficult for larger vessels by a flow up to 5km/h across the entrance on spring tides. The locks are built as staircases wherever possible and include two of the three longest staircases in Britain. They have horizontal capstan wheels with four ports to take posts for use as windlasses. Drum windlasses are also present by some locks, used until mechanisation in the 1960s. Sea Lock has the first of the lighthouses, small white cylindrical structures with black conical roofs. Indeed, the buildings on the canal are generally black and white, standing out smartly in the rugged scenery.

Corpach Tidal Basin has been enlarged to take 1,000t vessels with commercial wharfs for the Western Isles. The occasional Puffer may be seen here.

Corpach village has a Treasures of the Earth geological exhibition.

The staircase of two locks beyond the tidal basin takes the canal up to moorings and its first reach, one of the most dramatic in Britain. Directly ahead is Ben Nevis, the highest peak in the British Isles at 1,344m, 9km away with its summit just 7km from the sea although that summit is frequently shrouded in cloud. Fort William has one of the highest levels of rainfall in Britain, with an average of 239 days of precipitation spread right through the year.

A line of beech trees provide shelter from the prevailing wind as the canal curves round to the north-west at Caol. The water is wide and clear, usually edged with broken rock, which is covered with mesh in places.

The first two of the canal's 11 swing bridges are met

Quite large ships use the canal. This Swedish vessel is approaching Corpach top lock.

at Caol. A manually swung railway bridge carries the West Highland line from Fort William to Mallaig with steam trains and observation cars in operation in the summer; the railway line and bridge are now also associated with Harry Potter's Hogwarts Express. It is followed immediately by the A830, the current Road to the Isles.

Banavie Locks are the longest and widest staircase in Britain with eight locks taking the canal up 20m; walls collapsed here in 1829 and 1839. A rebuild was also required in 1929 when a vessel crashed through the top lock gates and fell into the lock below, the sudden water loading damaging the chamber walls. Indeed, the walls were built as the longest pieces of masonry on any canal, each 460m. The flight passes the Moorings Hotel & Restaurant, a craft shop, a post office and a quarry. From the top of Neptune's Staircase, as the flight is also known, the view back down Loch Linnhe is stupendous. Among those able to appreciate it are lock keepers whose houses have bay windows for better views down the flight, and those waiting to descend, as boats cannot pass on the flight. The 1845 proposal of a passing place in the centre was abandoned as impractical. To the south-east are penstocks down the side of Ben Nevis, to generate hydroelectric power for British Alcan's aluminium smelter. A different kind of power was invested in Inverlochy Castle, which rises above the trees to the east, almost as high as the low flying jets streaking up the glen.

For the next 10km the Western Reach is mostly sheltered by trees, to the extent that the B8004 on one side and the River Lochy on the other are not seen at all. The Upper Banavie aqueduct collapsed in 1843 and Shangan and Loy aqueducts were rebuilt in consequence.

Three sluices at Strone act as overflow weirs 1.2m wide and 3m high, the water falling 2.7m to the River Lochy. The turbulence of this resulted in Telford's becoming quite lyrical. Just before the canal passes over the River Loy, there are forest walks in a fragment of the old Caledonian pine forest. The crossings of rivers are usually only marked by groups of stone arches on the left bank or sluices on the right although aqueducts usually have side arches for farm access, not obvious from above. This one has three parallel 76m tunnels, the centre one 7.6m wide and the outer ones 3m wide, all of the order of 4m high. One is for access.

Moy swing bridge is the last remaining example of Telford's unusual cast-iron design for the canal. The bridge takes the form of two cantilevered halves, each 6m long x 3m wide, swung by hand and counterbalanced by containers of rocks. There is no land access between the two sides for the lock keeper if both sides need opening, so a boat has to be rowed across.

Allt Coire Chraoibhe joins under a set of stone arches on the left to provide more water. Beyond the River Lochy, boreholes extract additional water supplies for Fort William.

The two locks at Gairlochy are separated by the B8004, a telephone box and the Stable Tea Room.

From here, the canal opens out into Loch Lochy, 15km long and 162m deep, the very essence of Glen Albyn as it cuts dead straight between shores rising steeply for hundreds of metres. Canal construction included raising the top water level of the loch by 3.7m. As the canal enters the loch in the natural course of the

Looking down Neptune's Staircase towards Fort William and Loch Linnhe.

Moy swing bridge, the last remaining of Telford's design for this canal.

Looking up Loch Lochy on a wild day with large waves despite the relatively small size of the loch.

The Well of the Seven Heads by Loch Oich.

Invergarry Castle overlooks Loch Oich.

river, the River Lochy leaves in a new Mucomer Cut to pick up the River Spean at Bridge of Mucomir. A hydro-electric power station was added at the confluence with the Spean in the 1960s.

Because the B8004 was the line of a drove road, Telford and John Simpson had to construct a bridge over the new cut to takes herds of cattle. The sandstone masonry arch bridge includes mock arrow loops in the abutments and piers.

The only variation on the narrow loch comes near the beginning where the River Arkaig enters on the left at Bunarkaig from Loch Arkaig, opposite an obelisk in the trees.

Powerlines, the A82, the line of a dismantled railway and General Wade's Military Road all arrive beside Loch Lochy, about a third of the way up its length, via Glen Gloy. The River Gloy doubles back after incising a deep cleft parallel to Loch Lochy. The river looks as if it flowed into Loch Lochy some 3km further south-west.

Clunes Forest, on the north-west bank, becomes South Laggan Forest on both banks, providing conifers for Corpach paper mill. Above is the treeless Glengarry Forest, including 887m Sean Mheall, 910m Ben Tee and 935m Sron a' Choire Ghairbh, all parts of Glengarry's Bowling Green. Birdlife by the loch varies from sandpipers to raptors.

Wooden chalets precede the graveyard beside the Kilfinnan Burn, entering the Ceann Loch, the mausoleum of the Chiefs of Glengarry. In 1544, the Frasers fought the Macdonnels, Macdonalds and Camerons here in the Battle of the Shirts, so called because the clansmen stripped off in the sunshine before the combat commenced.

The two Laggan Locks take the canal up to its summit level and into the Laggan Avenue reach, a section of canal thickly lined with pines. This 2km reach, crossed by powerlines, is the only section of canal without a natural watercourse alongside.

Beyond the Laggan swing bridge, carrying the A82, the canal enters Loch Oich, the most attractive of the lochs used by the canal. The highest point on the canal at 32m and 6km long with several islands, it is only 50m deep at its deepest point, generally much less, so Jessops designed the earliest type of continuous bucket steam dredger to deepen the channel.

Chalets mark the Great Glen Water Park, which hires out boats and offers meals in the licensed restaurant long after other local restaurants have closed down for the evening.

Swans and seagulls, a picnic area and a grocery shop surround Tobar nan Ceann, the Well of Seven Heads. The well is topped by a pyramidal monument inscribed in English, French, Gaelic and Latin, the former towards the loch, surmounted by a hand holding seven heads. It was erected in 1812 by Macdonnel of Glengarry to recall the washing of the heads of seven members of the family of the 11th Chief Macdonnel, killed in the 1660s as a reprisal for their undertaking of the Keppoch Murders of their two other brothers.

Invergarry Castle was once the home of the Macdonnel Chiefs of Glengarry but it was burned by the Duke of Cumberland in 1746 after Culloden. Colonel Alexander Ranaldson MacDonnell opposed construction of the canal because he feared it would compromise his privacy, subsequently dying of injuries sustained when the *Stirling Castle* grounded in Loch Linnhe in 1828, six years after the canal was completed. Now the L-shaped tower ruins gaze over the Glengarry Castle Hotel, itself having interesting architecture. Islands are studded around the loch, thick with rhododendrons and attractive to swooping swallows and motionless herons. Unlike most other British canals, the Caledonian Canal has limitless water supplies. At Invergarry it is the River Garry that enters from Loch Garry past fragments of the old Caledonian pine forest. Beyond a boathouse, the Calder Burn enters at Aberchalder, opposite a rocky outcrop with heather and pines. Such scenery could hardly be less canal-like.

The River Oich leaves at the start of the Coiltry Reach.

The A82 crosses a swing bridge. Beside it are the

Loch Oich is dotted with islands which provide refuge for a variety of birds. Buoys mark the fairway for larger craft.

The shore of Loch Oich near Bridge of Oich, a beautiful spot which could hardly be less like the bank of a canal.

The lock flight is the central feature of Fort Augustus.

The Torr Dhùin cliff at the edge of Inchnacardoch Forest. The River Oich runs at the foot of the cliff.

abutments of a former swing bridge and a rusty suspension bridge across the River Oich.

Cullochy Lock starts the descent to the Moray Firth and has been laid out with a rockery, all the plants neatly labelled.

At the end of the next straight, a spur passes off to the right among the birches and alders. Secluded and with moss-covered banks, it looks the ideal spot to camp but the midges are waiting.

Canal and river together pass the foot of Torr Dhùin cliff with its walks on the edge of Inchnacardoch Forest. The canal widens into a lagoon before Kyltra Lock and a weir permits overflow into the river, the only time the river is seen from the canal.

Kyltra Lock also has a rockery, this time with a stream cascading through the centre.

The Fort Augustus Reach includes a lagoon crossed by powerlines and flanked by a forest walk. Some of the larch trees opposite the golf course are fire blackened.

The village of Kilcumin was established by St Chumein, a follower of St Columba, who set up a church. After the 1715 rebellion, a Hanoverian fort was built to help control the Highlands from this strategically important point, remaining in military use until 1854. It was sold in 1867 to the 14th Lord Lovat, whose son presented it to a Catholic Benedictine order. St Benedict's Abbey boys' public school was established in it at the end of the 19th Century. The school and clocktower were designed by Joseph Hansom. The Abbey Church was largely designed by Pugin in English Gothic style around the same time. The village was renamed **Fort Augustus** after William Augustus, the Duke of Cumberland – probably the most hated Englishman in Scottish history – following Culloden. (Fort William was named after William III.)

The central feature today is the five-lock staircase 12m down to the A82 swing bridge and Loch Ness, the grass neatly mown all around although there is little to stop a customer staggering out of the Lock Inn in

the dark after a dram too many and falling into one of the large lock chambers. Construction had to be undertaken with the stone at the base laid on moss to prevent sand being blown upwards, the base of the bottom lock being 7.3m below Loch Ness water level. Robert Southey was particularly impressed with the scale of the construction when he visited it with Telford. In 1857 the bottom lock wall failed. The flight is faced by a restaurant, several shops and the old canal workshop at the foot of the flight. The Caledonian Canal Visitor Centre is in a former lock keeper's cottage. The Clansman Centre in the village includes a copy of the inside of a turf house. Behind the shops on the south side is the Lovat Arms Hotel, where the non-resident canal user may take a bath.

Landing pontoons are adjacent to the entrance to Loch Ness. Loch Ness is Scotland's most famous loch. It is 35km long and has a catchment of 1,777km^2. It contains the greatest volume of freshwater in Britain, is deeper than the North Sea at 250m and has the greatest mean depth of any British lake. The shores are steep with underwater cliffs and the sea rises quickly with fierce squalls, 1.5m waves not being unusual. Thus, it needs to be treated with great respect in bad weather, particularly when the wind is the prevailing southwesterly that blows straight up the loch. The wave action and steep bed and sides mean that wildlife is not as extensive as on the smaller lochs but the water is clearer and it has never been known to freeze, the temperature remaining fairly constant at 5–7°C, especially at the bottom. What it lacks in wildlife, it makes up for in scenery, particularly in fine weather. Some of the cloud effects can produce majestic or threatening skies. In 1970 it was used for filming part of the *Private Life of Sherlock Holmes*.

As the loch leaves Fort Augustus it is rejoined by the River Oich on the left. The River Tarff flows in on the right.

There is a crannog on Cherry Island and Roman material has been found. There is also a Cherry Island Walk nature trail.

The Allt Doe descends over a waterfall to enter the loch on the right side. A sheet of scree slides down from 555m Beinn a' Bhacaidh, facing across to 607m high Burach. The latter is surrounded by Portclair Forest, covered with deciduous trees on its lower slopes, interwoven with bluebells in the spring. Portclair itself is the smallest of hamlets.

Caravans at Rubha Bàn precede Invermoriston. General Wade's bridge over the River Moriston was built after the 1715 Jacobite uprising. The Seven Men of Moriston sheltered Prince Charlie for a month in 1746. Jacobite Roderick MacKenzie, who looked like the Prince, was shot by Cumberland's men in Glen Moriston, allowing the Prince to escape. Opposite, the bed of Loch Ness reaches its steepest, dropping 199m in 110m.

Above Alltsigh, tucked away in Creag-nan-Eun Forest,

Waterfall near Alltsigh on Loch Ness.

the glen sides reach their highest with Meall Fuar-mhonaidh at 696m.

Conspicuous on the east side is an aerial before Foyers. Burns recorded in verse the Fall of Foyers with its 9m and 27m drops on the River Foyers. In 1895 it was used to drive the first large hydroelectric power station in Britain for the British Aluminium Company, now converted to a 300MW pumped storage scheme, using Loch Mhor, 179m above Loch Ness. An aluminium smelter was added to use the power at source, subsequently closing in the late 1960s. A woodland walk is echoed by a forest trail at Inverfarigaig. First comes Easter Boleskine with its pre-1777 church and many burial enclosures. The Farigaig Forest trail has an interpretation centre in a converted stone stable, showing the development of the forest, and has a picnic area and toilets. Near the River Farigaig there are the Inverfarigaig vitrified forts. There is another fort site on the west side of the loch. Goosanders might be seen on the loch.

The Cobb Memorial is dedicated to John Cobb, who was killed in 1952 while trying to set a new world water speed record on the loch in his jet speedboat, *Crusader*, which exploded.

Another vitrified fort at Strone Point was used as the site for the Grants' Urquhart Castle. At one time it was one of the largest castles in Scotland, commanding the junction of Glen Urquhart with Glen Albyn. Mostly post-1509, it has additions by John Grant of Freuchie, who received it as a gift from James IV. A royal castle in the days of William the Lion, it was fought over in the Scottish Wars of Independence in the 13th Century, Edward I capturing it and then losing it to Robert the Bruce. It was sacked in 1689 and, in 1692, blown up by the Royalists to keep it out of Jacobite hands. Despite its ruined state, it is still a popular tourist attraction and at times it is unusual to hear English being spoken because there are so many foreign visitors. Toilets are near the water.

It is hard to think of Loch Ness without also thinking of Nessie, the Loch Ness monster. Traditionally shown as a huge serpent with a series of coils looping out of the water, recent drawings suggest a more spherical body with a long neck and small head, perhaps also with four flippers. Recorded sightings go back to a kelpie seen by St Columba in 565. St Adamnan, monk of St Columba and Bishop of Iona, and the monks of Fort Augustus Abbey, the successors of St Columba, also recorded sightings.

Until 1935, General Wade's Military Road up the east side of the glen was the route through but the opening of the A82 brought tourists with many more sightings and photographs. The Japanese have even searched the depths of the loch with a mini submarine. As yet, conclusive proof has not been forthcoming. Allocation of the scientific name *Nessiteras rhomboideus* seems premature.

Perhaps some of the sightings can be put down to the waves that form momentarily on the loch which can look very black when seen against the light, as is the case for people on the A82.

The Loch Ness 2000 and Original Loch Ness Monster visitor centres in Drumnadrochit put the case for the monster. The findings of a 20-year scientific study said there are only 20–30t of fish in the loch, so they would not be able to support more than 2–3t of monsters. The surface temperature only rises above 12°C for four months of the year, so the suggestion is that people are probably seeing an occasional cold water fish migrating in from the sea. The preferred option is the sturgeon. It can grow to over 4m long and weigh in at 100kg, possibly getting lost from time to time while trying to find suitable breeding waters.

Drumnadrochit lies between the River Coiltie and the River Enrick. Below is an alder swamp, where beaver were present until at least the 15th Century.

Picnic sites line the east shore. The west shore has the Clansman hotel with its own harbour. The dipper is present along the shoreline although this seems totally unlike its usual territory of secluded rapids.

The west shore still rises to 501m at scree-coated Carn a'Bhodaich but the east side of the loch is now lower. The width halves abruptly at Dores with a ridge bearing the Kinchyle of Dores stone circle and then Aldourie Castle, nearly at the end of the loch at Bona Ferry.

Loch Dochfour leads gently back into canal dimensions after passing Alban Water, which is landlocked between the two lochs as Darroch Islands are dotted about.

Dochfour Gardens are seen at their best from Loch Dochfour and surely few other canalside gardens can equal them. Various shades of pink rhododendrons are set against yellow gorse and broom and a variety of magnificent trees on 6ha of terraces. Perhaps the daffodils, water garden, yew topiary and soft fruit are lost from this distance but the view is enhanced by the gardens, which come as a complete contrast with Loch Ness. On the east bank, oakwoods are brightened by wild rhododendron bushes. The wrecks of ferry and fishing vessels rot gently and herons and divers go quietly about their businesses.

The River Ness leaves to the right over the long sloping Dochfour Weir and then is followed by the canal for some distance. After Dochgarroch Lock, two sets of powerlines cross. A stone circle and Ness Castle are before and after them, respectively, but hidden in the trees on the far side of the river. Trees also hide Dunain Hill with its aerial and the fort site before the next set of powerlines. A gravel pit, which might be expected to be hidden, is conspicuous high on the side of a hill. Opposite it and across the river as it parts company from the canal are Holm Mills, home of James Pringle. Their former

Urquhart Castle is one of Scotland's most visited castles, now and historically.

Dochfour Gardens rise above the water as it tapers back to canal dimensions.

Wooden hulls quietly rot along the side of Loch Dochfour.

The Scot II moored in Muirtown Basin.

spinning and weaving factory for tweeds and tartans for over two centuries, it is now their retail and Clan Tartan Centre. Founded in 1798, it is the oldest woollen factory in northern Scotland, water- and steam-powered. Its cast-iron columns do not line up with the timber roof trusses, giving a decidedly rustic appearance.

Arrival in **Inverness**, Gaelic for mouth of the noisy river, is announced by the Inverness Rowing Club on the canal.

The tight Torvean bend, after Torvean Hill (where a navvy found a solid silver chain in 1808, now in the National Museum of Antiquities in Edinburgh), is dangerous for larger craft as they pass moorings.

Whin Park has the relocated 30m-span Dredge-style General's Well suspension footbridge. This now supports a people-carrying model railway. Tomnahurich Swing Bridge or Bught Bridge carries the A82 across for the last time. An isolated, steep, tree-covered hill on the east bank forms the attractive Tomnahurich cemetery, opposite a golf course. Seventeenth-century prophet Brahan Seer had predicted that fully rigged ships would sail inland this way. Ahead, Craig Phadrig has a forest trail near the site of the 4th-century BC double-walled vitrified fort that was the stronghold of the Pict King Brude. The 172m-high hill commands wide views of the Moray Firth and Beauly Firth. The heart of Inverness, the Highland Capital, lies only 2km from this ancient fort.

Caley Marina is the largest boatyard on the canal and has everything from narrowboats to drifters moored.

Muirtown Locks form a four-chamber staircase, the steps being hidden on the right in front of the Whitecross Restaurant. The 9.8m drop gives a view across Inverness to the Kessock Bridge, which crosses the mouth of the Beauly Firth directly ahead, carrying the A9.

The A862, the former route of the A9, crosses on Turn Bridge. Muirtown Basin follows, a wide basin with commercial wharves. Craft vary from herring boats to the *Scot II*, an icebreaker and former tug that takes passengers from here to Loch Ness in the summer. Despite the commercial nature of the basin, the west side consists of neatly mown lawns and forms a pleasant environment.

Clachnaharry Lock has a variety of interesting features, a derrick crane, decorated cast-iron lamp standards and a delicate belltower. Directly beyond it is Clachnaharry swing railway bridge, carrying the main Inverness to Wick line, painted white to reduce temperature effects.

Traditional architecture at Clachnaharry Lock.

The old light at Sea Lock by the Beauly Firth.

It was built to carry the Highland Railway across at 65° to the canal. Mounted on the east bank, its 38m hog-backed wrought-iron girders span 24m over the canal. There is a monument to the 1454 battle between the Mackintoshes and the Munroes.

The gently shelving shoreline would have given access to Clachnaharry only at high tide and the muddy bed of the loch was too soft to support structures. Telford solved the problem by running out twin 400m clay embankments that he surcharged with rocks to displace the mud over the following six months, during which it sank 3.4m. He then excavated the navigation channel between them, building Sea Lock at the far end.

Sea Lock operates within four hours of high water and has a light on a post. This supersedes the little cast-iron lighthouse that formerly served to aid navigation. The flows across the end of the canal are 9km/h on spring tides but there is a slack after the flood from an hour after high water at Dover for an hour. There is another after the ebb from five and three-quarter hours after high water at Dover for two and a quarter hours.

The firth is good for winter wildfowl and waders, including 2,000 red breasted mergansers and 1,000 goosanders in December, and is a moult migration ground for Canada geese.

Looking down the Muirtown flight and across the roofs of Inverness towards the Kessock Bridge and the Moray Firth.

Distance
96km from Loch Linnhe to the Beauly Firth

Navigation Authority
British Waterways

OS 1:50,000 Sheets
*26 Inverness
& Loch Ness
34 Fort Augustus
41 Ben Nevis, Fort
William & Glen Coe*

Index

Main coverage is listed in **bold** print.